COMPUTERS, COMMUNICATION AND USABILITY:
Design Issues, Research and Methods
for Integrated Services

NORTH-HOLLAND
STUDIES IN TELECOMMUNICATION
VOLUME 19

NORTH-HOLLAND – AMSTERDAM • LONDON • NEW YORK • TOKYO

COMPUTERS, COMMUNICATION AND USABILITY:

Design Issues, Research and Methods for Integrated Services

Edited by

Paul F. BYERLEY
DeTeBerkom GmbH
Berlin, Germany

Philip J. BARNARD
Jon MAY
Medical Research Council
Applied Psychology Unit
Cambridge, UK

1993

NORTH-HOLLAND – AMSTERDAM • LONDON • NEW YORK • TOKYO

ELSEVIER SCIENCE PUBLISHERS B.V.
Sara Burgerhartstraat 25
P.O. Box 211, 1000 AE Amsterdam, The Netherlands

ISBN: 0 444 89974 X

This book is printed on acid-free paper.

Printed in The Netherlands

Contents

Prefaces

This book grew out of contacts between research groups funded by the ESPRIT and RACE programmes the Commision of the European Communities. We are grateful to DG XIII for providing the following prefaces:

A Synthesis for Future Design Practice

Jakub Wejchert

ESPRIT, Commission for the European Communities DG XIII, Brussels, Belgium.

As we move towards the end of the twentieth century, humans have come to be exposed to huge amounts of information through various media—including, of course, computers and information related technologies. As a consequence, one necessary and timely challenge of information technology is how to best present and represent such information within the interface between humans and machines, so as to transform it into readily usable forms. This involves asking (and trying to answer) what tasks users typically need to accomplish and what tools have to be designed that are most appropriate. Solutions in this domain necessarily involve elements of industrial and graphical design, models of interaction, psychology and cognitive science. By creating generalisable and predictive models of the interaction between humans and machines one can at least lay the basis for certain design principles and gain insight into why some interfaces are successful and others are not.

It is at this level that the domain of basic research comes into play, that is, research that is fundamental but applicable, and is an appropriate point at which to introduce the ESPRIT programme. The principal aim of ESPRIT as a whole, is to coordinate and focus pre-competitive research in information technology throughout Europe. ESPRIT Basic Research is an 'upstream' component whose results feed into several of the more industrially oriented activities of the programme. Its aims are to replenish the reservoir of new knowledge and expertise from which industrial research can draw; its objectives being the generation of ideas, concepts and knowledge ultimately needed for future technological progress.

In this book we find an interesting interplay between the basic research of cognitive modelling and formal approaches to design rational on one hand, and technological developments and what designers currently do, on the other. The book strikes a healthy balance between the basic and the practical and it is encouraging that many of the contributors to the basic research sections are involved in ESPRIT Basic Research projects and activities. Indeed, it is from an interplay between the basic and the practical that the hope comes for a synthesis for future design practice and innovation.

The Future of Usability Engineering in Telecommunications R&D

Leon van Noorden

RACE Central Office, Commission for the European Communities DG XIII, Brussels, Belgium.

This book presents an overview of Usability Engineering as it is being practised in RACE and ESPRIT. Usability Engineering is in development as a discipline. Developing user interfaces can still be considered very much as a craft or an art, c.f. Dowell and Long (1987).

Every artist who is in the process of producing an extensive work of art has to take a few steps back every now and then to oversee their growing masterpiece. They have to see whether what they are doing close up to the work—at a distance that is dictated by the limitations of their arms and eyes—still fits in with the broad picture in their imagination.

This book appears at the end of RACE I and it therefore seems fit to take the few steps back to oversee the purpose of Usability Engineering in the total picture of developing Telecommunication Technologies and Services.

The hectic developments in this area have very much to do with improving usability, as can be concluded from the following three important tendencies:

Increasing freedom in bridging distance, time and scale

The very fast advance in the possibility of becoming independent from a fixed network connection by cordless and mobile services gives the subscribers the opportunity to move freely around and still be able to receive and make calls. Time shift recording, pay per view, etc. increase the opportunity to bridge time differences and the massive transfer of computer images gives the opportunity to see the microscopically small or the galactically large images at places far removed from the special and expensive equipment necessary for collecting these images.

Enlarging the sensorial richness of transmitted information

The growing capacity of telecommunication channels and the advances in the efficient coding of transmitted information enables communication via increasingly rich perceptual communication channels: increased audio bandwidth, surround sound, high resolution images and stereoscopic vision are all in development

Enlarging the power to act

Integrated Service Engineering will improve the ease of establishing information exchange with other people or information systems and the ease with which information can be manipulated and transferred between different services. Electronic signatures are an enabling technology for establishing trust between the different parties connected to the network and improve the power to act at a distance.

However, the promises made by these technological advances only come true if this complex machinery can be moulded in such a way that it fits 'naturally' to the skills, capacities and habits of the user. The human-machine interface has to be as transparent as possible, i.e. invisible, to the user. In order to produce these invisible interfaces their designers need to understand the behaviour of people as well as the technological possibilities which can be used to realise these.

Probably because of the large, perhaps fundamental, difference between the behaviour of technical systems and people there are not very many researchers and designers around who have a deep understanding of both realms. The obvious way out is the formation of multi-disciplinary teams for such tasks.

The complex infrastructure of the future telecommunication services will consist of many layers and as a consequence many interfaces between these layers. Examples of these layers are the chip technology, subsystem hardware integration, software and display technology. It is a sheer impossibility to design new systems that encompass all these layers at once. Developments happen at many levels at the same time and unavoidably are rather independent from each other. All these developments must have their own techno-economic justification. They must each have a reasonable chance of profit in their local environment.

The complexity of this process of evolution and revolution makes it impossible to predict what human-machine interfaces and services should be developed at some distance in the future. Not only the techno-economic environment changes, but also the skill and expectations of users. What seems to be sure, however, is that a continuous effort is needed to 'diffuse' at the 'local' level (i.e. taking only two or three layers into consideration). Designers should be made clear what the customers need and the customers must be made aware of what the technology can offer.

It is in this process that this book fulfils an essential role and contributes to the effectiveness of the recent work in the Community programmes on Usability Engineering.

In the first paragraph I established the link between Usability Engineering and art. I did this of course primarily to be able to develop the metaphor of the artist in the second paragraph. It permits me now to consider the development of Usability Engineering during the period of RACE I as an art critic. My conclusion is that the art has developed itself very much in the direction of an engineering discipline. Both the tasks of the end users and the designers have been analysed and systematised. The methods developed on these grounds will certainly help to avoid making badly engineered user interfaces and to build a continually growing understanding of the important issues. I see this as substantial progress and I am happy that I could watch this closely. However, I hope that these technical tools for designers do not scare away the more imaginative ones. The human-machine interface should not only be functional but also be able to stimulate the imagination.

Reference

Long, J.B., & Dowell, J. (1989). Conceptions of the discipline of HCI: Craft, applied science and engineering. In A. Sutcliffe & L. Macaulay (Eds.), *People and computers V* (*pp.* 9-32). Cambridge: Cambridge University Press.

Foreword

In the last half century, research on usability has been dominated by three main areas, military hardware, telecommunications equipment and information technology. Characteristically, research was carried out on local issues as they affected a particular class of technology and its design. In just the last decade there has been a massive convergence both in the technological bases of these areas of concern and in the human factors issues they raise .

The actual research traditions in each of these areas have developed with slightly different characteristics. The original emphasis of much of the early military and telecommunications work dealt with the constraints of human sensori-motor skill. There was a strong emphasis on the development of empirical paradigms and the evaluation of alternative designs. The lead times for the development and introduction of the technologies were substantial, as indeed was the cost of failure. As computer technology emerged to provide a dominant set of design issues, the lead times for development and introduction of new technologies has decreased. The changes have not only been structural, the nature of the issues at the heart of a particular generation of design now also change rapidly. As Grudin (19xx) has pointed out, research on human-computer interaction started out with a direct concern with hardware aspects, moved very rapidly onto more cognitive aspects of interaction with command lanaguages and menu systems and equally rapidly made the next transition to organisatonal and social issues raised as computers were increasing harnessed to support "co-operative work".

As information and communications technology increases in power and undergoes further integration, profound issues are raised not only with respect to usability itself but also with respect to the kinds of analyses and methodologies that will need to be called upon to support product development. First, integration and increased power means that information and communications technology will be able to produce with high fidelity all sorts of images and support a very broad range of tasks. Techniques of broad scope will be required that are not restricted to a specific set of issues or technologies. These will need to be grounded in a relatively full understanding of human capabilities in all their diverse aspects. It will no longer be practicable to rely on simple, checklists, constrastive methods or analyses. Secondly, design spaces will become increasing complex and interdependent. Designers will not simply want purely evalutive input, they will require support for the definition of novel options and how to understand the broader implications of adopting a specific design decision. This they will undoubtedly want yesterday, with close to perfect accuracy, and at marginal cost.

This book is set against a background in which there is terrific pressure towards convergence, considerable flux and dynamic change in the real world of system development; and diverse views about how best to move forward. It is also a book in which the research described is set in a specifically European context. In developing a strategy for the books coverage and content, we felt that it would be most useful to settle on a number of clear themes concerning aspects of requirements specification. As a result the particular collection here does not seek to provide a comprehensive view of all European research on related topics in usability research. Rather , we decided to draw on a limited number of projects and to juxtapose contributions concerned with practical issues in method development and more basic research. We have done this with three general objectives in mind:

1) To provide a description of the product design context for usability research and practice, focusing in particular on the problems and opportunities arising from the advanced communication services which are expected to be in business and domestic markets within 20 years.

2) To draw upon some illustrative basic research in usability, system modellling and design behaviour which could lead to significant improvements in design practice in the long term.

3) To provide an integrated guide to some recently introduced usability design methods which are particularly appropriate to the design problems of advanced communication services.

Specifically, this book should also viewed as a demonstration of the results of international collaborative research supported by the European Community. Researchers from most of the western European countries have jointly engaged in a series of projects in telecommunications and information etchnology. Support has been provided for tool development and the piloting of new services, for example within the RACE programme, and for basic research within the Esprit programme.

RACE (Research and Development in Advanced Communiations in Europe) was established in 1987 and its current phase will complete in 1994. The European Community provides support in programme administration via RACE Central Office and contributes up to 50% of the cost of each project. The overall cost of the 160 projects is approximately 2500 million ECU. The RACE programme is divided into Integrated broadband Communication (IBC) developments, Intelligence in Networks, Mobile Communications, Advanced Communication Experiments, and Test Infrastructures. An Overview of the technical results of the programme up to 1992 can be obtained from Bonatti et al (1991) and Byerley & Connell (1992).

The RACE projects which have contributed to this book are those that have at least partly addressed the issues of the usability of advanced telecommunication services.

ESPRIT (European Strategic Programme for Research and development in Information Technology), initiated a programme of Basic Research in 1988. This programme recognised that fundamental research, even when it does not have immediate industrial applications, was needed for technological developments with major industrial, economic and social impacts. Its goals included the provision of new technological expertise and to ensure the future availability within Europe of high calibre scientists and engineers. Community funding of 77 Million ECU covered a 30 month initial programme for a total of 74 actions and working groups. Most of the actions were started in 1989 and concluded in 1992. In 1992, a second call led to the funding of a further 108 proposals for projects, working groups and networks of excellence. In both of these calls research in human-computer interactions was regarded as a priority theme.

This book therefore not only represents a singificant sample of collaborative research within the RACE and Esprit programmes. It is also an attempt to bring together the results of the two programmes around the general theme of usability requirements.

Usability is a prime candidate for collaborative research and Community support as long as this is through a systematic, long-term programme. The properties of a system which make it usable are not constant, but depend upon the tasks and the users which the system has been designed to support. Thus usability should be defined for the whole range of a population, in the same way as accessibility has come to be defined with respect to people with special needs. Systematic research of this kind is not ususally within the resources or interests of any single company or research department, requires widespread dissemination for validation, and long-term objectives with consistent planning. If Cognitive Ergonomics is presented as a discipline which can offer some immediate demonstrable benefits from within a larger framework that provides the organising principles for, say, a twenty year collaborative research programme, then it can avoid the disappointments that arise from promising too much too soon, and it can start building on current achievements, many of which are reported in this volume.

We hope that this book will be of interest to those developing future research programmes, to reserchers in industry and university, and to psychologists and engineers designing new computer and service products.

Paul Byerley*, Phil Barnard† and Jon May†

*Alcatel—SEL AG, Lorenzstraße 10, D-7000 Stuttgart 40, Germany
(*now at:* DeTeBeKom GmbH, Voltastraße 5, D-1000 Berlin 65)

†Medical Research Council Applied Psychology Unit,
15 Chaucer Road, Cambridge, UK—CB2 2EF.

References

M. Bonatti, F. Casali, G. Popple, Editors, (1991): Integrated Broadband Communications: Views from RACE—Network and Engineering Aspects; Volume 16 in the North Holland Series in Telecommunications. Elsevier Sceince Publishers: Amsterdam.

P.F. Byerley & S. Connell, Editors, (1992): Integrated Broadband Communications: Views from RACE—Usage Aspects; Volume 18 in the North Holland Series in Telecommunications. Elsevier Sceince Publishers: Amsterdam.

Grudin. J, (1990) The computer reaches out: the historical continuiyt of interface design. In Proceedings of CHI'90 (Seattle Washington, April 1-5, 1990) ACM: NewYork, pp 261-268.

Acknowledgements

I would like to express my gratitude and respect to John McEwan who had the foresight to set up a cognitive ergonomics group within ITT Europe in the early 1980s. This group was managed by Marco De Alberdi to whom I also owe a debt of gratitude. This group was disbanded upon the formation of Alcatel but members of it continue within Alcatel and elsewhere to promote and conduct usability engineering of telecommunication services. I would also like to thank Leon Van Nordern, Emile De Hert, and John Nolan of Race Central Office for their support as project officers for the projects I managed.

Paul F. Byerley

We would like to thank all of the individuals and organisations who were members of the first Amodeus project. Their contributions at the project's 'matrix' meetings were an essential factor in the cohesiveness and productivity of the project, and working with them made the whole experience an enjoyable one.

Philip J. Barnard and Jon May

Section 1

Usability in a Service Engineering Context

Paul F. Byerley

Alcatel-SEL AG, Lorenzstraße 10, D-7000 Stuttgart 40, Germany

This book is based upon the conviction that user requirements can be systematically incorporated into design only when methods are available which allow them to be explicitly developed, recorded and communicated within a product development context.

The user requirements for a system refer to a description of the behaviour of that system to a user. However this description of the desired behaviour is not easily obtainable and often requires a process of decision making as complicated as that which provides the functional structure, or architecture of the system. In other words the development of user requirements shares many characteristics with later design stages, including the need for methods which are appropriate within the product development context. The diagram illustrates the importance of four criteria for the appropriateness of a method.

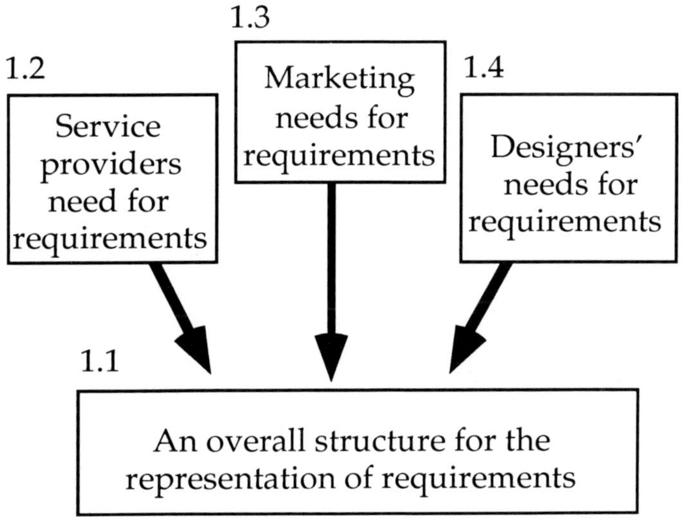

The method should be appropriate for the range of product opportunities and issues to which it will be applied. Chapter 1.2 describes a wide range of applications for advanced communication services and associated requirements issues, particularly those which are associated with the interface between end users and service providers. These applications and issues are at least a part of the currently known elements of the first criterion and they are used as examples throughout the book.

The second criterion is that the method should be appropriate to the development and sale of products. Usability engineering will finally be taken seriously when it can be demonstrated that it not only leads to usable products but also to increased sales. Chapter 1.3 reviews the relationship between user requirements work and marketing and sketches a proposed closer one.

The third criterion comes of course from the users of the requirements methods and the requirements themselves. Systematic use of user requirements methods may require designers to make major changes to their resource allocations and skills. Making design decisions explicit often initially appears to increase effort. Although the book assumes that designers should use user requirements methods, these should be designed to be appropriate to the decision making or problem solving characteristics of designers. Chapter 1.4 examines this criterion.

The final criterion is given by the notion of overall structure for user requirements. Usability is one of a number of desirable user requirements properties, some of which are incompatible. If the designer is provided with a number of methods which compete for the attainment of different properties within the requirements stage, then one or more of the properties will be discarded as an objective. Chapter 1.1 attempts to define the overall structure for user requirements methods which would enable the selection of appropriate properties at each particular step of an '8 Step Method'.

Computers, Communication and Usability:
Design Issues, Research and Methods for Integrated Services
P.F. Byerley, P.J. Barnard and J. May (Editors)

Chapter 1.1

Service structures

Paul F. Byerley

Alcatel-SEL AG, Lorenzstraße 10, D-7000 Stuttgart 40, Germany

1. Introduction

The scenario common to all the chapters of this book is of a user interacting with a work domain with the support of services. These services could be provided from the user's terminal, from local facilities such as a private network, from public networks, from remote databases, and from remote human users. For example a hospital consultant discusses diagnoses with a colleague at another site while both view some medical files and x-rays which one of them has retrieved from a central database. Alternatively a tourist retrieves multimedia images from a distributed database which integrates information from thousands of independent holiday providers, the tourist then plans his holiday with a travel advisor and purchases it with all the various transactions being initiated from the tourists site. Integrated broadband communication will enable a wide range of business and domestic tasks to be supported by telecommunication services. The attributes of these services will not be fixed but will have values which can be set to give maximum support to a particular task, for example the hospital consultants will require high quality video while viewing the x-rays but not while viewing each other during conversation. In every case the user will be relying both on the designers of terminals, databases, private and public networks and services, and on their own negotiation for and manipulation of the available services. Therefore the finally used integrated services can be considered to be the result of a complex sequence of design decisions and user-service provider negotiations many of which can affect the usability of the service. Thus the analysis of the usability of integrated services requires also an integration of previous research in for example human computer interaction and service and network design.

The objective of design is to produce a system which has properties such that the performance parameters of one or more applications can be met. An application is defined here as one or more goals and tasks with one or more users (human or machine). The performance parameters of an application concern the desired quality of the products of the application, and the acceptable costs of producing those products. By definition then the performance parameters are the expression of an interest. There may be many interests with respect to an application and thus to the outcome of a design activity, however the proposal reported in this chapter is that network and service design requires responding to two types of interest, those external to an application and those internal. It will be argued that usability is a system property relevant to meeting performance parameters from an internal perspective, and that the results of design for an external perspective can supply the context in which design for usability should take place.

This chapter will describe an eight-step method which consolidates design for the two perspectives and will provide a definition of usability which can be mapped directly to the objective of design.

2. External and internal design perspectives.

This chapter will only consider those perspectives which are appropriate to the definition of the operational requirements for a system, and in particular those operational requirements which are based on the needs of users. Thus perspectives for other stages of design such as, functional specification, architectural design and implementation are not considered.

The distinction between external and internal perspectives, reported in Byerley and Bruins (1992) was developed to replace a distinction, often used but wrong, which suggests that the major requirements for design are derived from market related notions of need and want and that usability is less interesting and entails only ease of use.

An *external perspective,* with respect to an application, enables a view to be taken of the whole application for the purpose of the design of services which will support that application. If an application is defined as a set of goals and tasks, potential users (human and machine), with performance parameters (desired quality and acceptable costs) then an external perspective would motivate the design of a network of users and tasks with services which support the operation of that network. This design would attempt to provide a system with properties that enable the attainment of the performance parameters of the application. These performance parameters might be those that represent the interests of a customer for the network of services, for example a communications manager of a large company. From an external perspective the internal structure of each of the users in the network is not determined. The operational requirements from an external perspective simply define the responsibilities of each of the users (or communicating entities) the relationships between them and the services which can support these relationships.

An *internal perspective* with respect to an application enables a view to be taken from within the network that was designed to support the application (from the external perspective). The view is taken from the perspective of one of the users, human or machine. The objective from this perspective is also to design the system but in this case that includes the user and the users interface to the system. If the chosen user is a human then the design process can be one of usability engineering. If the chosen user is a machine then the design process concerns the internal structure of, for example, a database. The context for design is partly given by the results of the design from the external perspective. It includes the goals and tasks allocated to the user for the external perspective, the characteristics of the user, the task network and service environment for the user created from the external perspective, and finally again the performance parameters of both quality and cost, however these do not have the same values as the external perspective performance parameters. Just as with the external perspective there are quality parameters to be satisfied but these only concern the tasks allocated to the user and cost parameters, but these concern the costs to the particular user, e.g., cognitive load (or ease of use). Design from the internal perspective initially takes as given the results of design from the external perspective. This includes the decision on the tasks allocated to the user. Thus one of the givens for the internal perspective is that the user is the agent of these tasks, there is no question of their reallocation to alternative users other than by design iteration.

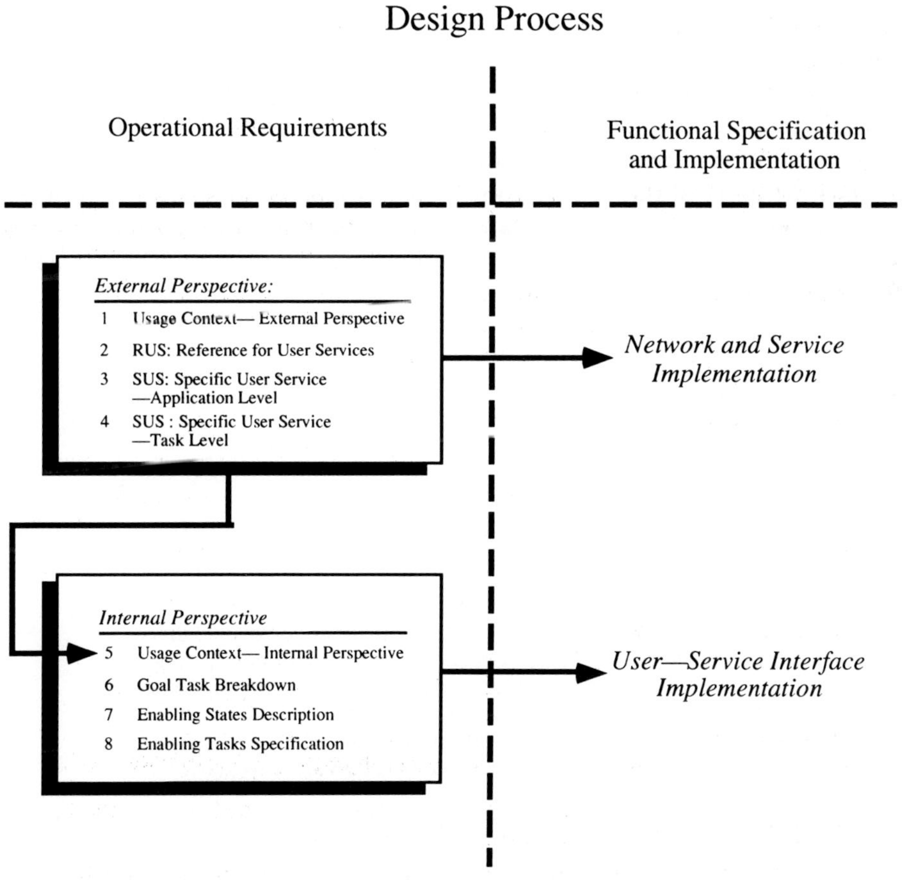

Figure 1: An Eight-step method for Integrated Service design

A clear conclusion from this approach is that usability engineering is not an occasionally dispensable process in the light of the greater importance of other design processes. It is in fact, along with the design of the machine entities of the system, a necessary process for the completion of the design of the system. Since the user requirements from the two perspectives may not be compatible and may need to be iteratively refined an eight-step method is proposed which sequentially combines them. The basis for the method is shown in figure 1. The Race Programme publishes results which have achieved an important degree of consensus as Common Functional Specifications (CFSs). The eight-step method is defined in CFS B310, B311, and P120.

3. Design from an external perspective.

Step 1: Usage Context—External Perspective

The four steps of the external perspective begin with a description of the application which provides the usage context for service design (the methods used prior to this step, e.g. business analysis methods, for the choice of the application are not covered here). The application is described with respect to goals and tasks, potential human and machine users, and performance parameters. Example parameters are given in Table 1. The use of these parameters to describe 15 of the applications being studied in the Race programme is given in 'Race Usage Research Review', a Usage Reference Model Project publication for Race.

Application and Task
- **application:**
  - content of application
  - application frequency
  - number of tasks
  - parties located at the same site
  - parties belong to the same organisation
  - level of user cooperation (informing, coordinating, collaborating, cooperating)
- **task:**
  - **task taxonomy:**
    - human to human, conversational communication
    - human to human communication, primarily send
    - human to human communication, primarily retrieval
    - human to machine / machine to human conversational communication
    - human to machine / machine to human communication, primarily send
    - human to machine / machine to human communication, primarily retrieval
    - machine to machine, conversational communication
    - machine to machine communication, primarily send
    - machine to machine communication, primarily retrieval
  - **information types:**
    - number of different source information types
    - amount of information communicated
    - type of information transfer required (audio, still / moving image, data) for each transfer:

Table 1 : External Perspective Usage Context Parameters and Variables (continued overleaf)

content of information communicated minimal / ideal quality of information type required
required transfer rate

communicating entities:

number and type (human / hardware / software)
type of communication (one to one, one to many)

locations:

number of user locations
distance between locations (same site, city, region, country, international)
mobility required within or between tasks

task temporal characteristics:

duration of task execution
task frequency and typical time of day and week
continuous or intermittent information transfer
maximum 'waiting time' per task
priority of task

task dependencies:

intertask dependencies
desired information processing (e.g. realtime, compressed format)
availability of other human communicating entities required
dependencies which are external to the network
required output of information (transient or permanent, format and content)
time control of information delivery

environmental conditions:

noise level (good, medium, poor)
lighting (good, medium, poor)
space (not limited, limited)

Communicating Entity Characteristics

user:

occupation
user location (business, domestic, public)
type of user (e.g. whether he executes a task regularly or casually)
special need categories (elderly, blind, deaf, etc.)
prior (task, technical, telecom) experience
required (task, technical, telecom) knowledge
whether support / help is required

customer:

sector
whether the customer is user or not
level of innovation (low, medium, high)
influence on policies
orientation towards change (positive, neutral, negative)

Performance Parameters and Cost

security:

user and participant identification
authentication
confidentiality
integrity

privacy:

anonymity
(environmental) privacy
repudiation control
availability of directory and x-directory option

cost:

cost to customer and user
usability costs

Table 1 (continued) : External Perspective Usage Context Parameters and Variables

Step 2: Reference for User Services (RUS).

For this step the major design decision is taken to distribute the tasks of the application to geographically separated users. Some of the tasks of the application may themselves be communication tasks with specified users, others will not be and distributing these tasks will impose communication requirements which will need to be deduced. All the communication requirements can be expressed as links in a network of users. An example RUS is illustrated in figure 2. The links in the RUS can be described using a simple communication task taxonomy shown in Table 2 (see also the 'task' parameters in Table 1). The taxonomy is based on the communicating entity type (Human or Machine), and the communication type (Conversation, Send, or Retrieval). The definitions of Conversation, Send, and Retrieval concern the relationship between any pair of users. A pair of users could have more than one relationship of more than one kind, which varies over time with the execution of the tasks of the application. Thus the RUS must contain a description of the task interdependencies and flow of work.

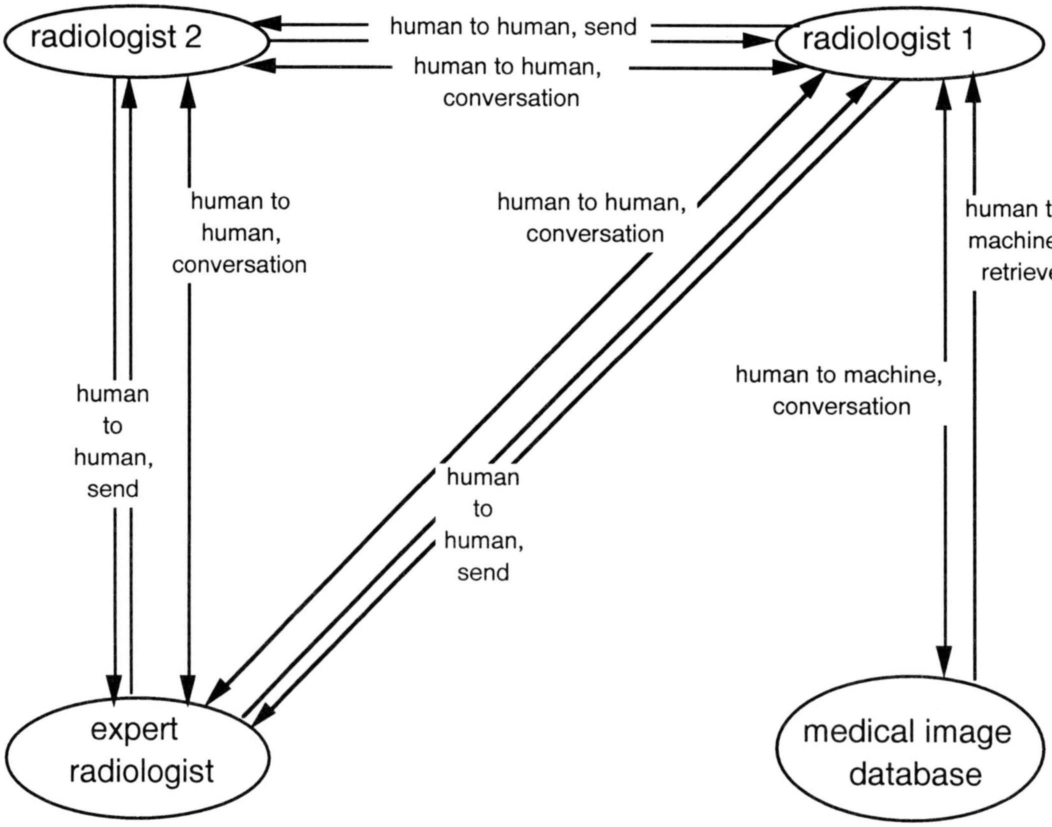

Figure 2: An example Reference User Service diagram for the application 'collaborative medical diagnosis'

The definitions of the communication tasks are directly relevant to service design. The reasons for the inclusion of the categories Human and Machine include the findings that the characteristics of services needed to support a link of the RUS are strongly determined by this factor, and in addition it enables the avoidance of a confusing use of the term 'Real-Time'. 'Non real time' is sometimes used to describe 'human to machine send' in the sense that some time later a human appears and retrieves the information from the machine. Thus the correct use of the descriptor 'Real time' in this case is to describe the relationship between the two tasks of send and retrieval and in general it should be kept for the description of task interdependencies.

- **human to human conversation**
  two doctors exchange information about a patient in order to make a diagnosis.
- **human to human send**
  a trainer provides students with instructions.
- **human to human retrieval**
  personnel in a social service centre watch over an elderly person at home.
- **human to machine conversation**
  a designer using a remote CAD/CAM program.
- **human to machine (or machine to human) send**
  a technician sends NC data to robots.
- **human to machine (or machine to human) retrieval**
  a salesman retrieves planning data from the workstation in their company.
- **machine to machine conversation**
  a workstation exchanges information with a remote printer in order to identify the right typesetter.
- **machine to machine send**
  a server sends information to a remote workstation.
- **machine to machine retrieval**
  the workstation in the financial department retrieves data from robots.

Table 2: The 11 communication tasks

Conversation, Send, and Retrieval are defined as follows :

Conversation occurs when the communicating entities share the roles of source and sink and both entities have control of the relationship. Communication can occur in both directions simultaneously. In broadband communication the service is not necessarily symmetrical, for example there might be high quality video in one direction but low quality in the other. Thus negotiations between users on media, media quality, and who pays can become complicated.

Send occurs when the source of some information has control and sets up a relationship without negotiation with the other user (the sink). Communication is unidirectional. In broadband communication messages can be multimedia and thus feedback to the source, of what media, of what quality, got to which point, becomes complicated.

Retrieval occurs when the sink initiates a relationship with the source in order to obtain information from the source. Retrieval is the most complex of the basic relationships because even though the major information flow is unidirectional sequential information flows in both directions are required. After the initiation of the relationship the sink requests information from the source. The source then sends the information without having to initiate a further relationship. Obviously information retrieval can occur using other relationships, e.g. conversation, or with complex combinations of relationships, e.g. conversation followed by send. However 'Retrieval' is a basic form with different consequences for service design than the other relationships.

The three communication task relationships are illustrated in figure 3.

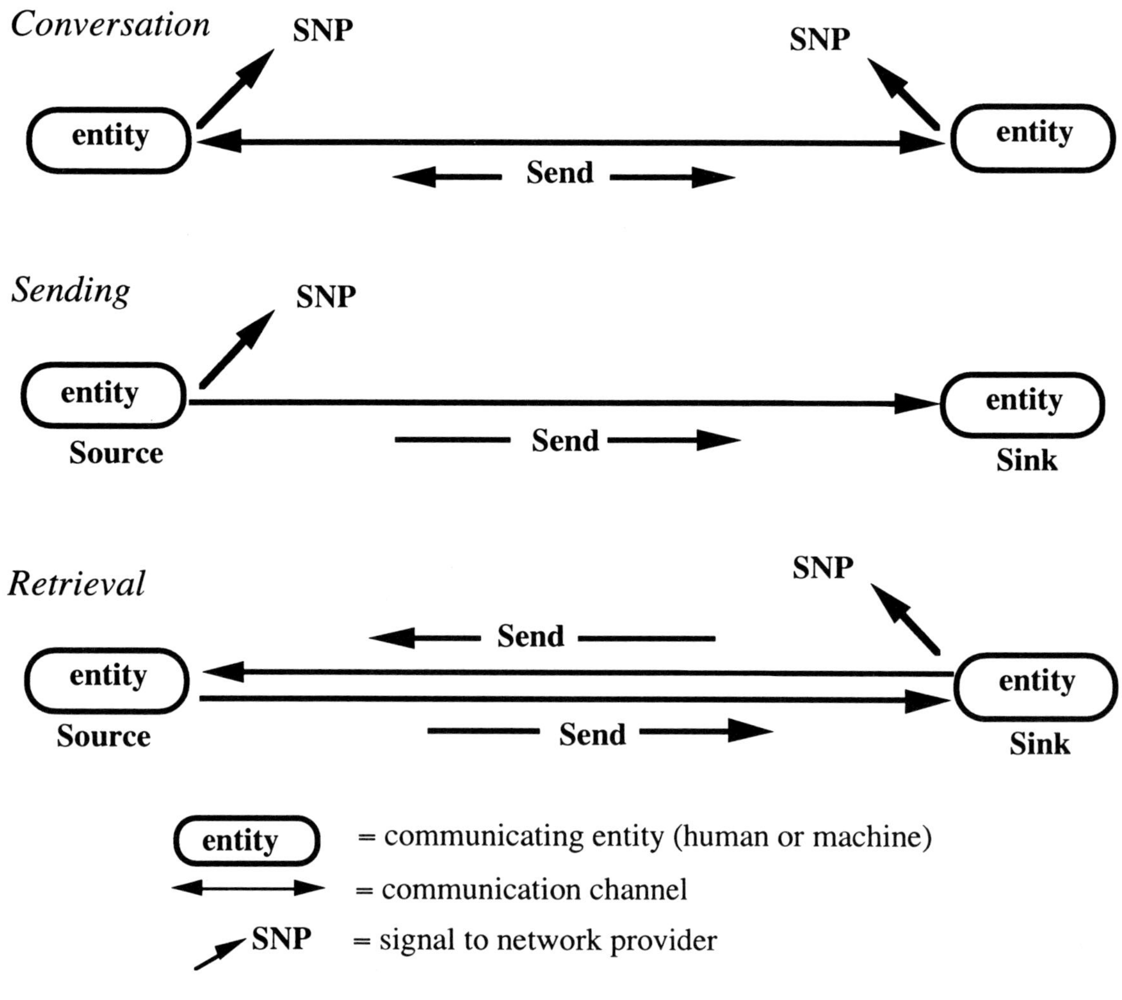

Figure 3: The three communication task relationships

Introduction to Steps 3 and 4.

In the next two steps specific services are recommended which can enable the task network described by the RUS to function. These service descriptions are made as far as possible in implementation free terms. The service descriptions use a two level structure which corresponds to the two steps. That structure will be introduced here.

Integrated broadband communication services can penetrate far more deeply into both domestic and business applications than conventional services. However if they are to do this they need to enable the attainment of the desired performance of the application better than conventional services and methods. The Reference for User Services description results from a design decision that the performance of the application can best be met by a particular task distribution amongst the user resources. That particular task network will require a particular set of services in order to function. If the variance of services follows that of task networks then there are a potentially very large number of different specific services that have to be offered. For example even for a simple point to point human to human conversational task the service needs of an elderly domestic caller may be quite different to those of a young office worker. When combinations of tasks and complex RUSs are considered the burden on the service provider appears even more enormous. However two traditional modelling techniques can be used to organise the requirements of users into manageable descriptions suitable for the design of service provision.

Conferencing	C	one or more conversation tasks
Distribution	D	one or more send tasks
Retrieval	R	one or more retrieval tasks
Conferencing with Distribution	CD	mixture of conversation and send tasks
Conferencing with Retrieval	CR	mixture of conversation and retrieval tasks
Distribution with Retrieval	DR	mixture of send and retrieval tasks
Conferencing with Distribution and Retrieval	CDR	mixture of conversation, send and retrieval tasks

Table 3: Seven application level services

Firstly, services for task combinations can be modelled as aggregates of services for single point to point tasks. Since there are only 11 types of point to point task, 11 types of point to point service can be proposed. For task combinations, if we leave the category of Human v Machine to the task level, then there are only 7 types of task combination (user-user relationship combination). Therefore 7 types of service would be needed if based on the task combination types and each of these services would be formed from one or more point to point task level services. A task combination can be described as an application therefore services for task combinations are referred to as application level services to distinguish them from task level services. The set of application level services is given in Table 3. At the task level there are appropriate service attributes, and at the application level there are appropriate emergent service attributes Tables 4 and 5 provide examples of these two types of attribute. These attributes can be used to describe each specific service.

Secondly, each specifically appropriate service can be considered as an example or instantiation of a generic service. Thus for each task type, or task combination type, the appropriate service type is defined as a generic service. Each generic service consists of a set of specific services which vary one from another on the values of the service attributes. The derivation of an appropriate specific service for a particular application can be modelled as the selection of the specific service from a generic one on the basis of the usage context. The Usage Reference Model project(R1077) has consolidated the results of all the usage projects in the first phase of the RACE programme. There are 28 projects of 3 to 4 years with an average size of approximately 7 Mecu. These projects have for the most part been evaluating advanced services for a wide range of usage contexts. The consolidated results provide indications of the variance of the usage context parameters, and of the service attributes, and the relationships between the usage contexts and the service attributes for each generic service. Obviously at this stage only indications can be given since the results do not cleanly reflect design to meet user need but are clouded by issues such as what advanced services the particular project was able to implement. However rules are being generated from regression analyses of the data for both task level and application level service attributes, for example :

- Where task duration is long or tasks are executed frequently, and where the source information type is video, a high image quality is normally required.
- In a conference where the source information type is multimedia and there are more than two participants, then the value of the dialogue management attribute for first and third party control of service components needs to be known and agreed by participants prior to call set up.

Step 3: Specific User Service—Application level

For this step the appropriate service attributes for each application taken as a whole are defined. This requires firstly a selection of the task combinations which will be defined as the applications. These task combinations, which could be the complete RUS, are selected on the basis of the intertask dependencies described in stage 2. The task combination can include more than two communicating entities e.g. a 'send' to a database by one communicating entity, followed by a 'retrieval' of the same information by another. Alternatively the task combination could involve only two entities on the same link, e.g. a database browse with a conversational task, followed by a 'send' from the database to the browsing entity. For each task combination the appropriate application level generic user service is selected, and a choice made from the available attribute values for each service attribute, using where possible design guidance from data or 'rules' such as those provided by the URM project.

Static characteristics

Channel level characteristics:

CPN relation (only CPN, internal & external network, only external network)
channel complexity (number of networks supporting a channel)
channel usage (number of tasks supported by the same channel)
number of calls per task
location of application functionality (e.g. terminal, CPN, public network)
directionality (bidirectional, unidirectional)
type of communication (one to one, one to many, many to one, many to many)
call control (source, sink)
mode of establishment (on demand, reserved or permanent connection)
interchannel dependencies (realtime, not realtime)

Service components:

service components (audio, moving images, still images, data)
for each of these service components:
format preservation features (e.g. readable, scannable)
content of information
quality level
transfer rate
symmetry of information transfer (symmetric, asymmetric)
mode of traffic (bursty or non-bursty)
mandatory or optional service component
type of usage (contribution, distribution)
requirements by combination with other service components (e.g. lip synchronism)

Terminal (hardware and software characteristics):

terminal type
(dedicated telecom terminal, dumb or intelligent terminal)
(dedicated or multi-purpose terminal)
(own or shared terminal)
location of terminal (workshop, office, library, booth etc)
number of terminals and screens per user
software types (operating systems, application software)
peripheral, input, output, pointing devices
dialogue style
specific acoustic, lighting, space requirements

Network characteristics:

minimum transfer rate of the network
network access
fixed / mobile (degree of mobility)
immediate / delayed
transfer medium (cable, satellite)

Nested services:

(e.g. directory service, call forwarding, multiparty call)

Dynamic characteristics

Control of channel level characteristics:

service request and payment
change of partners, including redirection
change of location
change of terminal
change of terminal quality
change of interservice integration (e.g. holding, background, simultaneous)

Control of service component characteristics:

allocate / deallocate service component
change of quality level
change of transfer rate
change of symmetry

Control for communicating entity states:

nested service control
control of other communicating entities (e.g. value added service provider, database, terminal)

Table 4 : Task Level Service Attributes

Application Level Service Type
(see Table 3)
Service Interdependencies
(both within the system and external)
Dialogue Management
(for example conference management procedures)
Nested Services
(for example aggregate service cost feedback)

Table 5 : Application Level Service Attributes.

Step 4: Specific User Service—Task Level

For this step the service attribute values needed for each point to point link in the RUS. are described. All the point to point links are examined, whether they form part of a task combination or not. Similarly to step 3 the appropriate task level generic user service is selected for each link and service attribute values chosen from the available ranges for each task level attribute. This can again benefit from consolidated data from the URM project.

Results of the external perspective

Thus the result of the first 4 steps of the method is a proposal for allocation of the tasks of the 'customers' application amongst a set of users, the description of the communication relationships which result, and a specification in customer terms of the services, at both application (task combination) level and task level, which would be needed to enable the distributed application to function. These results can be directly taken by the next stages of the design process for global functional and architectural specifications of the system as shown in figure 1.

Considering functional specification, the RUS contains a description of the user-user relationships of the system and from these the functions can be defined within each user which would provide these relationships. For example a particular retrieval relationship would require functions in the source for interpreting and responding to requests for information and thus a first level functional specification of a database can be made which would be consistent with the task network in which it resides.

Considering specification of the architecture of a system, the RUS contains a task topology with which the architecture topology should be consistent. The architecture topology can be based on the RUS topology but with the addition of components invisible to the users, such as service provision systems.

The RACE project R2078 TIM (Tourism Information Marketing) has used the eight-step method to describe the requirements for a European wide system which is being implemented to enable holiday providers (hotels, tourist boards) to market their products directly at point of sale sites in origin countries. The project has fully documented each design stage. It has found that the eight-step method interfaces naturally to the subsequent use of object oriented functional specification methods such as that described in Coad and Yourdon (1991), and to architectural specification.

However before functional specification of the human user interfaces can be made additional user requirements analysis is needed. This further work for the human communicating entities is described under the next 4 steps.

4. Design from an internal perspective

The four steps for the internal perspective are similar to those of the external in that the first two concern a detailed description of the application, in this case that part of the application given to the user in question, and the second two concern the design of the system which would support the application.

Step 5: Usage Context—the Internal Perspective

The usage context for the chosen human communicating entity (user) consists of the goals and application tasks allocated in step 2, the characteristics of the user, the service environment from steps 3 & 4, and the performance parameters of quality and cost. Examples of the parameters are given in table 6. It should be clear that the performance parameters are different from those of the external perspective. Of particular importance is the cost category which sets the constraint for the accumulated costs of two different kinds of task, Goal Tasks, and Enabling Tasks. Goal Tasks will be discussed now, Enabling Tasks in Step 8.

Users: how many? simultaneous use? co-operative use?
gender? age range? demographic factors? special abilities? special needs?
technical experience? technical background? attitude to use of system?

Work goal: what is the overall reason for the use of the system?

Work domain: what objects exist to be used, created or modified in reaching the work goal?

Performance parameters: what is the required task quality?
what are the acceptable human and machine resource costs (i.e. time, number of errors, effort, cognitive demands,...)

System: what are the technological constraints upon the design?

Organisation: what sort of organisation will the system be used in (e.g., size, accountability, management styles, permanence, investment potential, etc)

Service Environment: Within what service environment will the system operate? These service environment assumptions are provided at the application level analysis and should include a Reference for User Services diagram.

Table 6: Internal Perspective Usage Context

Step 6: Goal Task Breakdown.

The user has two types of task to perform. There are those tasks by which the users intend to move closer to their goal (Goal Tasks), and there are tasks which create or modify the state of the system which is used to enable the user to execute their goal tasks. In IBC these Enabling Tasks include accessing or modifying services by interacting with service providers. The goal tasks for the user are those that have been allocated to him as a consequence of the design of the RUS in Step 2.

The objective of Step 6 is to analyse the goal tasks in more detail. It is important to note that this goal task analysis is taken from the perspective of the particular user. It is an analysis of their tasks not of the application in general (as in steps 1 to 4). In addition this task analysis is declarative in that it provides a more detailed definition of the goal tasks. It does not provide a complete analysis of the sequence of interactions in which the user will be engaged. Preconditions for goal tasks and the tasks which create preconditions are not examined in this step. For example acquiring information about or being informed how to do a task is not part of the definition of that task (this contrasts with the approach described in chapters 2.2 and 2.3, which has been derived in order to define system properties). A simple example of a task definition is given in Table 7.

The task analysis requires taking design decisions, since there is more than one way to define most tasks. Clearly therefore, like the RUS Step in step 2, this step is a critical reference point for later design stages and therefore for iteration. If later steps in design provide unsatisfactory solutions, then the results of these two 'reference points' need re-examining.

Description: a brief textual description of the goal task

Sources

a list of the goals that refer to this subgoal

Task parameters:

Numerousness:

users: an estimate of the number of human-end users who might be involved in each instance of this subgoal.

other entities: the number of other entities (eg databases) that may be involved in each instance of this subgoal

Temporality:

frequency: how often will this subgoal be performed

duration: how long will the performance of this subgoal take, with respect to the total use of the application

Intertask dependencies:

necessary pre-tasks—list any other subgoals that must have been performed for this subgoal to make sense

task interleaving—list any other subgoal tasks that are likely to be performed concurrently with this subgoal

Design aspects:

Functional requirements:

A list of what this subgoal should achieve

Interaction requirements:

A list of constraints upon the way the interaction should be performed

References: a list of the subgoals that this subgoal makes reference to

Table 7: the description of a subgoal in a goal task breakdown

Introduction to Steps 7 and 8.

Unlike other approaches to task analysis, the task analysis in the eight-step method is now halted and begins again only after another step in system design. Step 7 consists of a description, for each component of each goal task, of the state of the system which is needed in order for that goal task to be successfully executed. Step 8 then contains an analysis of those tasks, Enabling Tasks, which can create these 'Enabling States' of the system. Thus there is a logical break in the task analysis, with Enabling State design intervening.

Implicit in steps 7 and 8 of the method is a two component definition of the system property *usability* which corresponds to the two kinds of performance parameter, quality and cost, which the system must assist the users in attaining.

Step 7 is concerned with the parameter quality, and the first component of the meaning of usability is as follows:

> *A Usable system is one which contains all the enabling states for the envisaged users and tasks.*

The break in task analysis is also reflected in the consideration given to cost, the other performance parameter stated in Step 5. The costs associated with goal tasks and enabling task need to be separately analysed. Goal tasks, in this approach, have the user as the agent. The costs associated with them to the user therefore depend directly on the structure of the work domain. Enabling tasks, in contrast, are not necessarily allocated to the user, indeed in many cases they can be allocated to some other entity. In general, if the cost performance parameter sets the total acceptable costs of the application, maximum resources can be devoted to the goal tasks if the costs associated with the establishment of enabling states are kept to a minimum. This provides the second component of the meaning of the system property 'usability':

> *A usable system is one where the enabling states can be achieved with minimum cost to the user*

In order for this definition of usability to be useful it is necessary to describe in more detail what is meant by the term Enabling State.

An enabling state describes the preconditions for a goal task. A goal task is executed by a user on an object. The task may require the support of a service in order to be executed. Therefore an enabling state description must consist of a three tuple as follows :

> Enabling State=(User, Service Relationship, Objects)
> *where:*
> User = required characteristics of the user,
> Service Relationship = the interface between the user and the object
> Object= required characteristics of the object.

The inclusion of the User in the enabling states description allows the actual and required states of the user to be compared. Discrepancies between the two will allow the designer to propose enabling tasks in Step 8 such as the provision of user manuals, tutorial services, etc, which will reduce these discrepancies.

The Service Relationship is a pre-implementation user requirements concept whose purpose is to describe the bridging of the gap between the user and the object. The term is preferred to 'user interface' since this would create confusion with implementation concepts, for example, pc user interfaces. The described Service Relationship could eventually be implemented on a combination of, terminals, public and private networks, and value added service providers.

The Service Relationship provides a view of the object, for the planning of actions by the user, and it transmits to the object the effect of the actions of the user. In both cases the Service Relationship can act either as a channel between the user and the object or it can impose a model of the user or the object between the user and the object. For example, a task of conversation between two humans who are geographically separated can be supported by a video telephone. In this case the object of the task 'converse' is another human, and the service relationship required is one which provides a view of him/her. Speech and visual signals are transmitted without being subject to analysis, the service relationship providing a channel between the two. In contrast, a user remotely controlling a complex physical process may require the state of that process to be modelled, and represented to the them, and the user will plan their actions on the basis of the model. This issue is important because the different cases involve different kinds of design issue, e.g. the veridicality of transmission or in contrast the adequacy of the representational techniques. The distinction between represented and non represented objects, (and interpreted and non interpreted user actions), is dependant on the definition of the goal task and needs to take into account the perceivability of the object attribute.

For each goal task a relevant object attribute is defined. That object attribute may be perceivable by the user. If so the appropriate peripheral device, e.g. a camera, can respond to the attribute, and the attribute will then be coded in a way appropriate to the transmission medium, transmitted, and eventually decoded and presented. Design issues in this case concern the capacity of the service to preserve the information in the object attribute, for example during image compression. In contrast the object attribute may not be perceivable and the attribute values must be transformed, and the transformed information presented. The design issues in this case concern the preservation of the mapping between the object attribute and its representation. This mapping implies that the service relationship between the object attribute and the user contains a model of the object attribute and that it is to this model that the user responds, as illustrated in figure 4.

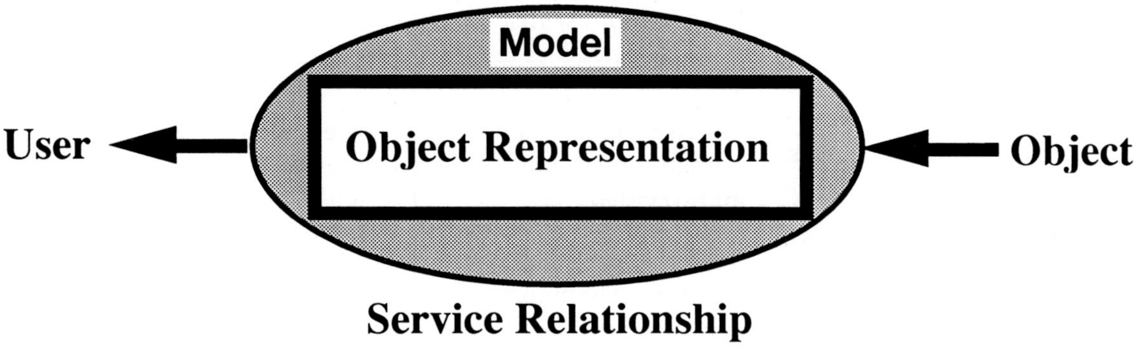

Figure 4: Object representation

It is important to bear in mind that the definition of the goal task determines what is the appropriate object attribute, and that the concept of a service relationship is a pre-implementation one. The example of editing text should make this clearer. A text editing goal task has the relevant object attribute, say, letters. This attribute is perceivable and does not require modelling and representation. The design issues therefore concern maintenance of the attribute information, e.g. through size and layout, given the remote or non-remote location of the object (which is described in Step 2). Some objects such as text will also require implementation decisions which will in turn affect the implementation of the service relationship. However the appropriate object attribute is still, for example letters, whether the object is implemented as a file in a computer, or on a piece of paper.

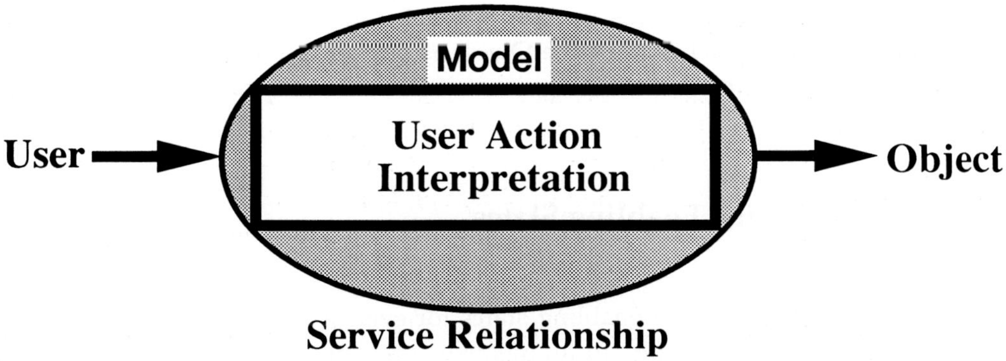

Figure 5: Action Interpretation

In a symmetrical way as indicated in figure 5 an analysis can be made of the users action as to whether the object is affected by the action or by a model of it. The distinctions for representation and interpretation can be allocated to a 2x2 table as shown in table 8, which can be used to broadly categorise design issues. Chapter 3.2 contains an analysis of the use of this table for design principles.

		Goal Task Object	
		unrepresented	*represented*
User's Goal	*uninterpreted*	1. e.g. real time human to human conversation using audio/video	2. voice activated database retrieval
Task Action	*interpreted*	3. e.g. human to human conversation using text to speech synthesis for a dumb person	4. e.g. automatic translation of voice mail with an icon providing confirmation of receiver

Table 8: Examples of the four basic service relationships

The goal task analysis of Step 6 includes a description of the flow of work. With task sequences and task interleaving, more than one service relationship may be extant at any one time. Thus the method requires the designer to consider and revise the enabling states descriptions for the occasions when more than one service relationship can be present.

The Enabling States approach includes the Object of goal tasks as part of the system to be designed since unless the object is in a state appropriate to a goal task that task cannot be executed. For example, a golf ball must be 'teed up', or a human user needs to be alerted to a telephone call. Inclusion of the object of the goal task in the enabling state description allows the designer to define enabling tasks in Step 8 which can create the appropriate state e.g. making the telephone ring. Inclusion of the Object as part of the design is another feature of the Enabling States approach which distinguishes it from other approaches.

Step 7: Description of the Enabling States

This consists of descriptions of the enabling states for every sub goal task and for every case where service relationships can be simultaneously present. Example parameters for each element of the three tuple are given in table 9.

Sources: A list of the goal tasks that require this enabling state

User Description

A list of the required attributes of the user, eg knowledge, capabilities.

User Principles:

A record of any principles that are relevant to this enabling state

Service Relationship:
Are the user's goal task actions interpreted or uninterpreted?
Is the object represented or unrepresented?
Description of the User Task Interpretation :
How shall this be done?
constraints and requirements on the performance of the model
User Task Interpretation Principles: a record of any principles that were used
Description of the Object Representation:
How shall this be done?
constraints and requirements on the performance of the model
Object Representation Principles: a record of any principles that were used
Service Attributes:
(If the service relationship is both interpreted and represented, this section can be omitted).
Involved Entities: which entities are linked by this relationship? eg user, workstation, recipient
Communication Type: The type of communication involved in this relationship (eg Human-Human Conversation; Machine-Machine Distribution)
Channel characteristics:
Information Content: A description of content of the information that is to be communicated by this relationship, (eg head-and-shoulders view with sound)
Service Components: What service components (eg audio, video) are involved in this relationship, and whether they are mandatory or optional.
Synchronisation requirements: a description of any synchronisation required between service components and the temporal tolerations.
Format Preservation: What degree of loss is tolerable in the information communicated by this service relationship (ie what are the constraints upon compression)
Quality of Audio Service Component: Constraints upon the quality of any audio components (eg telephone quality, CD/HiFi quality)
Quality of Video Service Component: Constraints upon the quality of any audio components (eg frame rate, colour depth, etc)
Quality of Service: What attributes of the service relationship are most important in ensuring the acceptability of the service relationship to the user?
Service Attribute Principles: A record of any principles used in completing any of the Service Attribute information
Object Description
A list of the required attributes of the object, including any knowledge it must have and any actions that can be performed upon it (ie changes that can be made to it).
Object Principles:
A record of any principles that are relevant to this enabling state
References: Enabling Tasks required by this enabling state

Table 9: An enabling state description

Step 8: Description of the Enabling Tasks.

This final step consists in describing the enabling tasks for each element of the three tuple for every enabling state description. Unlike the goal tasks for which the user is the agent, the agent of the enabling tasks is not given and therefore major function allocation decisions need to be made in this step. The second criteria of usability requires that enabling states are achieved at minimum cost to the user This implies that where possible the agent of the enabling task should be an entity other than the user and that where the user has to be the agent the tasks are designed to be simple, and easily learnt

Since the enabling state description and enabling tasks apply to the user, service relationship and object, the Enabling States approach should lead to the development of user interfaces which are goal relevant and complete.

The eight-step design method provides a way of structuring and sequencing design decisions. But a designer may require support in the form of relevant principles and criteria for each design decision. These principles can be developed from system properties, in particular the system property 'usability'.

5. System properties, principles, and criteria

The objective of design is to produce a system with the properties with which the performance parameters of the application can be met. The performance parameters of an application are described in terms of the desired *quality* of the product of the application, and the acceptable *cost* of production of the application products. In order for design to be successful therefore the properties of the system need to be directly relatable to these two performance parameters such that attaining some desired system properties during design ensures that the performance parameters of the application can be met with the system in operation.

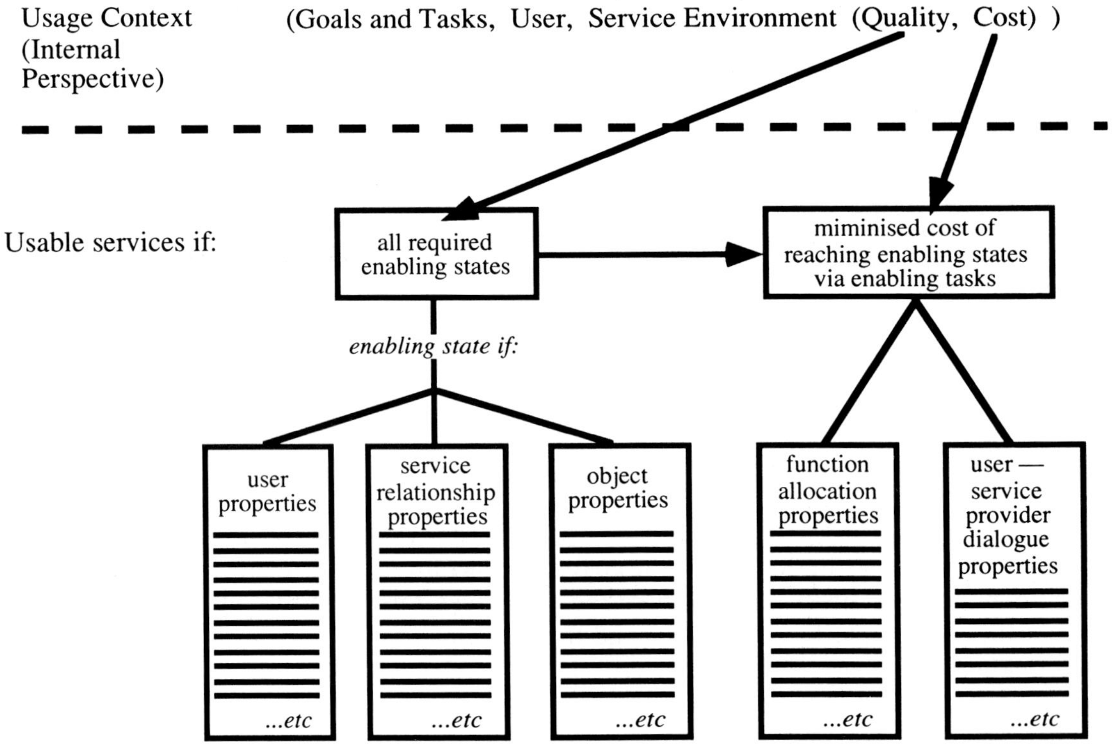

Figure 6: relationships between performance parameters and usability properties

In the last four steps of the method the system property 'usability' was introduced. This was defined in terms of, the range of enabling states that were available i.e. the goal tasks which the system could support, and the cost of reaching the enabling states. Thus the first level breakdown of the meaning of the system property 'usability ', within the Enabling States approach, can be mapped directly to the two performance parameters of the application, where the first component of the meaning maps to the quality performance parameter, and the second component of the meaning to the cost parameter. This mapping can be maintained as the meaning of 'usability' is further broken down. An enabling state consists of a description of the user, service relationship and object. An enabling state can be defined as a state which possesses a set of system properties relevant to each of these three elements. For example if the service relationship contains a representation of the object then that representation can have the property 'compatibility`. Thus usability can be defined as a hierarchy of properties which are derived from the two properties at the first level breakdown which in turn map to the two application performance parameters. This is illustrated in figure 6.

The two components of the first level breakdown of the meaning of 'usability' express relationships between the user and their tasks and the attributes of the system (which includes the user). Enabling states are defined with respect to users and their tasks, and costs are minimised for a particular user. This applies also to the lower level properties.

The system properties are, at any level, expressible as relationships between usage context and attributes of the system. This is illustrated in figure 7. Since the properties can be broken down until they are specific to particular design issues and since they are expressible as relationships between usage context and system attributes, they can be used as principles for design and as criteria for evaluating designs.

IF	<usage context>
THEN	<system attributes>
BECAUSE	<cognitive psychology rationale>

Figure 7: structure of a system property

Within the field of usability engineering good system properties are becoming well known and defined. Chapter 3.2 of this book reports on system properties relevant to service design and this builds on work reported in Byerley, Denley et al (1993). Interacting Cognitive Subsystems theory, described in chapter 2.1, is a major advance in predicting system behaviour from the attributes of users and their tasks, and system attributes via a model of cognitive processing. This work can provide the rationales for the definitions of the system properties. In particular this approach should enable in future the generation of specific instances of properties from generic ones, which will provide specific advice to designers for particular users and tasks.

The lack of a theory enabling the routine generation of specific instances of system properties for design is even more the case for the first four steps of the method.

The URM project has consolidated a large number of specific instances of usage context x design decision for service design. Assuming that these associations are partly based on the desire to meet the needs of users then these associations can form the first empirical basis for a set of 'external perspective' system properties. At the moment they are categorised only by type of communication. In order to become the basis for design principles they need to be categorised in terms of system properties. Examples might be, flexibility, manageability, enhanceability etc. Then it would be possible to build a structure of system properties, and show how these can be mapped to the external perspective performance parameters as is being done for the internal perspective.

6. Design methods and user service provider negotiations.

The eight steps of the design method with the associated design guidance from the principles could be followed from start to finish by one designer or design team who wish to express the external and internal operational requirements for a system. This system could then be functionally specified and implemented. However the model of design which is more likely to be relevant to advanced communication applications is where a sequence of negotiations and design decisions take place between a number of users and service providers, the end result of which, at some point in time, is a service or set of services appropriate to the application from both an external and internal perspective. These negotiations can take place prior to or during a call.

At any stage in the negotiation the decisions will include both the selection of service attributes and the selection of the design decisions to leave to the next phase of negotiation. For example a hospital communication network manager may decide, in negotiation with public network operators and value added service providers, to install a set of services with the capacity for the transmission of high quality video images from a medical database, but will leave the decision on bandwidth allocated during a call to the consultants who will be using it.

Negotiation can be first party, e.g. negotiation by a user before or during a call for an alteration of his/her service, or third party, e.g. negotiation before or during a call for a change in services by someone who will not be using the services. The hospital communications manager is one example of a third party negotiator, a video conference manager is another. In both cases the eight-step requirements method can be used to structure the service negotiation dialogues. For example a third party negotiator can extract requirements from the application users in order to design a network, and, applying the enabling states approach, decide which design tasks (enabling tasks) to give to each of the end users.

The concept of third party negotiation implies possible chains of service creation, where the process driving the chain is a 'configuration' of the service options left open by the service provider next lower in the chain. More precisely, a service provider provides service options to a user. These service options must be created in order to be provided. Therefore for a particular user and service provider, service creation is the entire process whereby the service provider has service options to offer to the user. These service options then form the basis for the negotiation process between that user and service provider. The user may then use those configured services (first party configuration) or may then offer them to another address (third party configuration). In the latter case the user is acting as a service provider and the configured services become the next level of created services which are offered to a user at the new address. The service creation chain is primarily generated through the process of service configuration though of course it must have as its starting a set of service building blocks and configuration rules.

The application of the eight-step method to the design of first party configuration for an internal perspective is illustrated in figure 8, where the negotiation dialogue is broken down into 3 main phases as recommended by the R2089 ASCOT project. The ASCOT project is implementing a toolset which will mediate between the user and a range of service providers such that service configuration causes minimum cost to the user. Essentially therefore the toolset will have the responsibility of designing and undertaking many of the enabling tasks required for the service relationship. As shown in the figure, the requirements phase of the dialogue results in the design of the service configuration enabling tasks, of which pre-call tasks are executed first in the binding phase followed by call set-up and in-call tasks in the dynamic phase.

The dialogue within the requirements phase is undertaken in 3 subphases, the first of which is termed *Defining the User's Goal Tasks.* In this subphase the objective of the toolset is to obtain from the user sufficient information so that it can define results for the first 6 steps of the eight-step method. This subphase may itself be viewed as being composed of two phases, Network Context, and User Goals. In the Network Context phase the toolset obtains information from the user on ; the application of which the user's tasks are a part (step 1), the distribution of tasks amongst the other users which define the particular user's cooperative responsibilities (step 2), application level features of the services such as overall dialogue management procedures which will constrain the particular user's freedom to configure the services (step 3), and task level features (step 4). In the User Goals phase the toolset extracts information about the particular user, their overall goals, their special needs, and desired performance parameters, (step 5) and more detailed goal task information (step 6). The major part of the dialogue with the toolset is therefore in the Defining the User's Goals phase.

The second subphase is termed *Defining the Enabling States.* From the user's input in the previous phase the toolset will define the required service relationships for each of the user's goal tasks. and it will then search through its service and network databases for available services and appropriate service providers. It will collate the predefined service configuration procedures offered by the service providers which can be used as enabling tasks for pre-call, and in-call configuration. This subphase is equivalent to step 7 of the eight-step method. It will probably require little or no information from the user.

The final subphase of the requirements phase is termed *Defining the Enabling Tasks.* During this phase the toolset examines the collated service configuration procedures that have been predefined by the service providers. The toolset will decide whether to pass on the procedure, as defined, to the user for them to execute, or whether to modify the procedure into a simpler or more consistent form, this would include combining elements of procedures from more than one service provider, or whether to perform the procedure entirely itself. Once these calculations have been made the toolset will inform the user of the service configuration actions that they will have to perform to be able to undertake their goal tasks. Thus the requirements phase ends by defining the dialogue that the user will operate in the binding and dynamic phases.

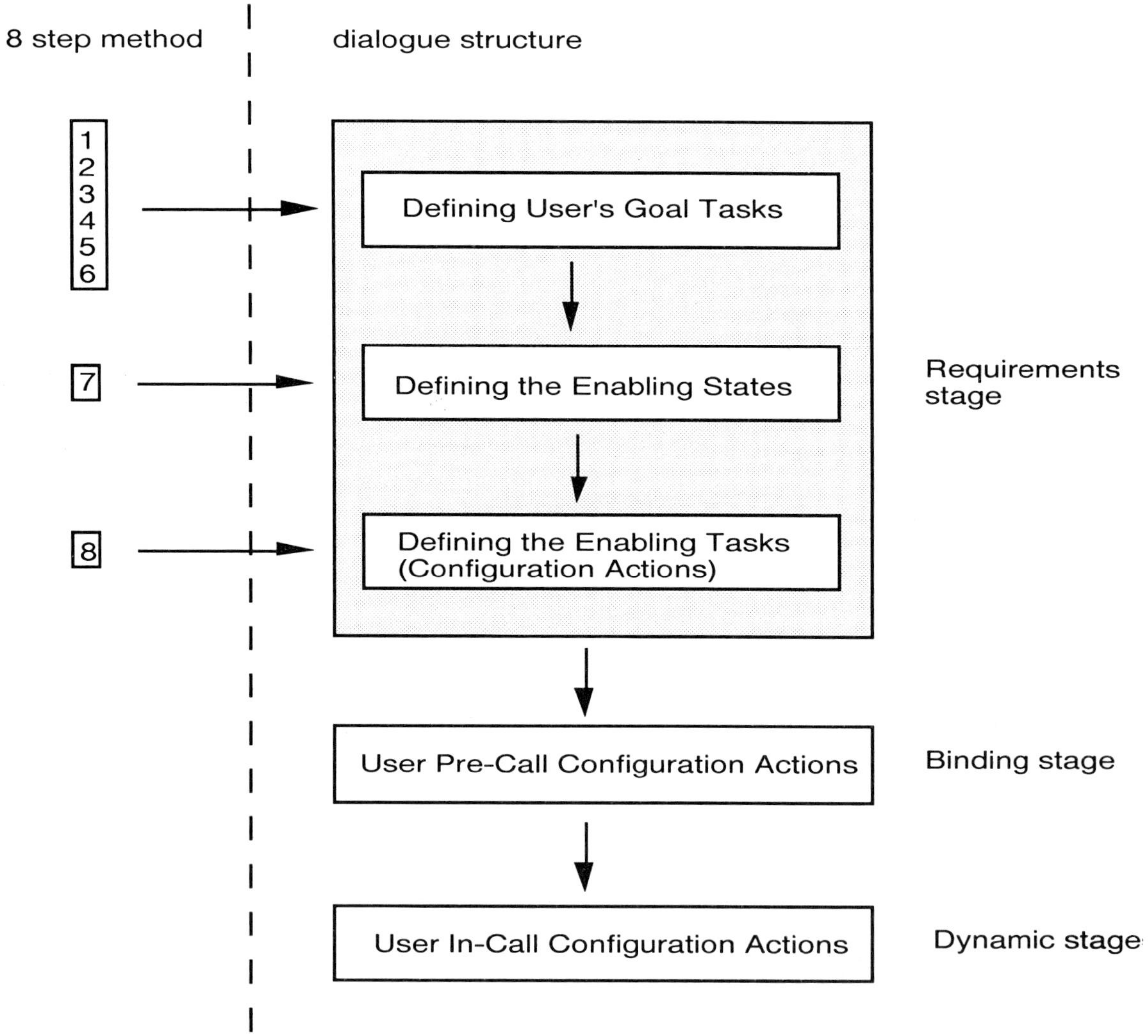

Figure 8 Mapping the eight-step method to First Party Service configuration dialogue

7. Conclusions

The objective of this chapter is to provide a unifying structure with which user requirements for services can be defined and then smoothly input to later stages of design such as functional specification and architectural design.

The method derived for the last four steps of the eight-step method is described in detail in chapter 3.1

Chapter 3.2 presents some of the latest work, from a number of projects, on the definition of system properties.

References

Byerley, P. and Bruins, R. (1992) *Conceptual Framework for Service Definition* in P.F. Byerley and S. Connell (eds) *Integrated Broadband Communication : Views from Race : Usage Aspects*. North Holland Studies in Telecommunication. Volume 18.

Byerley P., Denley I., Melchior E-M., Voigt U-B., Floruß S., and Hermann S., (1993). A unified Method for the design of Telecommunication Services. *Ergonomics. Special Issue on Telecommunication.*

Coad P., and Yourdon E., (1991) *Object Oriented Analysis*. 2nd Edition. Yourdon Press. NJ.

The Usage Reference Model Project Publications for Race cover in addition to the topics in this book:

application pilot design
a guide to Integrated Broadband Communication terminology
and usage implications of the Race Common Functional Specifications.

They can be obtained from : IFC Research, 99 High Street, Guildford, Surrey, GU1 3DP, U.K.

The Race Common Functional Publications documents cover all aspects of telecommunication system design They can be obtained from: Race Central Office, European Commission, DG XIII/B, 200 Rue de la Loi, BU-9, 1040, Brussels, Belgium.

Acknowledgements

The work reported in this chapter was carried out as part of the Usage Reference Model (R1077), Guidance (R1067), Ascot (R2089), and TIM (R2078) projects supported by the Race Programme of the European Commission.

Computers, Communication and Usability:
Design Issues, Research and Methods for Integrated Services
P.F. Byerley, P.J. Barnard and J. May (Editors)

Chapter 1.2

The potential and the problems of IBC

Jon May*, Ian Denley†, Becky Hill† and Michael Jeffries‡

*MRC—APU, 15 Chaucer Road, Cambridge, UK—CB2 2EF

†Ergonomics Unit, University College London, 26 Bedford Way, London WC1H 0AP

‡Octacon Ltd, 9 Petch Street, Stockton-on-Tees, Cleveland, UK—TS18 1PD

1. Introduction

A little over a century ago the invention of the telegraph and telephone brought about a revolution in human communication. For the first time it became possible for people to communicate almost instantaneously over long distances, and the effects upon everyday life, upon trade and commerce, and even upon systems of government and social organisation followed quickly. Since then the technological infrastructure supporting telecommunication has continued to evolve, adding transcontinental networks, direct dialling and satellite links to spread the coverage and enlarge the capacity of the basic service, but until about ten years ago the way that the average person used the service had changed very little. The vast majority of traffic over the growing networks was conducted by a plain old telephone that might have had its casing redesigned, and a dial replaced by a keypad, but which was essentially the same as it had been for decades.

The situation is now changing rapidly. A proliferation of new services such as fax, teletext and computer communication use the existing networks, but require additional terminal equipment in the user's home or office. At the same time devices resembling the telephone have become mobile, requiring additional network development. These new services have been aggressively and successfully marketed, to the extent that high-street shops are selling domestic users comparatively cheap communication devices that were expensive business executives' luxuries only a few years before. One factor constrains all these developments, however: the lack of bandwidth.

Over twenty years ago, long before the current burst of change began, a character in Stanley Kubrik's film *2001* made a call to his daughter from an orbital space-station using a Bell videophone pay-booth. Kubrik did not include instantaneous facsimile transmission or personal computers in his vision of the future. Instead he showed an enhancement to the existing communication service that everyone could immediately understand, could see the benefits of and would like to use. Despite its appeal, videotelephony is not yet widely available, while the less obvious applications are. There have been many prototypes developed since then, but until now all have come up against the unavoidable problem of fitting all of the information required to construct a real-time video picture into the 64 kilobits per second capacity of existing phone lines. Ingenious algorithms, faster chips, and smaller videoscreens have brought partial solutions, but for a real videophone service to be practical more bandwidth is needed than is currently available.

In the next few years this bandwidth will become available. Where the standard phonelines can carry 64 kilobits per second, the newer ISDN networks provide 2 megabits per second channels, and B-ISDN 140 megabits per second. These networks are being used as the testbed for the development of new forms of communication service. At the moment the record communication rate for a single channel is somewhere between 20 and 32 gigabits per second, using an experimental network based on 'solitons', pulses of laser light a few millionths of a second long that can be carried by an optical fibre without chromatic dispersion or non-linearity due to variations in light intensity. This bandwidth is around half a million times larger than existing phonelines, a difference so large as to practically remove the upper limit on the amount of information a channel can carry. All the changes in personal communication that have occurred over the last ten years or so will become insignificant by the end of the millennium. We are unlikely to see orbital space-stations open to the public by 2001, and Bell no longer exists as a single institution, but we will be able to use public videophones.

The change from an electro-mechanically switched communications network, carried over a physical network of metal wires, to one in which the medium is light and the switching digital, will not happen overnight though, and it will not be cheap. It has been estimated that more will be spent on the global network infrastructure in the 1990s than in the rest of the century put together. Investment on this scale clearly requires careful planning, and the involvement of national and supranational government bodies. It is inevitable that in the time taken for a given level of technology to be developed, its commercial viability proven, and the appropriate investment decisions made, a newer, cheaper, better and completely different technology will be on the horizon. It is therefore vitally important that any services are designed to be flexible in terms of the bandwidth they can handle, so that they do not become defunct before they are completed.

For these reasons the development of the services that will be provided over the new networks is progressing independently from the research into the technology that will provide the infrastructure. Services are being planned in a modular way so that they can be combined with whatever other services are available, and configured to suit any available network. The collected name for these services is Integrated Broadband Communication, or IBC. The Commission of the European Communities has established the RACE programmes to carry out 'Research into Advanced Communication in Europe', with a budget of around 2.6

billion ecu, or £2 billion, 60% of which is provided by industry and 40% by the Commission. Phase 1 of RACE began in June 1987, with the objective of preparing for the 'introduction of IBC taking into account the evolving ISDN and national introduction strategies, progressing to Community-wide services by 1995'.

The first phase of RACE concentrated on the conceptual and technological basis of IBC, and with the development of some testbed applications that have provided the groundwork for the second phase, which began in January 1992. In this second phase the emphasis is much more on the introduction of IBC and the development of 'product lines', with the aim being the establishment of a European IBC network by the interconnection of 'island' testbeds and demonstration systems before the end of 1995. In the first part of this chapter we will present an overview of the application pilot work from the first phase of RACE to explore the potential that the new technologies hold, and examine the problems that are raised for the designers and providers of new services. A review of the developing methods and approaches to the solution of these problems will lead into a set of recommendations for the support that must be given to designers, service providers and end-users.

2. Some potential applications of IBC

The application pilots in the first phase of RACE dealt with a number of different sectors of economic life and study and were intended to demonstrate the potential of and the user reaction to IBC. The breadth of the RACE programme can be seen from the range of sectors covered: pilots were run in the fields of health, manufacturing, publishing, entertainment and culture, transport, social service provision, finance and insurance. Without the data collected by these pilots on the expected level of use of the various services and of users' preferences for particular combinations or variations of services it would be impossible to plan for the full scale introduction of IBC, and the Usage Reference Model project (URM) was charged with collating and analysing the results of the pilots. Across the sectors listed above, the pilots provided a number of different services, and the URM identified the following applications:

- distributed case handling
- interpersonal communication
- remote delivery of expertise
- distributed collaborative decision making
- distributed learning and training
- monitoring and surveillance
- telemarketing
- entertainment and leisure distribution
- multimedia interpersonal messaging
- multimedia information assembly and access
- distributed collaborative design

To give an idea of the potential services that will be offered by IBC, we will give a brief review of the RACE 1 pilots under these application headings (for brevity, only the names of the projects will be used here —their RACE numbers are listed at the end of the chapter).

2.1. Distributed case handling

Distributed Case Handling relates to the handling, management and collaborative processing of a series of multimedia documents or separate documents in a variety of media, interrelated by a common theme, whose origin and location may be distributed across one or more organisations and locations. In an office environment, for example, documents circulate for processing, and several people perform a variety of tasks, approving, correcting and editing documents. Tasks have to be performed within certain time limits, and frequently according to a particular, pre-defined sequence or process. It is essential to be able to follow the document in order to locate it and to determine which preparation steps have actually been performed.

IBC will allow the locations of the people involved in case handling and the nature of the operations to be controlled by factors other than the need to be close to the information—the need to physically move paper or other media will be replaced by the ability to access records stored centrally or distributed over a number of sites. Indeed, a single 'case' may consist of separate records stored at different sites. The information that can be managed will also change, to include still images, audio and video. Multiple editing will become possible, with workers at different sites working on the same case documents simultaneously.

This is the main application of the ECHO pilot. The services involved in this application are typical of those required by many business applications, across many sectors. The specific cases handled in ECHO are those dealing with new life insurance business. Insurance companies operate in a highly competitive market place where customer relations are of primary importance. Managers are aware that existing methods are far from optimal; cases are often misfiled and information is not always recorded in a consistent manner. Furthermore, case handling in its paper based form is slow and involves large volumes of material.

In order to make case handling more efficient and therefore improve the quality of customer service a computerised version of case handling is being implemented. This is achieved by:

- allowing fast and easy access to any case folder by any case worker in any location.
- reducing the volume of paper required and also the need for photocopying.
- controlling workflow by activity planning.

Remote handling of cases can be performed throughout the life of the case. Most of the information will be scanned documents. During the initial stages of case preparation a quote must be produced and printed out to be sent to the client. After the case has been set up access may be required at any time for a variety of reasons: case folders are continually updated as new information comes in, and information may need printing out at any stage. It may also be necessary to find out how a case is progressing.

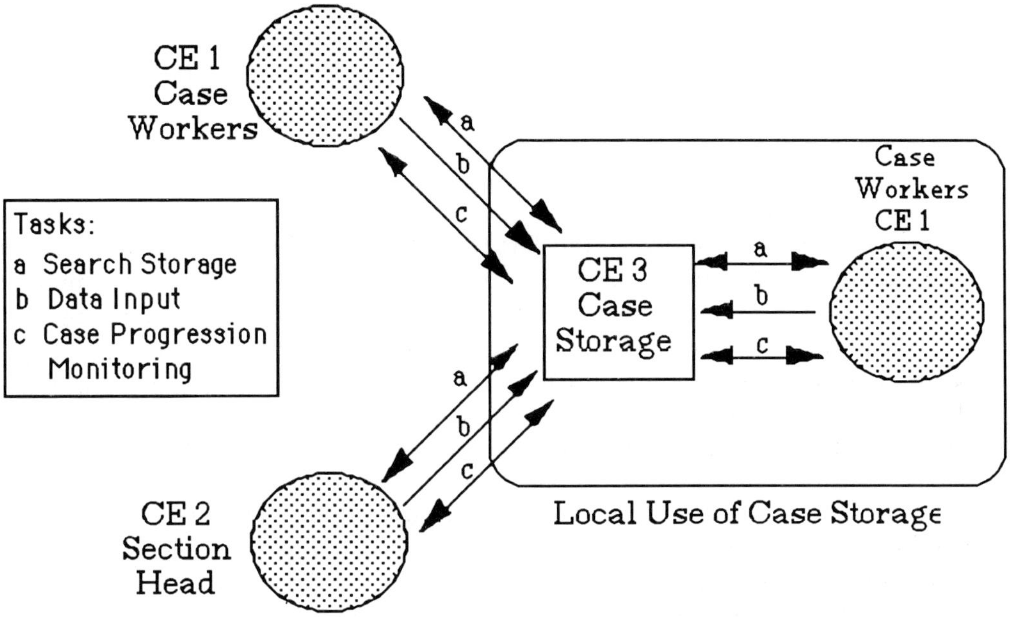

Figure 1. Tasks studied by the Distributed Case Handling pilot ECHO

A key benefit that the system provides is the guidance of case handlers through the correct procedures in order to attain thoroughness and organisational consistency. This facility eliminates the major problems of losing files, storage and inconsistent processing. In order to perform these activities the system requires a central store of information accessible to many parties and a means for rapid transmission of this information around the organisation. The services provided by ECHO must be robust, high quality and provide a high level of security. More specifically, speed is critical. The time required to produce a document on screen must not exceed 10 seconds. The printer must be capable of printing an A4 page in 5 seconds and an image page in 5 minutes. Response time after key depression must not exceed 2 seconds. If the response time during mainframe sessions is slower then the user must be made aware that the system is still active. Other requirements include reliability (minimum downtime), easy to use interface and peripherals, access control, and regular back up. These requirements cannot be provided by the narrow bandwidth of existing telephone services, but are well within the capacity of IBC.

2.2. Interpersonal communication

A central feature of IBC is that it will enable people to communicate with each other in a way that is as close as possible to their being in the same room, using communication channels that are a great enhancement of currently available services. High fidelity stereo digital sound, real-time video and many-to-many connections offer a significant degree of realism. In practice, the requirement is met by the provision of a multimedia workplace or domestic environment with a user interface that facilitates easy multimedia communications to another site. The process that the IBC services support can be communication for any number of reasons, from business to purely social contact. The component services required may vary according to the purpose of the communication, and so the user will be able to select the most appropriate form of communication for their task. A simple audio service like an existing telephone system may be sufficient for a simple factual enquiry, while higher quality sound, video and colour printing may be required for more demanding tasks.

One of the applications supported by the DIVIDEND pilot aims to provide interpersonal communication to support effective dealing in international banking's currency market. File transfer, or electronic data interchange (EDI), or document image processing (DIP), is supported in parallel with audiovisual communication for the exchange of a range of documents. These include scanned documents, graphical information, text files, forms and so on.

Senior managers in the banking arena have a coordinating and strategic decision making role. Their objective is to ensure that their departments are working with the same strategy to achieve the organisations business aims. Audiovisual communication takes place between two senior managers, or one senior manager and a customers. DIVIDEND supports and enhances the communication between those parties.

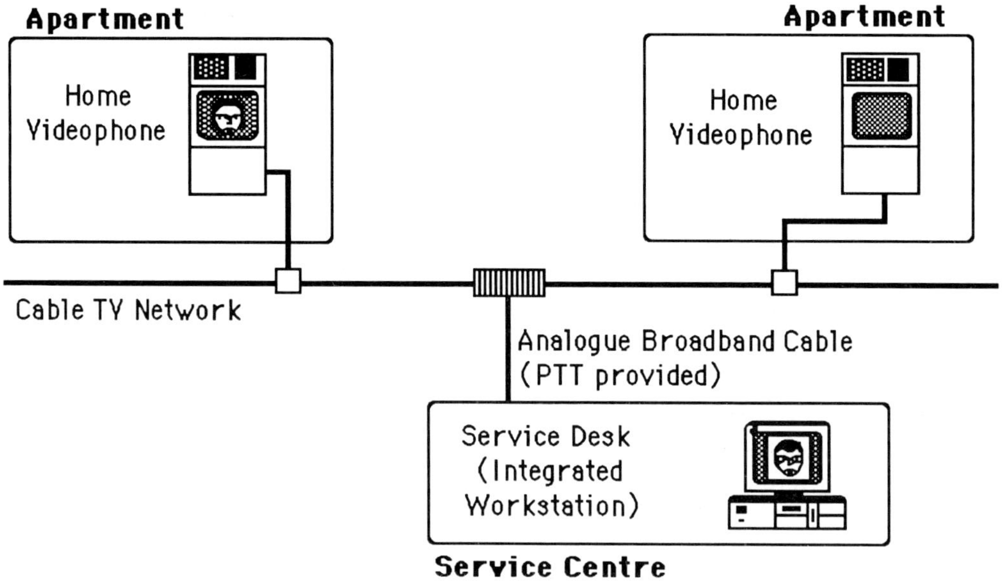

Figure 2: Interpersonal communication is supported by the APPSN pilot

The APPSN pilot covers a number of applications for people with special needs, such as the elderly, including interpersonal communication. In this pilot the application supports a social 'chat' between the elderly person and social service staff such as nurses and carers. In the introductory stages it is being used to encourage usage of the service. A call may be initiated by either party.

At this stage, there are two main circumstances under which such calls occur:

- early morning calls from the centre to encourage use and, perhaps, arrange an evening call-back
- evening calls from elderly persons (perhaps due to loneliness).

The users in this project are elderly people living in their own apartments in a complex in Frankfurt. Four main performance requirements need to be met:

- service to clients must be accurately directed (given to the right person), especially for this remote care application
- the service must be reliable—always available and properly operating during service hours
- adequate communication of the remote scene/person
- complete confidentiality and privacy of the service.

This means that the implemented service should provide:

- switching reliability (no wrongly directed calls/no calls lost)
- technical reliability (complete back-up or fall-back system)
- undistorted good quality pictures (TV standard)
- security (confidentiality and privacy).

2.3. Remote delivery of expertise

Many IBC applications will allow people to consult with human or machine sources of expertise, with several providers of expertise being consulted in a single session. IBC will enable access to expertise to be interactive and efficiently shared by many geographically remote users. This is of particular relevance as demographic changes increase the need for skilled personnel while the proportion of young people in society falls.

The operational requirements for this application include conference session management, two-way multimedia communication and perhaps expert-system based machine support. Examples of its use range from the ability to refer medical problems between specialists without needing to move either the patient or the specialists to the ability of individuals to consult a lawyer specialised in the area of law relevant to their specific situation.

This application is of particular relevance to the medical field, where the number of people with expertise in a particular area may be in single figures. The MULTIMED pilot, for example, allows generalists to send images (X-ray pictures) to radiologists for consultation and advice. With medical information, the main performance requirements are:

- high quality of images
- high reliability (working all the time)
- high security (private network)

This pilot also involves videoconferencing support for interpersonal communication and remote delivery of expertise between the gynaecology and histopathology departments in two London hospitals (Hammersmith and Queen Charlotte). Here the main performance requirements are:

- confidentiality and security of the system
- user friendliness
- acceptable costs.

Slow scan rates (512 x 512) are acceptable, but users want colour rendition and ability to change records quickly.

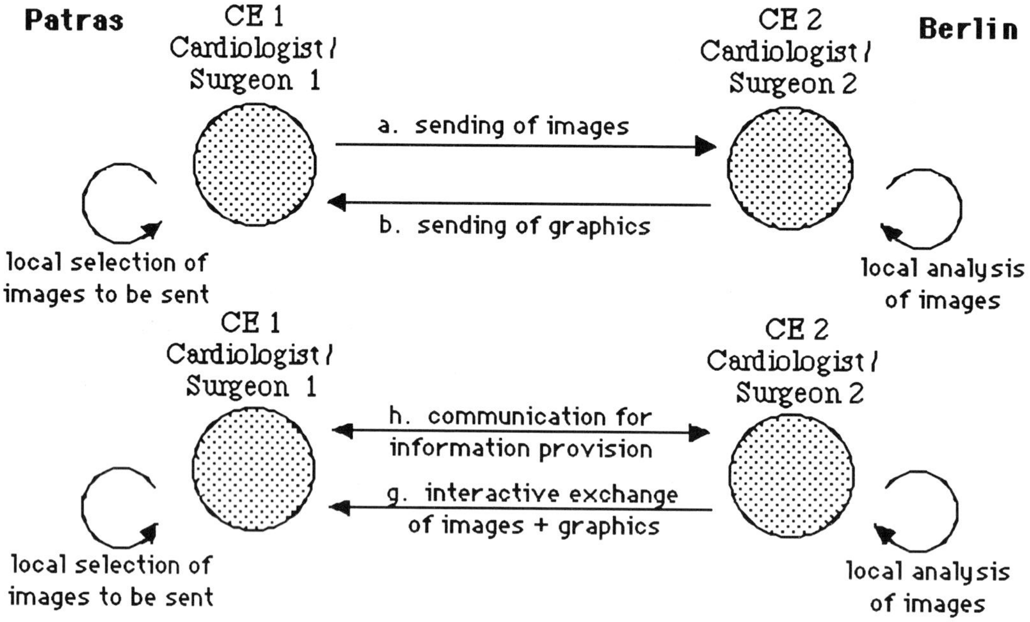

Figure 3: Examples of the IBC support provided in TELEMED

In the TELEMED pilot, medical images are sent from Berlin to Rotterdam where they are analysed, specific hardware and physicians' knowledge being available there; some numerical values are then computed in Rotterdam and sent back in the form of graphics to Berlin. These graphics can then be used by cardiologists in Berlin. High data integrity for transmission of digitised images is also important. The user group in this trial consists of cardiologists and heart surgeons.

Medical images can also be sent from Berlin to Leeds in order to be technically evaluated there. The result is some numerical values which are transmitted back and allow the technicians in Berlin to assess the quality of the imaging devices. This may indicate the need for some maintenance on the imaging devices, in case they are not at a high enough level of quality. Alternatively, the results may be more generally taken into account in the later image analysis made by the doctors that such correction should be made on images originating from these devices.

Finally, medical images are sent from Patras to Berlin where they are analysed, specific hardware and physician's knowledge being there. Some numerical values are then computed in Berlin and sent back to Patras in the form of graphics. These may be sent either in interactive or batch mode.

Other areas in which remote expertise is being supported include the assessment of structural damage and/or repair of an aircraft In RESAM, all the expertise from a central technical service centre can be made available where and when it is needed. A conferencing facility allows the technician to be part of a communication group including an expert from the aircraft manufacturer, a trouble shooter and engineer. Particular performance requirements include:

- high quality video
- high quality video printing
- videoconferencing
- image resolution at level of 35mm film
- colour images are preferred
- camera equipment must be portable, small and lightweight; with operation so easy that no instructions are required; operable at all temperatures; operable whilst wearing gloves; automatic focusing and fast stop; own light source
- special camera equipment for boroscopic inspection and able to work behind structures through relatively small apertures
- remote directional guidance of video camera
- sending and receiving calibrated high quality still pictures (e.g. for crack estimation purposes)
- storage capacity for such images
- access to on-line documentation, with multi-screen display
- have documentation available for active conferencing, with voice and paint facilities
- simple addressing codes etc.
- short waiting times per display

The users are generally high profile engineers on main site in offices. They deal essentially with the troubleshooters after a case has been processed. They update the system and need authoring tools.

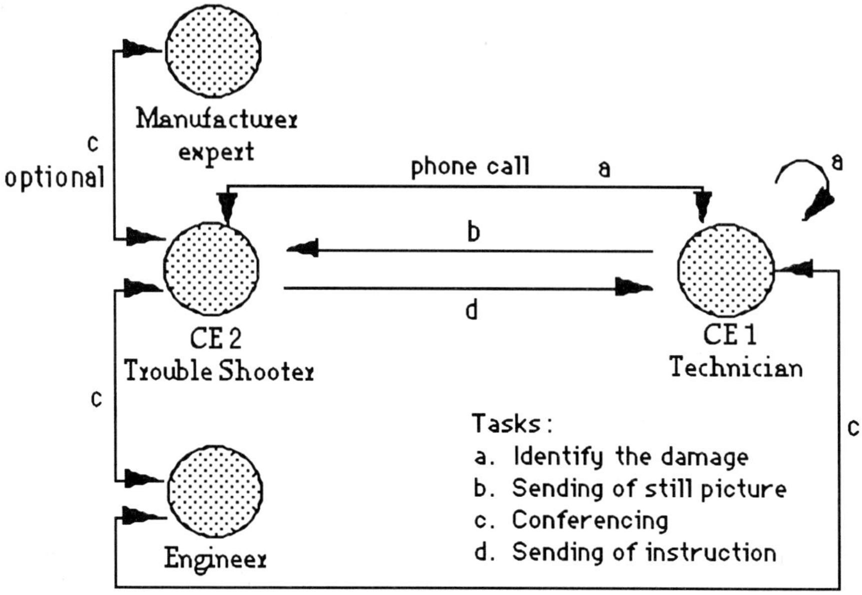

Figure 4: Aircraft maintenance tasks supported by the RESAM pilot

Another pilot providing information in this area is MARIN, which provides shore-based expertise to a ship while it is at sea. In this 2-party application, the shore-based expert is able to view (and listen to) the engine remotely and engage audiovisual conversation with the technician on-ship to advise about repair work.

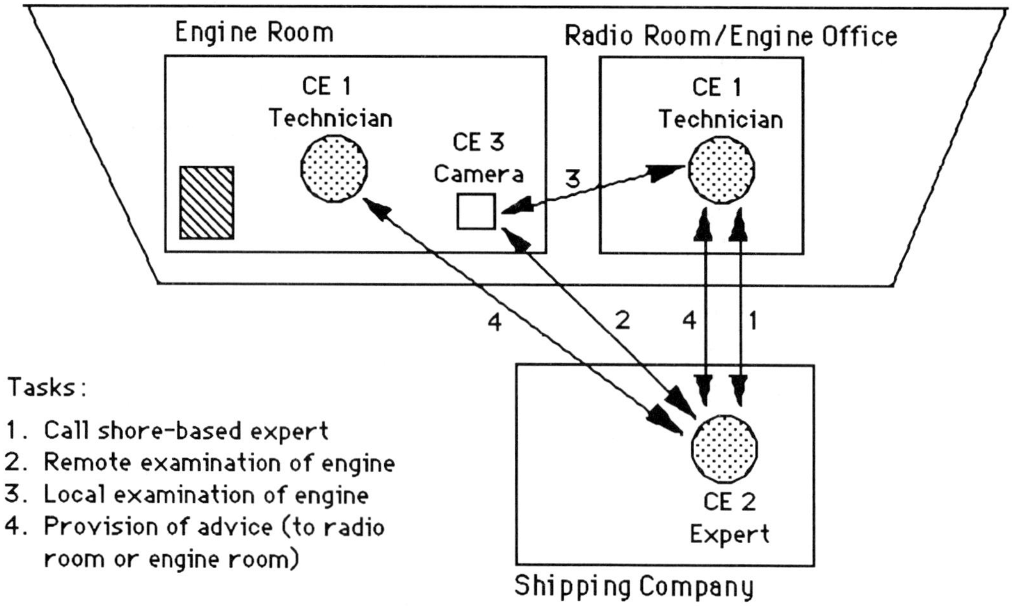

Figure 5: Shore-based expertise can be provided to ships at sea by MARIN

The main performance requirement is that the engine (or other) image or moving image sequence is of sufficient quality for remote diagnostics. If the video quality is not sufficient, back-up still image transmission should be available. For video, 64Kbits/s is at the limits of acceptability, 2Mbits/s is adequate and TV quality would be good. For audio, telephone quality is adequate.

The users are engineering staff employed by the ship owners. They provide the back-up expertise for the chief technician on board the ship. They are located at the shore head-quarters. On board the ship the main user is likely to be the chief technician.

2.4. Distributed collaborative decision making

During a typical decision making process many discussions take place, some between all those involved in the decision making and some that need to be kept private to particular groups or individuals. Complex issues or technical details may need to be clarified, and experts may unexpectedly need to be consulted to solve unanticipated difficulties. Until now the people involved have had to come together, demanding considerable amounts of travel, or the decision making process has had to be made in several stages. Both of these options are time consuming, expensive, and prone to error.

IBC will offer the potential for individuals to confer over long distances, at short notice, and with the opportunity for them to contact additional decision makers as the need arises. To do this it will require audio-visual conference facilities which are able to convey the facial expressions and gestures central to human interactions, and which allow people to present and explain detailed, possibly multimedia, information in an appropriate context.

The two pilots that have provided data in this area are both in the medical area, where the decision making (diagnosis) is linked to the need to consult remote expertise, as described in the previous application.

In MULTIMED, for example, the pilot has to provide the technology to support communications (multimedia information) between the professionals responsible for the treatment of cancer patients and between the geographical locations and the centre in charge of radiotherapy, as an integral part of the patients general treatment. Patient folders are exchanged between the hospital that sends its patients and the service in the Madrid hospital that receives the patients. The folders can be elaborated (local mode), exchanged (mail mode), and oncologists can have a videoconference with radiologists to obtain expertise on radio images (dialogue mode) and to come to a diagnosis. The system thus needs to work in three modes:

- local mode: where user deals with medical record
- mail mode: where the user can send messages
- dialogue mode: where two parties can use the same information.

The main performance requirement is for 'what I see is what you see' in the dialogue mode. Bandwidth must be high enough (2 Mbits/s) to ensure that the delay in information transmission is acceptable. Given the ethical aspect of the medical context, confidentiality and access control are also relevant.

TELEMED has also looked at decision making, linking psychiatrists with an international videolink; this videolink can be used for diagnostic exchanges in case a foreign psychiatrist is more able to treat the disease (either because the patient is from the same country as they are, which may happen more often with greater mobility of the population in Europe, or because one specific specialist for this patient might be in another country).

Another area TELEMED has examined is where a radiologist needs to have an expert's advice on some images; they can established link with the expert, through which they can discuss the images (and show the images to one another) and reach a joint diagnosis. The main performance requirement concerns image quality. This must be good enough for diagnostic purposes and both parties must have the same quality of image (whether from an analogue source in a videoconference or digitised).

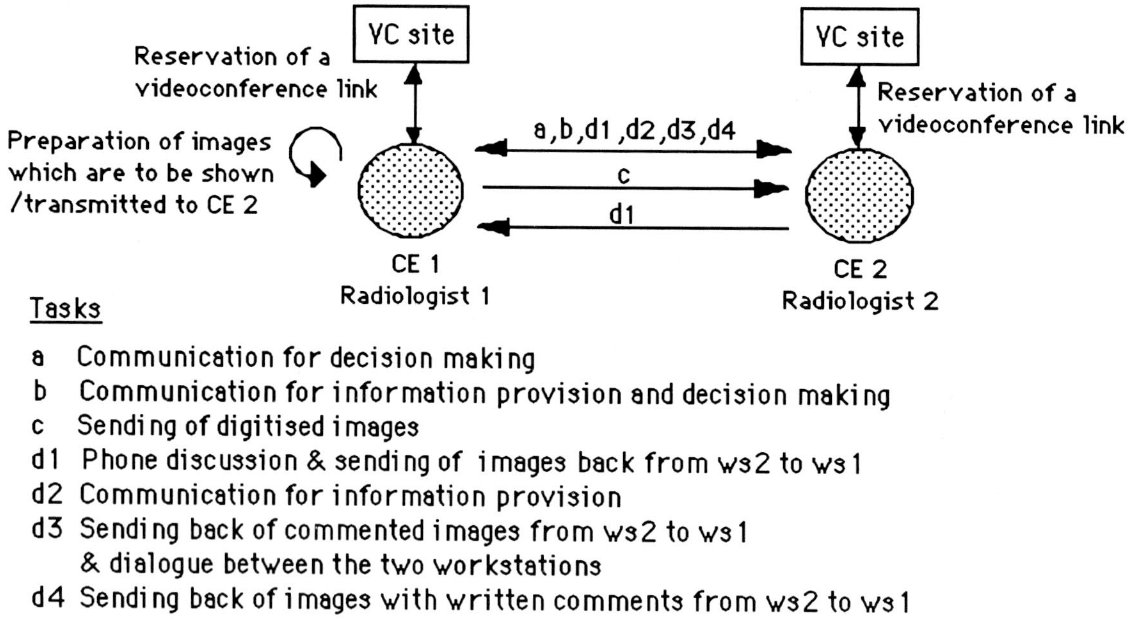

Figure 6: Tasks involved in the distributed decision making aspects of TELEMED

In another variation, angiogram films (that are moving image) and administrative documents are sent from Rotterdam to Berlin for a patient who is now in Berlin but was previously examined in Rotterdam. This transmission of images and documents takes place because a decision must be made quickly in order to know whether the patient in Berlin is to be operated on or not. Therefore, the application takes place in a context of urgency.

2.5. Distributed learning and training

In many areas, technological change has increased the need for people to be continually educated throughout their life. New developments mean that it is no longer feasible to learn the skills of a particular trade early in life and then to continue to apply them throughout a career. The availability of IBC will significantly improve the prospects for remote education and training on demand. Conferencing capabilities will enhance the potential for interaction between the teachers and students.

A typical classroom situation is an example of a conference in which only one person is presenting while the others are listening and watching. Direct visual contact helps to clarify terms by conveying gestural and facial expressions, and by showing objects and techniques directly to students.

IBC may change this pattern, by allowing a more interactive style of presentation and providing the teacher with access to more presentational facilities. If the teacher's presentation is stored on a database it would be possible for a student to control the pace of study, to study whenever is convenient, and to choose the particular information that is required.

RESAM, providing support to the aircraft industry, is typical of this type of application. This pilot allows a remote technician to request tutorial assistance for a particular job from a central video library provided by the aircraft operator or manufacturer.

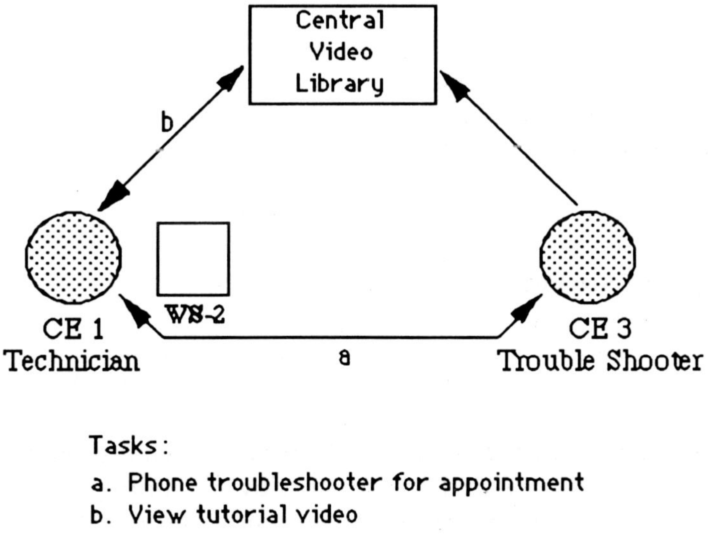

Figure 7: The distributed training support offered by RESAM

In the medical sector, TELEMED's connections between internationally separated psychiatric and radiology departments also offers the possibility of training to be carried out in a number of ways:

- through videoconferencing, while discussing methods, through multimedia documents which will have previously been prepared
- during an operation or consulting session
- educational exchange : teaching hospitals may have very different specialisation and it would be interesting to train one another through such a link

The APPSN pilot uses its extended safety telephone system (which includes a video link) to provide access to advice and guidance for elderly people in their own homes. The aim of this service is to help clients with everyday problems dealing with household, daily living skills and general need for information. The client calls the servicer to seek information, advice or guidance. The elderly person can also access a remote counsellor at the service centre for information or guidance on particular issues (e.g. on diet for a diabetic, or how to fill in a complicated benefits form).

This is primarily an availability service, i.e. the elderly persons are informed about the times during which the expert is available and they then call the expert at the appropriate time. A therapeutic service can also be provided from the service centre to the elderly person in their own home. The application is characterised by:

- regular sessions with the therapist
- more formal and scheduled interaction.

These features distinguish it from the other counselling applications. Therapeutic areas planned for coverage include interpersonal communication, physiotherapy, caring skills and general living skills. Although both partners can initiate the call, it is expected that in most cases the therapist will do this.

2.6. Monitoring and surveillance

This application covers all situations where knowledge of what is happening at a situation remote from the enquirer is made available over IBC. Examples include security surveillance, monitoring of unmanned industrial sites, observation and control of motor traffic, remote monitoring of medical conditions and the support by monitoring of the housebound. The application involves little or no human intervention at the remote situation, although in applications which involve 'telepresence' this may not be the case. The sensors may include microphones, cameras and static or dynamic measuring devices, and the transfer of the information to the enquirer is accompanied by the flow of control information back to the sensors.

In the APPSN pilot, for example, the local carer of the elderly person can arrange with the service centre for remote checks to be made on the elderly person's well-being while the carer is out of the home. The local carer indicates the period of time to be covered and the frequency of checking calls to be made. In this case the application is interactive. This means that the elderly person being cared for is aware that the checking call is being made and can interact with the caller.

The application can also be non-interactive. This means that the elderly person being cared for is not aware that the checking call is being made and cannot interact with the caller. The idea here is that the elderly person is not interrupted from sleep, or whatever.

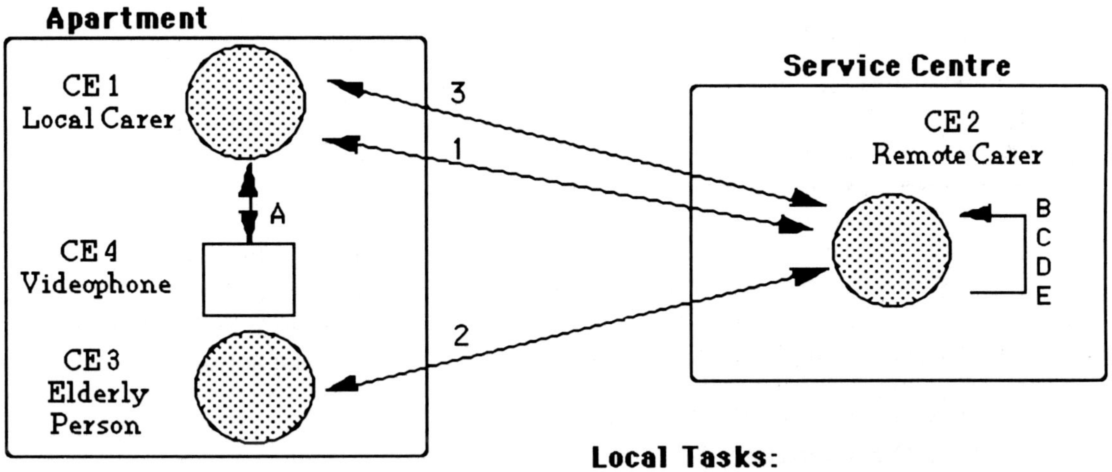

Communication Tasks:
1. Request Remote Care
2. Check Elderly Person
3. Cancel Arrangement

Local Tasks:
A. Select interactive Option
B. Set up the checking Arrangement
C. Receive Prompt that Call is Due
D. Make Checking Call
E. Cancel the Service Arrangement

Figure 8: Support for the remote monitoring of elderly clients' well-being is provided by APPSN

The APPSN pilot also provides an emergency response system which essentially involves adding a video link to an existing 'safety telephone system'. The old system involved a remote control button (usually worn on the wrist) which the elderly person could use in an emergency to trigger a telephone call to a service centre. In the event of an emergency, help was despatched to the home or apartment of the elderly person.

This system was operated over the telephone network. The new system adds a video link supported by CATV. In this case, when an alarm call is triggered it automatically makes a video connection between the service centre and the home. This allows more detailed appraisals of the situation to be made so that the response can be appropriately targeted.

Particular performance requirements include:

- information quality: TV quality for video and telephone quality for voice
- security: PSTN level for audio; privacy must be guaranteed for the video
- reliability: PSTN levels are adequate for audio and video

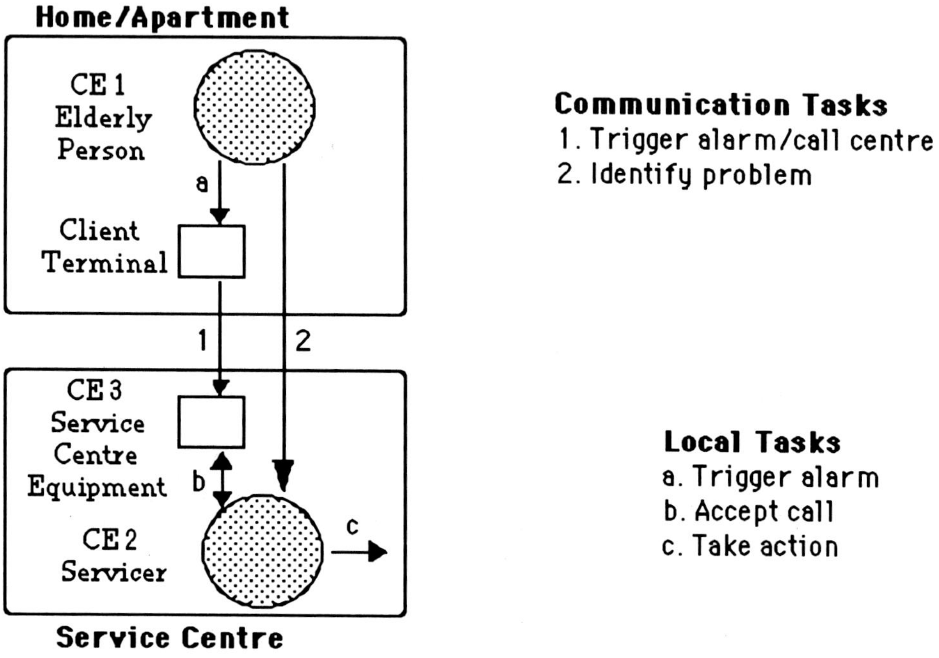

Figure 9: The emergency response system provided by APPSN

A more traditional variation of this application is represented by the CAR pilot, which allows for the transmission of live video frames, and which could be used in the a range of different CAD/CAM scenarios. A typical example of this would be the remote supervision of manufacturing processes, ie the transmission of full motion video from a manufacturing cell to a remote process engineer.

2.7. Telemarketing

The concept of telemarketing covers the remote offering, ordering and sale of goods or services together with payment, authorisation and authentication. Telemarketing systems using IBC would be found in quite different applications areas and would involve a broad spectrum of functions. The business areas could range from retail sales to real estate, with the provision of services such as banking, healthcare and insurance. The process may require broadcast multimedia information about the goods or services on offer, and purchasers would be provided with the means to place orders and authorise the transfer of funds from bank accounts or credit facilities. Auctions could be held via conferencing services, and audiences could be targeted for promotional purposes with increased accuracy and effectiveness.

The DIMUN pilot allows wholesalers of data cables to transmit their orders direct from a PC on their site, via a modem link to the host computer of the supply company. All relevant order and delivery information is kept there, so speeding up both the order and the delivery processes.

On a more specialised front, DIVIDEND supports the conversational interaction between international currency dealers that includes negotiating the deal, then the deal is carried out and verified using the EDI facility. This application supports trading both within a single bank and between different banks. While the quality of the communication need not be particularly high, the accuracy and reliability of the electronic data interchange is vital.

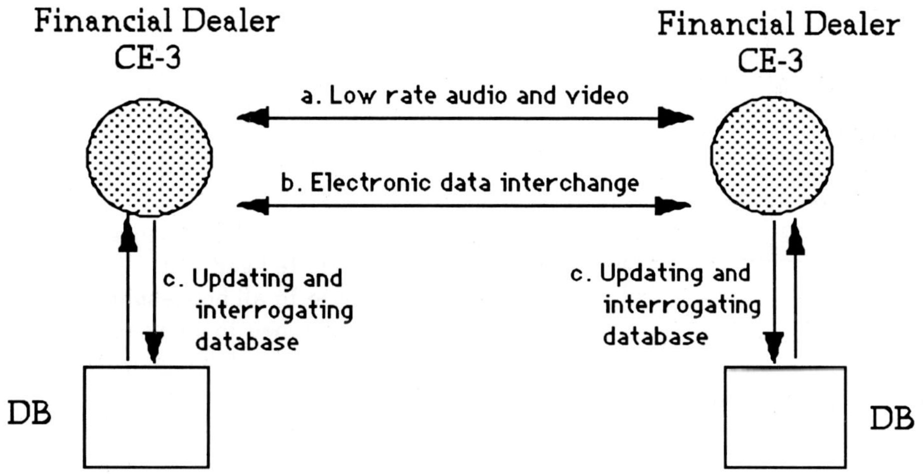

Figure 10: Support for telemarketing provided by DIVIDEND

2.8. Entertainment and leisure distribution

The IBC environment will bring with it the potential to distribute a wide range of entertainment material, with the main difference to existing services being the degree of control that the user can exert. Rather than being broadcast, material could be ordered specifically, when required. The range of subject matter on offer could be increased to serve specialist audiences, avoiding the problem of needing to appeal to the 'lowest common denominator'. Broadcasting would become more versatile, with the possibility of facilities such as pay-per-view for funding, which is being examined by the PPV pilot. The receiver of the material would also be able to participate directly in competitions, games, public affairs and interview programmes, for example.

In the PPV pilot, prior to the viewing of movies, the subscriber must consult the database provided by Totaalnet and completed by Cinema TV. The data base gives the list of available movies (6 different movies each week which are advertised at least four weeks before the schedule), their synopsis and references (cast, producer, and so on) and the cost of viewing.

The service also has an application in the business community. In this case, the application is the same as for the residential sector except that here the user is the manager of a TV shop who uses PPV programmes in order to sell TV sets. The customer who buys a TV set in such a premise gets a free subscription to Cinema TV for one year.

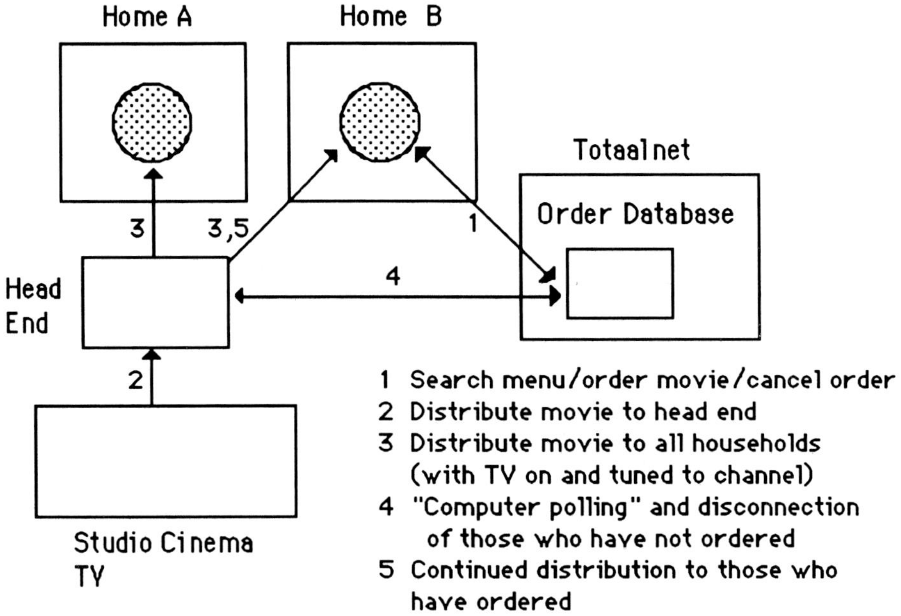

Figure 11: The tasks supported by the 'Pay Per View' pilot

2.9. Multimedia interpersonal messaging

In its most basic form this application represents a 'store and forward' service that can handle any combination of media in a single document, or group of documents. Messages would be in two parts: an 'envelope' carrying various forms of identification about the content and sender of the information, and the actual message. The application requires the writing and reading of all forms of information, with the reading likely to be by a real-time link between the recipient and the distributed storage service used by the originator of the message. It may be necessary for the storage device to transform the message to suit a variety of quality levels as specified by the recipient rather than the originator, and also to provide the originator with information about the use or non-use of the message.

The CAR pilot demonstrates multi-media messaging between two workstations. A multi-media mail facility has been designed and implemented that enables the user of the system to use a large part of the usual components of a messaging system. Facilities have been added to support the entry of text, sound, and pictorial forms. The trial task involves the user receiving a multi-media message, reading it and then composing a simple reply using still images, text and an audio overlay.

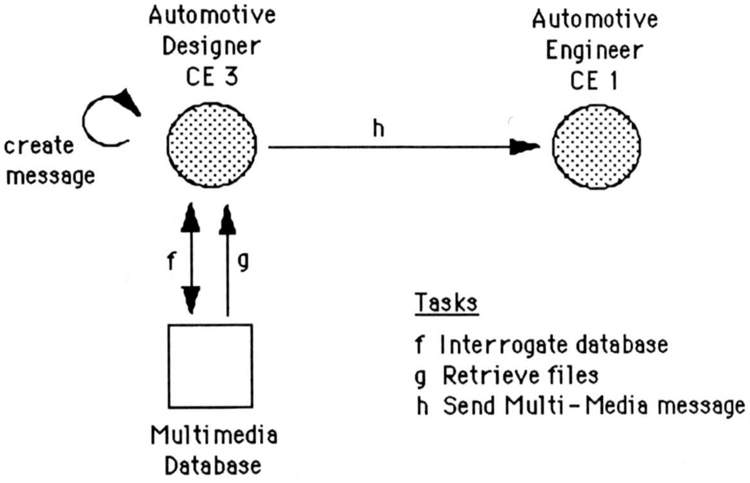

Figure 12: Multimedia messaging tasks supported by CAR

2.10. Multimedia information assembly and access

IBC will enable the maintenance of multimedia databases, and must also provide the facilities for the creation, editing, searching, retrieval and manipulation of the information that they hold. As the amount of information proliferates, it will become increasingly important for accurate and rapid location of relevant information to satisfy a user's inquiry. Browsing of databases will also be an important feature of the services. The charging basis for use of the material, and the permission for re-use of proprietary information, will also need to be controlled by the database provider.

This is a key feature of many IBC pilots. MULTIMED combines the constitution of a medical records database, access to these records by oncology specialists and remote consultation with a dosimetry expert. The application is centred at the Clinique du Mail in Grenoble and is based on the need of the oncology specialist to access all forms of medical data and images from patient files. The constitution of a database enables the centralisation of all information dealing with one patient in a folder. At Clinique du Mail there is a radiophysicist with dosimetry expertise. Remote clinics, such as Chambery, can send their patient images and data in order to have parameters for cancer treatment. This allows better treatment and time savings.

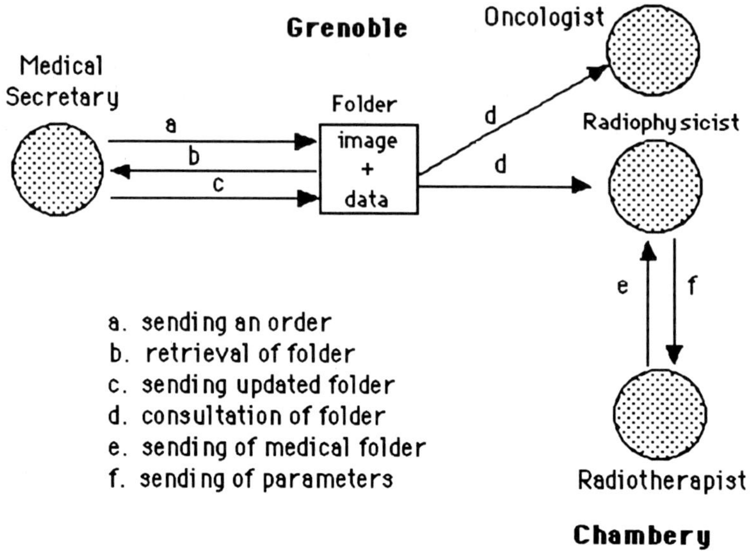

Figure 13: The use of multimedia databases in MULTIMED

The TELEMED pilot also includes a multimedia database of reference radiology images which can be used by radiologists to obtain some reference images; these reference images would primarily be used for teaching or research purposes, but can also be used for diagnostic purposes. Radiologists have to supply the data base with the images on which they are specialised; these images must be processed (made anonymous, commented upon, and so on) before being accepted in the data base for other radiologists to be able to consult them.

In the RESAM pilot, aircraft technicians have remote access to up-to-date information on the current aircraft status and recent history. The application makes most hard copied technical documentation superfluous. Regarding performance requirements, security is especially important. Authentication, encryption and other procedures are judged to be mandatory. More general requirements include:

- all documentation must be very easy to access
- text and images must be integrated
- searching must be much better than existing microfilm systems
- browsing should be possible
- editing facilities are needed, including revisions of both text and images
- response time must be fast
- multi-screen facilities
- at least 3 images simultaneously
- transmission of documentation in original or editor form
- image quality as good or better than existing manual quality

Once the technician has carried out any repair work, this must be logged in the aircraft history database. Multimedia records here can be more precise than textual descriptions.

The targeted customers of the GEOTEL pilot are essentially the french oil and chemicals industry. These firms often have their head offices in Paris, with some provincial offices or units, and can be considered as rather decentralised. These firms are willing to restructure their office automation systems in these remote services, and more generally to reconsider the office automation relationships between headquarters and remote units. The users are people who either belong to a patent department or need to work on standards. These kinds of services can be documentation, keeping up to date with technological developments, or patents service. Managers or technicians working in these industries need to have access to technical data, particularly concerning patents and standards; these data are updated very regularly by national as well as by international organisations.

Individual electronic newspapers (IENs) are built up of articles on topics the reader has previously expressed an interest in, and are being examined by the TELEPUBS pilot, which is aimed at the publishing industry. This is represented sometimes by completely integrated large groups, and sometimes networks of independent subcontractors linked to a publishing house. This pilot involves the composition of IENs and provision of access to these for customers. Two main types of session are involved. One type is a cooperative session, involving shared access to documents and interpersonal communications to agree modifications, discuss layout etc. The other type involves transfer of whole documents to the next step of the process.

The IEN composition process results in individualised electronic newspapers. This includes

- interactive identification of readers interests
- appropriate composition of IEN on the basis of available articles, including text graphics and pictures
- cooperation with and communication to other components like archives, publishing house, IEN access module

The IEN access module enables readers to get their IEN. This includes:

- accounting aspects
- communication to readers via public networks like IBC
- delivery of IENs via public networks (eg high quality fax)

Each user first defines their preferred content. They then choose either hardcopy on paper or direct display on screen. In the latter case the IEN will be a multimedia stack, including moving pictures. A knowledge based system is an important aspect in the development of this scenario, in particular to optimise layout for adequate composition.

The editor is responsible for composition and integration of text, images, graphics etc into a designed layout, and has close contact with one or several authors and selects in co-operation with the orderer the images and acquires this information from databases, archives and the like. The editor needs internal communication links to several colleagues who are working on specialised tasks like image recording, image editing, mixing of images, text and graphics etc and making up columns and mounting them into pages. These activities are today done manually by several persons. They will be integrated in specialised and/or multifunctional workstations to ease and shorten the design, realisation and co-operation phase.

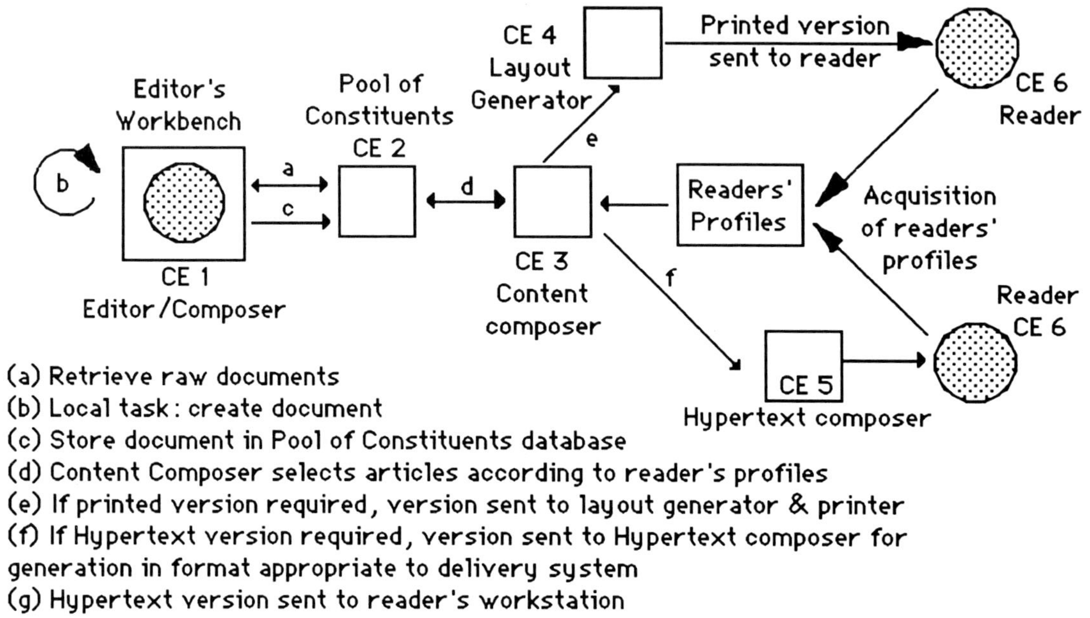

Figure 14: The tasks involved in producing an IEN in the TELEPUBS pilot

Particular requirements include:

- selection of pictures (>1500 per month) by on-line access to a databank/archive in digital high quality form
- remote preview of such pictures using a workstation or TV equipment
- digital pictures will be edited at a workstation to fit into the layout
- information should be actual up to the last minute
- file transfer (bulk data) and remote job entry
- joint editing functionality
- proof prints in b/w, grey scale or colour
- easy to learn user interfaces
- multi-windowing.

Regarding bandwidth, broadband is needed for fast response times. 2 Mbits/s may be acceptable for tasks such as image selection but not for sending real print. For the catalogue production, a document with a tree structure is sufficient. For IEN, a hypertext system is needed. True colours are very important; 24 bit encoding, at least. Thus one catalogue might require 100 Gbits of memory.

The MUSNET pilot is providing multimedia database support for museums. The museum curators are multi-disciplinary teams for the most part with no real IT culture. Their needs therefore appear progressively as their awareness of the available technology increases. Large data bases already exist worldwide but there is no means to access them remotely and although there is also much collaboration between museums traditional systems such as mail, telephone and travelling are relied upon. The MUSNET data base, in contrast, can be updated and modified by the curators at individual museums. Various input devices are used, keyboards for programming and keyword definition, scanner, voice recorder and other means to input the data to be displayed to the user. The database can then be used by any visitor to any of the connected sites.

Once the data base has been set up by the museum scientist, the museum visitor may decide to use the information in one of two ways. There is a guided tour option during which the user sees a series of objects and related information linked together by the scientist to form a logical sequence. Alternatively, the user may choose to navigate the system themselves in an "associative" manner, i.e., each set of information will have a list of key words. By selecting one of the key words the user may receive more information on a particular topic, period of time, genre of art etc and will have the option to look at associated objects. Thus the user may navigate the system according to their own tastes or interests. At any given time the user may look at the route by which they have reached a certain object, so tracing back the associations.

All links established during the visitor's sessions are recorded and a statistical analysis is regularly made. Results of this information recorded and stored by the system, enable the development teams to continue to ensure that the system really does meet the needs of the user. Particular performance requirements include:

- easy to use:
  both curators and visitors will usually be novice users
- fast processing:
  new information should be easily added and editing should be fast
- immediacy:
  visitors will require an immediate response to their database interrogations
- reliability:
  the service must be always available and operating during service hours.

2.11. Distributed collaborative design

Many design processes involve the participation of a number of different people with particular skills, who may be located in different places according to their specialities. IBC will enable multiple, iterative and remote contributions to a design process that has a single object as its focus, whether a physical object or another product, such as a software-package. In decision making processes such as those described earlier, the accurate communication of information plays a key role, while in design a focus on the object is more important, but the need will still arise for individuals to discuss design problems and their solutions. This application therefore has a number of similarities with collaborative decision making, although the emphasis upon a particular 'end-product' places additional requirements upon the services that must be provided.

In the DIMUN pilot, the users are electrical and mechanical designers located at a transformer company head office in Norway and at the component manufacturing site in Reinhausen, in Germany, where the most important component of the transformer is manufactured to order. Previously, communications between the design offices in Norway and the factory were purely on paper and by telephone. The design process was unnecessarily long due to the dependence on couriers to firstly convey the design to Germany and then to transmit backwards and forwards any necessary modification. Exchanges are now through an interactive CAD system which allows the German design office to transmit their design via the PSTN and to receive any amendments from Norway.

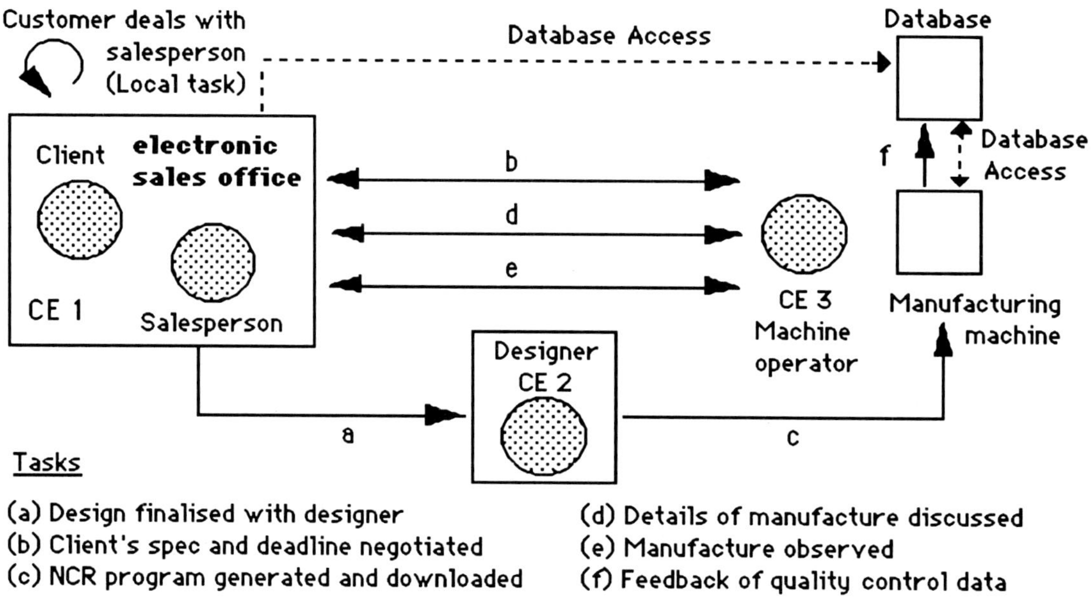

Figure 15: Design support tasks offered by the DIMUN pilot

Another part of the DIMUN pilot involves live interactive videoconferencing integrated to large volumes of data for chip design between manufacturing companies in Bremen, in Germany and Hemel Hempstead, in the UK. CAD is distributed but connected to the main management information system. The designers have the capability at their workstations to perform such tasks as design, analyses, testing, as well as determining product cost, providing project quotes and assessing the availability of component parts in inventory with the assistance of material and resource planning systems (MRP II), Computer Aided Planning systems (CAPP) and other industrial computer systems. Important aspects of the pilot include:

- Distributed Engineering:
  Engineers with the same qualifications, on different sites, have to work together to develop new products.
- Customer tailored products:
  Another result of the progressive interconnection of all the entities of the manufacturing chain is that the customers can be put in direct relation with the designers and influence the creation. Salespeople who are in direct contact with users needs are now in direct contact with designers.

The DIDAMES pilot involves a distributed CAD/CAM system to enable collaborative working supported by videoconferencing facilities at an integrated workstation. This improves efficiency by allowing the integration of technical and managerial activities into distributed meetings by applying an integrated videoconference system. The designers require advanced interpersonal communications to be integrated to the CAD CAM system because the process is distributed. They need it partly for discussing technical problems and partly as a management tool. For the pilot the users are exclusively engineers. In the future though, as technology costs and communication costs drop, the major user group may well be technicians who will use it to solve more practical problems. It is also envisaged that sales staff may become system users. Particular performance requirements include:

- variable picture quality levels should be provided for graphics and still images, from real time low quality to slower high quality; broadband is not needed continuously but bursts of data can occur when high quality graphics are required
- a default configuration of windows seems necessary although this should be modifiable by users; with multi-windowing the screen can quickly become submerged with windows
- needs easy switching on and off of audio and video (to avoid feeling of 'big brother' type of presence)
- shared access to documents is not easily managed without some form of token system; however it should remain as an option, but with each party owning their own information and only they can decide to make one of their documents visible to others
- several transmission modes should be available, allowing trade-off of one information type's quality level against another's (e.g. lower quality images for higher quality audio)
- the system should be accessible to different terminal types; where a user has no audio and/or video then this should be recognised by the system and taken into account
- at present up to 4 people can join a conference but it is planned to extend this to 8
- regarding image quality, it must be high for CAD/CAM although lower than TV quality is acceptable for videoconferencing

The CAR pilot is mainly aimed at supporting design in the automotive industry. Automobile manufacturers maintain highly qualified staff to support the complete production of automobiles from concept to physical realisation. The industry is geared toward improving efficiency and quality. Competition from the lower priced Japanese sector has meant that European manufacturers have to respond aggressively with large improvements in efficiency and quality to survive. CAR is a vital element of that as it facilitates the reduction of time required for the life-cycle of a design. The engineers referred to in the CAR project are employees of automotive manufacturers or an external supplier. Engineers are highly qualified users of CAD/CAM design systems and will have a wide knowledge of the automotive industry. It is likely however that their level of experience with communication tools will be limited.

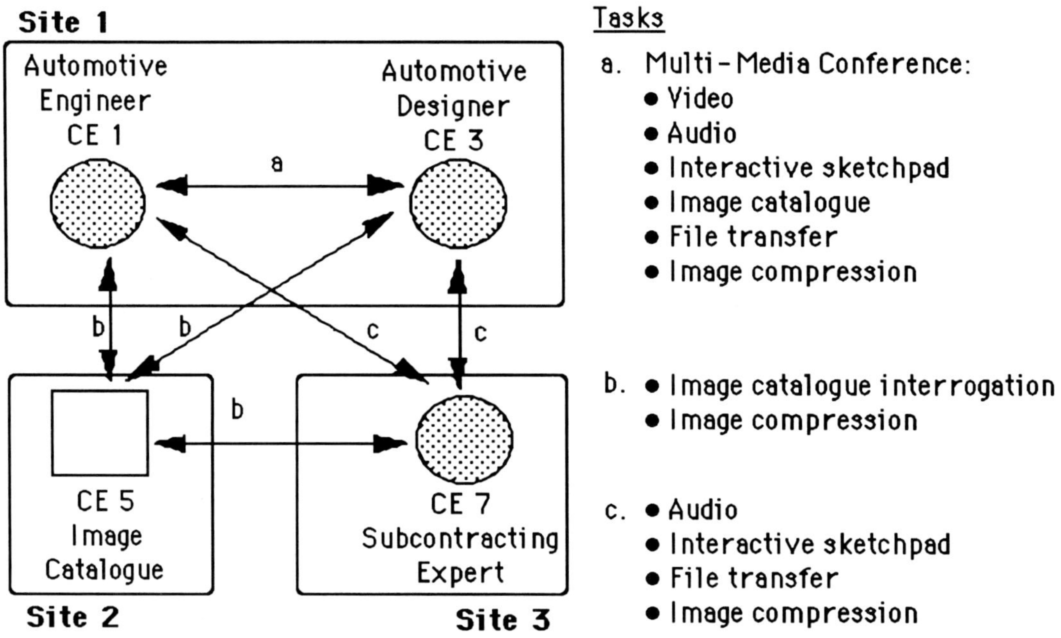

Figure 16: Design support provided by the CAR pilot

CAR brings together automotive designers, and automotive engineers into a single design environment. The application supports the initiation of a videoconference between two such users. With the aid of the conference controller software the partners can introduce distributed services to support their transaction, ie:

- Interactive sketchpad
- Interrogation of a remote parts suppliers' image catalogue
- File transfer

The interactive sketchpad equates to freehand sketching in terms of ease of use and supports most operations found in existing drawing packages. It permits multi-user use in different locations in real time, and can also be used as a stand-alone tool. The engineer and designer can call upon a remote sub-contracting expert through this demonstrator. This has the effect of linking the sub-contractors low-cost workstation to the conference via the ISDN. Image compression and decompression software supports the transfer of image files to the sub-contractor. A similar arrangement allows for the transfer of image files from the parts suppliers remote image catalogue. The main area of performance requirement is in support for three-way working.

CAD CAM is not the only area in which design support can be offered by IBC. The main objective of the DIMPE pilot was to demonstrate, on an international level, the possibility of implementing a low-cost operational distributed regional editorial office. This was set up while testing different elements in different countries. While keeping control at the headquarters over the production process itself, the regional editorial office must be able to communicate with the headquarters:

- with text transmission (but this is not very new)
- with image—and even high resolution image—transmission
- regarding complete page layouts

Two main aspects of performance were looked at:

- in production monitoring: speed of updating of the different processes must be sufficient for the production manager to have an accurate view of the state of the production of the newspaper
- speed of image transfer must be acceptable for the working context.

More generally, quality of transmitted objects (images, text etc.) must be sufficient to ensure that (costly) re-transmission is not necessary. Also, flexibility in quality levels is be important and this needs variable bandwidth on demand.

A similar variation was examined by the TELEPUBS pilot, involving the distributed production of catalogues. The areas addressed include establishing communication between existing technical equipment, changing conventional methods to a new cooperative form of catalogue production, and integration of an external image agency/archive. The make-up of copies which have to be produced as input for the printing house will no longer be necessary if direct 'computer-to-plate' data transfer to a printing and production house is possible. During these activities co-operation should take place internally as well as externally with the orderer whenever needed, for instance on the basis of a proof printout or a preview at a TV screen.

Within the reproduction office the tasks of image recording, editing and output have to be done on different systems, normally at different locations. During the co-operation phase and when producing final copies the systems have to interchange and integrate the different material electronically using LANs or an IBC network.

For an integrated editing of colour images special hard- and software is required for high speed block transfer between memory and the display processor and for short access time to storage media. Several image editing functions are available today on workstations but the interaction from the user's point of view is not always acceptable. Besides that, the systems need interconnection to scanner, recorder, databank or archive via high-speed LANs or an IBC network. Additional image editing features on a workstation are required.

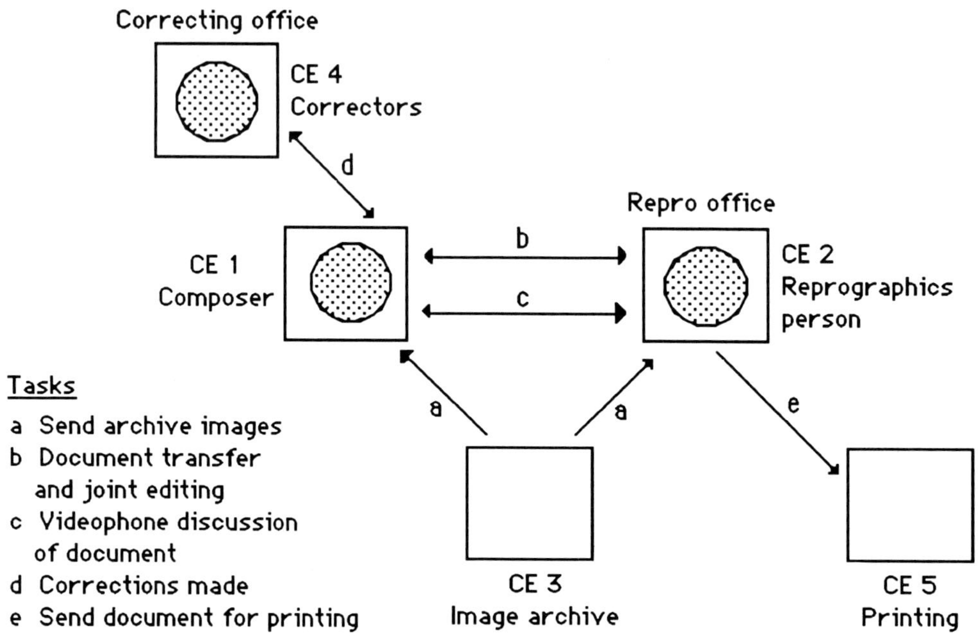

Figure 17: Design tasks supported by the TELEPUBS pilot

In addition to the requirements listed in section 2.10, this application requires:

- Reduction of time during co-ordination. Today large delays must be accepted while using delivery by PTT, rail or courier.
- Information should be actual 'up to last minute'.
- Printing in parallel at different printing locations and shorting delivery distances and time between printing house, dealer and subscriber.
- Connection between printing and publishing house and different customers like advertising agency, designer and reprostudios to shorten the process of co-operation and handle the process of correction based on proofs in an easy time-saving manner.
- Transmission of a high amount of data ca. 6Gb for 70 x 100 cm plate) for several plates which will be directly produced at the printing location within less than one minute
- Broadband connection between national and international locations
- Interconnection of a workstation and a highly sophisticated system for colour editing via a standardised LAN and interconnection to an IBC network using gateways

3. Configuring appropriate services for IBC applications

The applications described in the previous section show how varied the uses to which IBC can be put. Whatever one's image of the broadband future, one aspect is almost certain to obtain: the age of IBC will be the age of personalised communications. It will occur at a number of levels and affect us in a number of ways. We shall no longer be unobtainable, for example. Intelligence in the network will be able to redirect our calls to whichever terminal, anywhere in the world, that we happen to be near, our proximity and personal identity number signalled by our personal local transmitter. When we are engaged, or do not want to be disturbed, incoming calls will be answered by a range of devices, using different media depending upon the type of incoming call, and with different content depending upon the category or individual identity of the caller.

At an even more fundamental level, however, every call we make could in principle involve our personal selection of the communication task or tasks, in various combinations, to be supported (i.e., conversation, send, and retrieval), the media needed for each task (i.e., audio, video, still image, text, graphics, or data), and the required media quality and performance parameters determining the bandwidth and connection type.

In addition, we may have the choice of bearer services offered by different network operators, perhaps with different tariffs for different services at different times during the day, and providing different value added services (intelligent searching within directory enquiries, mobile redirection, etc). In short, the age of IBC will be the age of configurable telecommunications services, delivered not by monopolistic national operators but by a number of competitive suppliers offering a variety of service extras and inducements.

The danger, however, is that, with such a wide range of possible services and an even wider rage of configurable attributes for each service, not to mention all the value added extras, we could find ourselves needing to become telecommunications experts just to use the telephone. No network operator or terminal manufacturer would try to foist such an unusable solution upon the market. Rather, what will almost certainly emerge is a set of compromises or trade-offs between a completely flexible system where all aspects are configurable by the end user and a constrained system where some aspects cannot be reconfigured, in the interests of maximising 'usability'.

What these trade-offs will settle on, in different circumstances and for different types of user, is almost impossible to predict at this stage. It will be the result of an evolutionary process by which the most common user requirements will emerge alongside the development of the technology and of the market. And, indeed, there are likely to be a number of generations, or iterations in the design cycle before the optimal solutions begin to stabilise. Whatever these optimal solutions are, however, they will not be appropriate for all end-users (i.e., those who require more than the 'lowest common denominator' service), and so it will still be necessary to provide tailorable services to those who require them.

In describing the selection of specific services to realise a specific integrated service out of the universe of possible options, we have already used the term 'configuration'. The user will 'configure' the service to meet their requirements, and will do so in two ways:

1. by aggregating a number of services and service components to create a complex service, as when one might combine retrieval from a on-line database with a desktop conferencing service in order that a geographically distributed group can work on the data together,
2 by selecting specific values for a variety of service attributes such as picture quality, screen refresh rate etc.

Figure 18 shows a simple case of two users communicating via a service, which is provided by a service provider.

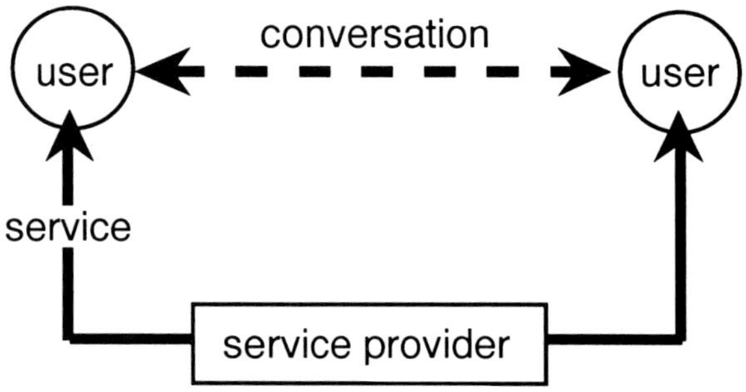

Figure 18: The relationship between users, services and service providers

The user will obtain a service by interacting with a service provider, as shown in figure 19. Since IBC services will be much more flexible than conventional services, users will be able to request and configure services to meet their personal requirements. This means that the interaction required between the user and the service provider, to choose (for example) the appropriate media, media quality or communication partners, may become complicated.

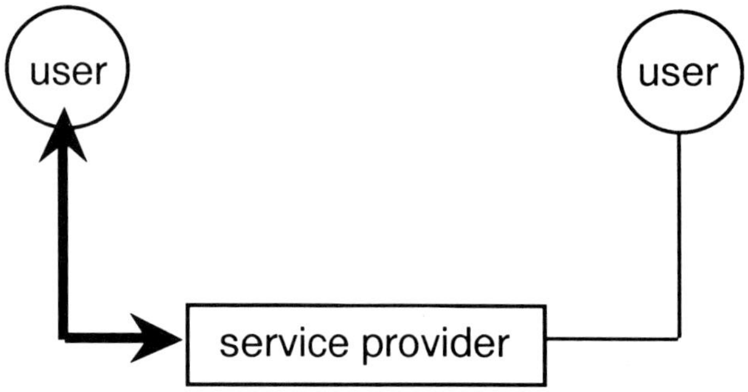

Figure 19: User—Service Provider interaction

A usable IBC system is one which people may use to achieve their goals easily. A usable IBC system must therefore provide services which are appropriate for all types of user and task, *and* these services must be accessible— for example, the interaction or dialogue between the user and the service provider must be simple.

If we examine the dialogue between the user and the service provider through which access to an existing service is currently negotiated it is somewhat limited, and is usually rendered as follows:

User:	'Off hook'	(Request for connectivity service)
SP:	'Dial Tone'	(Offer of connectivity—user has paid the bill, and there are no malfunctions at their end—request for further specification of connection required)
User:	'Dials'	(Specification of connection)
SP:	'Ringing Tone'	(Offer of specific connection—the line is available and the called user has paid their bill and there are no malfunctions at their end).
User:	'Waits'	(Accepts offer of service on the understanding that as soon as the called party picks up the phone the user will have contracted the service and will be liable for payment).

In this standard user/telephony service provider dialogue the only parameter to be specified by the user is the identity of the other user to be called. However, this is not the only configurable parameter inherent within the service. Another possible configurable option is the polarity of the payment liability, which can be changed so that the user *receiving* the call is the user billed (i.e. reverse charges).

It is not, however, a configurable option for the caller, who must first call the operator, not only to obtain acceptance of this liability from the user called, but because the caller actually has no way of implementing this configuration.

This is not, however, the way things necessarily are. It would have been technically quite possible for us all to have telephones with toggle switches for altering the polarity of payment liability, so that the standard telephony service would have contained this configurable parameter. It does not do so because this would almost certainly have resulted in the reverse charge setting becoming the default setting, with, as a consequence, users either refusing to pick up the phone to accept calls or refusing to pay bills for calls they had not requested and did not want. The service has therefore been configured by the service provider, the public network operator (PNO), to not allow the user this configuration option without first going to the service provider for permission for a specific instance.

The importance of this example is that it is a clear case where 'user options' have been deliberately restricted not for reasons of usability, but because of the wider demands of the communications enterprise. Indeed, there are almost certain to be many similar examples which will emerge with respect to IBC, many of them not determined by the PNOs but by end user organisations for economic and security reasons. Wherever configuration is allocated to an entity other than the end-user, the possibility arises that it will be undertaken as much for the purposes of control as for usability, with the consequence that the user will not be able to configure some aspects of the service with the tools and instruments at their disposal while actually using that service. This is a fact that must be borne in mind when design decisions remove configuration tasks from the end-user in the apparent interests of 'simplicity': there is no guarantee that the resulting configuration choices will be appropriate to the end-users' requirements, and so the services may actually become less 'usable' from their point of view.

This conflict of interest can be formalised using the concepts of external and internal perspective presented in chapter 1.1.

3.1. Service Configuration Actors

There are three logical places at which service configuration could occur, and three different actors who could take on the role of configurer. These are:

1. The end user, or their surrogate;
2. The platform provider;
3. The service provider.

The first option is, of course, that the end user will configure their own services to their own requirements. Indeed, there will always be pioneer end users who have the ability, the time and the enthusiasm to master the intricacies and potential of the most obscure option setting on their machine.

There are, however, a number of limitations to the possible extensiveness of this as an option. Not only are such pioneers likely to be few in number (like the number of people who use the full suite of tools within their word processor), but the scope for them to exercise their skills is also likely to be limited. The reason for this is that few initial end users will have single terminal connections to IBC services. Within major companies, who will almost certainly be the first to invest in this technology, it is far more likely that the services will be mediated via the customer premises network (CPN), which, in the short term at least, is likely to be a contentious local area network. Even if there were no issues of cost and security, which we shall consider later, it is therefore unlikely that such end users will have unrestricted freedom with respect to service configuration. Indeed, it is much more likely that just as the end user of standard telephony services has to ask the services operator to reconfigure the polarity of payment liability for a particular call, the individual end user on the customer premises network will have to ask the CPN manager for permission to use particularly high bandwidth services, which may be a configurable option granted to only a small number of key users.

There are still, of course, the home users, teleworkers or independent consultants, who, by their very nature, are going to be oriented to geographically distributed collaborative work and to telecommunications.

The second option would to be to build service configuration into their terminal, as an intelligent support system. The idea would be that the end user would answer a few simple questions about their requirements and the terminal would configure the communications services itself.

This, however, is almost certainly more easily said than done, and although some aspects of such a system are almost certain to emerge over the next few years, some of which will almost certainly be retained into the maturity of IBC, there are other factors which point against this becoming a general solution.

The most important consideration is that a configurable, all-purpose terminal is unlikely to be a low cost terminal, which is very important for both the home market and, therefore, the corporate market, where we should expect a great deal of the required functionality to be housed within the network.

A probable alternative approach from the platform providers is therefore to develop a limited number of fairly standard terminals, of limited configurability, designed to meet the specific requirements of defined subsets of users. This, for instance, is the approach currently being pursued by the project APTITUDE which is concerned with the telecommunications needs of professional teleworkers. Significantly, the main objective of APTITUDE is not to develop a configuration management system for teleworking, but a generic, integrated environment, i.e. one that integrates the most typically used services by the average professional teleworker, allows for easy configuration of those services within a necessarily limited set of options, and provides a standard user interface for interworking with those services with just the configuration options offered by the standard windowing systems. In short, APTITUDE is engineering a package and attempting to create a standard.

Moreover we can see that the combined forces of economics and usability always tend towards this form of solution. We might describe it as the tendency of all mass market products and services towards becoming commodities.

All mass products tend towards becoming commodities in that it is more cost effective to manufacture a standard product than to engineer infinite variations to order and far easier to sell a standard type about which there are few decision points rather than individual tokens about which there can be infinite pros and cons. The end user, of course, gets less choice, and may not find the packaged solutions appropriate to their needs.

A problem, however, is that it is likely to be some time before we see such packaged solutions emerging. The process is very much an evolutionary one, with partial solutions, either offering very limited functionality or alternatively requiring a considerable degree of configuration, and hence specialised skill in use, coming on to the market in an iterative process through a number of generations before just the right package is produced.

A considerable amount of work has already been carried out in the RACE programme aimed at prototyping tools to help service providers, in the form of public network operators (PNOs), gather requirements from and design networks for their corporate clients. Indeed, it is generally held that the PNOs are the leading contenders to extend their existing corporate business into this developing market. The position, however, is not without a certain degree of ambivalence due, in no small part, to the perspective which most PNOs have upon their services, which, as with standard manufactured goods, are typically regarded as commodities, standard market offerings out of which or upon which the customer builds his applications.

There are a number of reasons for this. Firstly, as mentioned above, a small, finite set of simple services is far more cost-effective and profitable than a large set of individually tailored services. Secondly, the telecommunications industry has a profound interest in inter-service-provider connectivity and standardisation which can only achieved on the basis of fairly low level, commodity-like, components. Finally, the history of most PNOs as public utilities and national monopolies, has tended to create corporate structures and an ethos in which a move away from secure, non-competitive trading in simple commodities is immensely difficult to engineer. The messiness of the market into which they would be entering is by no means attractive. It would be the same as national electricity power generating companies or water boards moving away from their mass market supply into the ever changing world of kitchen appliances.

That the PNOs are going to have to enter this market in order to survive commercially very far into the twenty-first century is something upon which most analysts would agree. The alternative would be to retreat into the supply of low-value-added bearer services while new value added service providers skimmed off more and more of the cream, which would make it very difficult for the PNOs to maintain their high level of investment in the infrastructure.

Moreover the PNOs are being inexorably drawn into higher level value added services. The clearest example is multi-user conferencing, which has to be hosted somewhere in the network if each user is to avoid having to have multiple open channels and multiplexing at each terminal. The host does not have to be the PNO, of course, and one of the options for owners of the large customer premises networks will be to host their conferencing themselves. But this would be an expensive option for all but the most high volume users, and would certainly deprive the PNOs of a great deal of revenue.

Mobile redirection is another value added service which almost has to be undertaken by the PNO. It is not impossible for other organisations to provide such a service, but it would be far more difficult in engineering terms and far more expensive. From all this, we can therefore quite reasonably deduce that a considerable amount of service configuration will be undertaken by the PNOs.

It is not, however, the only option, for there are two major reasons why large corporate customers in particular may wish to undertake their own service provider configuration:

- in order for themselves to assess the options with respect to hosting various value added network services within their own customer premises network, as opposed to buying them in as a packaged service from the PNO;
- in order to remain independent of any one network provider, and hence be in a position to purchase bearer services in an open market.

3.2. Design issues of Service Integration Tasks

The problems are not limited to the configuration of services. Consider the independent communication services, and associated items of equipment, that can be found today in a typical office. A simple scenario would be for an office worker to use, on a daily basis:

- a person-to-person audio communications service (e.g. for speech conversation) involving a standard telephone
- a person-to-machine realtime text data communications service (e.g. for remote database access) involving a character-based display
- a person-to-machine realtime numeric data manipulation service (e.g. for spreadsheet calculations) involving a bit-mapped monochrome display

It is clear that the isolated nature of these services will cause some problems for end-users in interacting with the services efficiently and effectively. Examples of problems of different types are:

- problems in physically moving between and accessing equipment
- problems of alternative input devices e.g. the layouts of the numeric keypads on the telephone and computer keyboard (for use with the spreadsheet) will differ
- problems of alternative command language syntax and semantics e.g. the spreadsheet and database services will probably run under different operating systems (such as MS-DOS and UNIX); they may also have common items (such as a 'SEARCH' command) with different meanings and formats
- problems of alternative task structures e.g. the discrete 'set terms' and 'search' components of the database access task contrast with the unitary and instantaneous 'set value and recalculate' task of the spreadsheet

These problems all concern some aspect of the integration of the various services, above and beyond any usability problem that may be specific within the individual services. Each of these integration problems will tend to have a negative effect on the usability of the set of services. This effect might manifest itself in various ways, for example by increases in learning times, or in data input times, or in command language syntax errors, or in user dissatisfaction, and so on. Such problems can of course only increase with the addition of other common office services and equipment e.g. word processors, fax machines, videotelephones, electronic mail, videotext, etc.

End-users therefore face problems in using current services that are caused by the low level of integration between services, and these negatively affect usability.

The principal reason for the existence of usage problems such as those described above, is the technical incompatibility of separate services due to the historical development of the relevant technologies. These technical problems of integrating different services will largely be solved in IBC systems, which will merge computing, networking and telecommunications technologies. IBC systems ought, therefore, to offer users many advantages: the delivery of services at a single terminal over a single network; better capabilities for transfer between services; and an increased range of services, especially for collaborative and distributed working.

However, the service integration problems faced by the users will mostly not be solved by technical integration into a single terminal. Although the physical problems of moving between and accessing separate items of equipment will largely disappear in IBC systems, the mental problems of learning, conceptualising and recalling how to use services will not disappear. End-users will still be faced with service integration problems affecting usability, such as:

- variations in the semantics of input device events (e.g. 'double-clicking' the mouse button may have different meanings in different services, and may not even be legal in some services);
- inconsistent screen layout conventions (e.g. different positions for common buttons in screens or windows for different services);
- confusion between operating procedures (e.g. similar database services may require very different access commands depending upon where they are physically stored);
- incompatibilities between the representations of objects (e.g. the nature of links between the component parts of different multimedia documents may well differ functionally);
- differences between representations to support collaborative working (e.g. two users working together to edit a drawing will require the same view of it, regardless of who performs which edits);
- difficulties in handling simultaneous conversational, retrieval and messaging services.

Indeed, the likelihood is that end-user service integration problems will increase in IBC systems, for two main reasons. Firstly, the number and variety of user services will certainly increase with IBC, thereby creating more opportunities for service integration problems to arise. Secondly, as mentioned in chapter 2.1, the psychological evidence suggests that increasing the similarities between tasks increases the negative transfer effects as well as the positive transfer effects. That is, if users do not know how to perform a task, they may well try to use knowledge they have of a similar task; some of this knowledge may work for the new task (a positive transfer effect) but some of it may be inappropriate and may hinder the performance of the new task (a negative transfer effect). Which effect will be more important depends upon the type and level of the similarities. Thus although delivering multiple services through a single terminal will increase similarities between them (e.g. in the input devices used) this may well have negative consequences for usability in some cases.

With IBC systems, therefore, end-users will still face problems caused by lack of integration between services, and these problems may make the integrated systems' usability even worse than existing independent services.

4. An analysis of service configuration in multimedia conferencing

So far the tone in this chapter has been a little negative, raising many potential problems in the provision of IBC services. It is not our intention to pour cold water on any enthusiasm for IBC, rather to point out the pitfalls that lie ahead in order that they might be avoided. Central to our argument is the belief that to introduce IBC services successfully, they must be planned in a methodical way that takes into account the needs and requirements of their users. This is the rationale behind the eight-step method described in chapter 1.1. While sections two and three of this book concentrate on the internal perspective of the latter four stages of the method, usability for the end-user of integrated services, the external perspective of the first four steps gives support to the service provider. To illustrate the sort of decisions that must be made in the design of a service, we will work through an example of the allocation of service configuration tasks for a video conferencing service.

The requirements of an end-user wishing to carry out a video conference need to be expressed in terms of an enabling states analysis (described in detail in chapter 3.1), so that the enabling tasks concerned with configuring the video conference can be identified. Thus, the overall goal of the user would be to communicate with more than one person using video, and all of the other requirements of the user would be concerned with enabling this goal to be performed in a particular way i.e. they would be enabling tasks (these enabling tasks would be made explicit during the enabling states analysis). Some of these enabling tasks will be concerned with the configuration of the video conferencing service; they are thus, configuration enabling tasks.

When using a particular service a user wants to have the flexibility to allow them to configure many of the options within a service themselves; this implies being able to predict how the service will operate and also being able to control the operation. The configuration options within a service will be directly related to the configuration enabling tasks of the end-user; e.g., if a user wants to be able to change partners during a conference call, which is a configuration enabling task, then one static configuration option would be whether the user wants to be able to change partners during a conference call. Ideally the end-user would like to have all these configuration enabling tasks under their own control, so that allocation of all the configuration enabling tasks should be to the user's worksystem. The user can then keep control of the tasks, but does not have to carry out all the behaviours associated with the tasks. The user has thus been relieved of the enabling tasks, increasing the usability of the system, and the enabling tasks have been re-allocated elsewhere within the worksystem. However, because of network/hardware constraints, policy or cost etc., when the configuration of the video conferencing service is being negotiated some of these configuration enabling tasks will have to be allocated to a different worksystem, that of the conference manager or public network operator.

Therefore when configuring a video conferencing service, distinctions must be made among the users of such services with regard to the role they play. The different roles in a conference will have different tasks allocated to them. The following example describes the possible roles in a video conference, identifying which configuration enabling tasks should be allocated to each role.

Originating a conference call, hosting the conference and being charged for the call can be compared with inviting people to a meeting, hosting the meeting and paying for it. There must be a clear separation of these three roles between the chairmanship of a conference, and the main user of a conference call. In this example the chairman of the conference is the conference manager, and the main user is the end-user of the service, for whom the conference is being configured.

What the main user of a conference is allowed to control will depend on the rights and duties of the chairman. The chairman of the conference wants to be able to plan actions in line with conference requirements, i.e. to have control of the access of new users to the conference and the nomination of users already participating to the floor, because this will allow the main user to plan their actions in line with their task demands. Therefore the chairman must negotiate with the main user about any configuration enabling tasks/requirements which are conference requirements. These configuration enabling tasks/conference requirements will then be re-allocated to the chairman e.g. to be able to promote or demote calls in a queue, identify incoming calls during current connection and indicate request for floor or nominate and remove floor holder. The chairman must then further negotiate with the service provider to make sure that the service is configured such that these conference requirements are under the chairman's control. The main user will keep control of all configuration enabling tasks relating to cost of call, identity of speakers and call ownership etc., for example, handing over control (and hence billing) of the call to another party; being able to drop out of the conference for a short time, and easily jump back into the loop when required; being able to change the ownership and the costing arrangements mid-call, being able to request a change of service from the service provider and being able to set the quality levels for each media type etc. The main user will also be provided with a clear indication of who is talking during the conference, who is listening to the call and what the cumulative cost of the call is.

Identifying the different roles, and understanding the negotiation process which must take place between these roles within a video conferencing service, will allow the end-user configuration enabling tasks to be optimally allocated across the different roles, to configure a service that best meets the end-user requirements.

4.1. The Video Conferencing Scenario

This scenario is only concerned with multi-point multimedia conferencing. The scenario concerns a main user who wishes to set up a conference with two other users via a conference manager (the system manager) and then to distribute to each participant a video clip which forms part of the multimedia document which is stored on the main user's private database. For the sake of simplicity we are assuming that this is a conference within the same organisation (but with geographically separated users) so that the role of the conference manager is not compromised if the main user drops out of the conference. i.e. the conference manager would still be fulfilling wider organisational goals by remaining in control of the remaining participants.

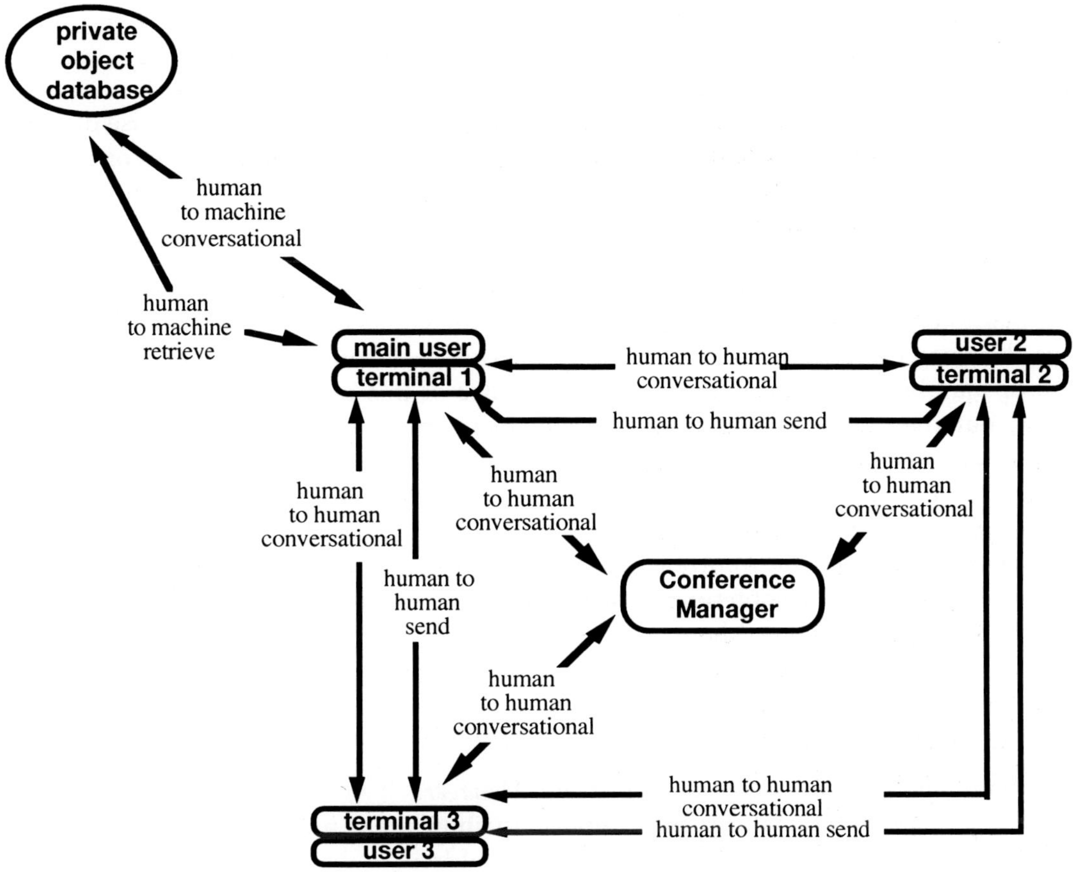

Figure 20: An Example Network for Multimedia Video conferencing

The task network which can support the tasks described in this scenario, part of the *Reference for User Services* (RUS) for this application, is shown in figure 20. This network describes the communicating entities and the channels that link them. The communicating entities involved, namely the participants and the private object database, are connected by an internal network. For human to human conversation tasks a multi-point connection is required which means that all users can be involved in any conference. Both audio and moving image service components are required. The audio service component must be mandatory whilst the moving image service component should be optional. Audio is needed for the transmission of speech, moving image for the transmission of the users' faces. For human to human distribution tasks the connection required is point to multi-point since only the main user has access to the database.

The specific scenario used here takes a snapshot of the group conferencing task to illustrate how the allocation of configuration enabling tasks affects the design of the conferencing service through its configuration by either the end-users (first party) or the system manager (or conference manager as a third party). Static and dynamic conference set up are described as there is a different set of configuration enabling tasks concerned with both types of service usage.

4.1.1. Static Conference Set Up

Static configuration tasks are those tasks which must be configured before the use of a service i.e. at the specification/binding phase of the service. Below are a few examples of these types of tasks which illustrate trade-offs in control of configuration between the main-user and the conference manger (systems manager).

Participant Directory: The main user wants to be able to browse a list of potential conference participants (by name and organisation/department, say). This requires a directory service listing their names/ addresses etc. which has to have been configured prior to call initiation. The user may be able to configure certain attributes of this service (e.g. their own personal details), but there are other attributes which may only be configured by the conference manager (or must be negotiated) e.g. authorisation levels may be specified by the conference manager that can be used in resolving network resource allocation conflicts involving channels requested by this user. It may also be used in determining whether the user is authorised to make bookings, use certain information services or request channel with quality of service above a certain limit. The conference manager will also need to map the participants to their appropriate communication address.

Configuration Enabling Tasks for main-User:

- Specify limited directory service attributes.

Configuration Enabling Tasks for Conference Manager:

- Specify critical directory service attributes (e.g. authorisation levels).
- Map participants to communications address.

Terminal Capabilities: The main user needs to be able to determine the terminal capabilities of other conference participants e.g. the media types permissible, file transfer capabilities, types of files that the terminal cannot receive successfully etc. i.e. the main user needs to know which users can participate in which aspects of the conference in order to plan their conference activities and reduce their costs in attempting to perform actions which ultimately will be unsuccessful. For instance, the main user may want to be assured that all the participants can use video if requested (that their video CODECS use the same standards) and can transfer data if requested (that they possess compatible protocols) etc. Terminal capabilities for all users are likely to be specified through negotiation with the conference manager. However, the conference manager will translate the user's requirements into information that is meaningful for terminal specification e.g. Video Capabilities in terms of coding standards (MPEG, say); maximum output bandwidth (2mbps, say); maximum usable input bandwidth (384 kbps, say), acceptable delays etc.

Configuration Enabling Tasks for main-User:

- Interrogate recipients' terminal capabilities.

Configuration Enabling Tasks for Conference Manager:

- Specify users' terminal capabilities (through negotiation if appropriate).

Quality of Service Request: The main user may want to specify in advance of the conference the quality of service that is required. The quality of service as requested by the user may be as simple as low/medium/high, but for the conference manager there are a number of configuration tasks to be performed such as specifying the minimum bandwidth required; the minimum end-end network delays and variations in these, the maximum acceptable bit error rate etc.

The user may also want to request that the agreed quality of service be guaranteed i.e. the user may want some assurance that the quality of service they have requested is what they receive. This could be agreed through negotiation with the conference manager or through default settings for particular services or channels. For the conference manager such guarantee agreements are important so that when the network is under stress lower priority channels can be dropped first. They may also be used to resolve conflicts in resource allocation when setting up conferences. This may be a fairly complicated task for the conference manager who has to manage bandwidth across a number of different channels and users. For instance, should all bandwidth be taken from channels with lowest priority or a little taken from several channels?

Configuration Enabling Tasks for main-User:

- Specify quality of service high/medium/low.
- Specify/negotiate guarantee of service quality

Configuration Enabling Tasks for Conference Manager:

- Specify quality of service in terms of bandwidth, network delays and variations, error rates etc.
- Specify/negotiate guarantee of service quality.

4.1.2. Dynamic Conference Control

When the call is in progress there are a number of call management configuration tasks which may be necessary. Below are a few examples of these types of tasks which illustrate trade-offs in control of configuration between the main-user and the conference manger (systems manager).

Add New Users: When setting up a conference call it is unlikely that end-users will be able to specify all participants in advance on every occasion i.e. the main-user will often want to add a new user to the conference during the call. For the end-user an ideal solution to this configuration task may simply entail selecting a new participant from the directory service. If there are available channels associated with the conferencing service this may not be problematic; however, if resources are limited or the network is under stress then the conference manager would require control of the allocation of channels, and hence the addition of a new user would require negotiation. Of course, adding a new user also has implications for the required quality of service for the conference as a whole as specified prior to the call.

Configuration Enabling Tasks for main-User:
- Add new user from directory

Configuration Enabling Tasks for Conference Manager:
- Add new user from directory
- Allocate channels and privileges etc.

Change Quality of Service: The main user may want to change the quality of service during a conference. If the main user is monitoring costs during the call they may for instance be prepared to go to a lower quality call in order to reduce costs and prolong call-time, or conversely may wish to increase call quality during difficult tasks. For the user, changing the quality of service might simply be from low to high or vice versa. For the conference manager, the capacity for such changes would either have to be specified in advance and given as default settings (in the case of high to low, say) or negotiated in real time (for low to high, say). As in the case of static configuration, the conference manager has a number of configuration tasks to be performed such as specifying the minimum bandwidth required; the minimum end-end network delays and variations in these, the maximum acceptable bit error rate etc.

Configuration Enabling Tasks for main-User:
- Specify quality of service high/medium/low
- Negotiate service changes with conference manager.

Configuration Enabling Tasks for Conference Manager:
- Specify quality of service in terms of bandwidth, network delays and variations, error rates etc.
- Set configuration defaults

Split Conference: During the course of a conference users may want to split the conference into smaller sub-groups of participants if, say, the conference becomes unmanageable because of numbers or if sub-groups want to increase the efficiency of the meeting by splitting into work-groups to discuss particular topics etc. From the users viewpoint, simply allocating users to users to sub-groups without losing the conference connection would be preferable to having to re-start two separate conferences. Establishing the second conference, however, would also require the users to specify the participants for each sub-group, designate a new main user for second conference, and negotiate ownership and charging of the services amongst themselves (or through negotiation with the conference manager). For the conference manager such tasks become complicated with respect to establishing and allocating appropriate channels for the two conferences whilst maintaining guarantees of service quality if possible and balancing the needs of the network as a whole.

Configuration Enabling Tasks for main-User:
- Specify participants for each sub-group
- Designate new main user for second conference
- Negotiate ownership and charging of the services
- Negotiate conference changes with conference manager

Configuration Enabling Tasks for Conference Manager:
- Allocate channels
- Maintain service quality guarantees.

In sum, the above description of static and dynamic configuration was designed to show how the negotiation which needs to take place between first and third party conference users, controls how the configuration enabling tasks can be allocated across these different users worksystems, which affects how the design of the video conferencing service.

5. The allocation of service configuration enabling tasks

The allocation of configuration enabling tasks to worksystem entities (human or machine) has implications for the ease of use of services and hence their uptake and use. Inevitably there will be trade-offs between desirable properties of the service for the end-user (e.g. simplicity and controllability) and design/operational constraints for other nested users (e.g. service providers) who are supporting the end-user. The purpose of this section is to consider *how* to allocate configuration enabling tasks and to propose some candidate criteria for deciding to *whom* (or to *what*) they should be allocated. Section 5.1 discusses allocation in the context of the Enabling States Approach and describes the distinction between fixed and contextual allocation. Section 5.2 describes points in the IBC network design process where allocation decisions have already been made and discusses their influence of configuration enabling tasks. Section 5.3 proposes some candidate criteria which can be used when making decisions about the allocation of configuration enabling tasks.

5.1. Types of Allocation

In general, the Enabling States Approach takes the position that end-users will not want to be burdened with enabling tasks, since they detract from the cognitive resources available for their goal tasks. The agents for Enabling Tasks should, wherever possible, be their terminal or some non-human communicating entity, and where an enabling task must have the human user as agent, then the task should be designed to be as automated as possible, requiring the minimum of effort on the part of the user. However, because of the complex interrelations between the nested users of an IBC system (as in the example of the video-conference) it may be the case that some enabling tasks identified for the end-user should not be allocated to the end-user's worksystem agents, but to a different user, e.g. the network manager. This allocation will depend the negotiation processes that need to be carried out between the different users, to configure a particular service in the best way to meet the end-user requirements—certain enabling tasks would need to be allocated to the network manager because of the negotiations which would have to take place between the network manager and the relevant value added and public service providers.

It is clear, then, that decisions about where and to whom to allocate configuration enabling tasks is a complex issue with respect to the usability of services for end-users. There are two basic types of allocation decisions that can be made which we will term *fixed allocation* and *contextual allocation* (this distinction is more commonly known as 'static allocation' and 'dynamic allocation', but the terms 'fixed' and 'contextual' are used here to avoid confusion with the distinction between 'static' and 'dynamic' configuration tasks made in sections 4.1.1 and 4.1.2). In the case of fixed allocation a fixed number of tasks are allocated either to the end-user or some other work system entity (including the system manager) with each being responsible for their allocated tasks. Fixed allocations are likely to be assigned during the set-up phase of a service.

A contrasting approach is that of contextual allocation where the end-user and other worksystem entities (e.g. the user's terminal or the network or the system manager) are jointly responsible for particular tasks. In this case responsibility for a given task at a given time (during the use of a service) can be assigned either to the end-user or some other system entity depending on who is best able at that moment to perform the task. This kind of contextual allocation might be specified during the set-up phase of a specific service using simple heuristics or may rely on some degree of 'intelligence' in the network or terminal.

For example, whilst using a single service the end-user may wish to maintain control of bandwidth directly; however, during the use of multiple services it may be more appropriate for the system manager or the user's terminal to monitor and control the use of bandwidth on behalf of the user, i.e. during the use of multiple services some enabling tasks would be taken away from the end-user. This heuristic could be specified as an option during set-up or might alternatively rely on suitable network functionality being available to monitor and control network load. Of course, a further example of contextual allocation is the negotiation process between end-user and system manager or service provider.

5.2. Function Allocation Decisions Points in The IBC Design Process

The service design process has been described in chapter 1.1 as consisting of a number of iterative stages:

Step 1) Usage Context -External Perspective

A formal description of the application is made in terms of goals and tasks, user and customer characteristics, and performance parameters. In the context of usability, performance parameters include individual variables such as desired levels of cognitive and subjective costs. This stage is the starting point, and does not include any function allocation.

Step 2) Reference for User Services (RUS)

For this step the major design decision is taken to distribute the tasks of the application to geographically separated users. The result of this stage is a Reference for User Services, or RUS. The network is described in terms of communicating entities and the channels that link them. The communicating entities are described in terms of the states that they need to be in to carry out the tasks allocated to them (a description of their required knowledge, skills, abilities, etc). The channels are described in terms of the tasks that they must support. This is clearly an important function allocation stage, which affects the complexity of the network for the user, the tasks the user is responsible for, the complexity and interdependence of the tasks, the number of different entities that the user needs to communicate with, etc. Relevant user characteristics are individual goals, roles, skills, and budget.

Steps 3 and 4) Specific User Service Analysis

The result of these steps is a Specific User Service—SUS—or a Generic User Service—GUS. This stage consists of the specification of service features for the channels necessary for the tasks. There are two components to this.

Firstly the required states of the communicating entities are compared with their actual characteristics (given in Step 1, above). Where there is a difference, Step 2 can be repeated to redescribe the network, or nested services can be proposed to support the communicating entities, or the entities can be left to acquire the necessary state themselves. This is again an important function allocation decision, determined by user characteristics. For example, the APPSN project has designed service centres that will provide remote audio-visual translation services for users with visual and hearing impairments. There are clearly some tasks that cannot be left to these users, and so at this stage additional entities and services could be proposed to support them.

Secondly, the infrastructural features of both nested services and the main channels are described. These features are determined by user, task and performance parameters, but do not, we think, require function allocation decisions.

Steps 5-8) User-Service Interface Design.

These steps (described in chapters 3.1, 3.2 and 3.3) begin with an Enabling States description and end with a functional specification of the interface. An Enabling State is a state of the network at a particular point in time needed for a communicating entity to perform a task or tasks. The state must therefore be appropriate for particular types of communicating entity, and consists of a detailed integrated description of nested services, main services, other communicating entities and communicating entities terminals. No function allocation decisions need to be made here.

Following the descriptions of Enabling States, the tasks needed to create the states need to be defined in an Enabling Tasks Description (e.g., initiating a call, asking for someone's attention, manipulating terminal equipment, etc), and the agent who is to perform these Enabling Tasks must be specified. This is the third important function allocation decision point. In general, users will not want to be burdened with Enabling Tasks, since they detract from the cognitive resources available for their goal tasks. The agents for Enabling Tasks should, wherever possible, be their terminal or some non-human communicating entity. In some cases, for example, the people with special needs of APPSN, the users may be physically incapable of acting as the agent of an enabling task, while in others, they may simply lack the knowledge or skills required (for example, to set up network parameters and connections). Where an enabling task must have the human user as agent, then the task should be designed to be as automated as possible, requiring the minimum of effort on the part of the user.

Subsequent design levels work from these task descriptions to provide a functional specification of the interface.

In sum, then, within the design process, then, there are three points at which function allocation decisions important to users occur:

1 Usage Context—External Perspective
2 Reference for User Services------------FUNCTION ALLOCATION
3 SUS—Application level---------------FUNCTION ALLOCATION
4 SUS—Task Level
5 Usage Context-Internal Perspective
6 Goal Task Design
7 Enabling state descriptions
8 Enabling task descriptions -------------FUNCTION ALLOCATION

5.3. Guidelines for the Allocation of Service Configuration Enabling Tasks

The purpose of this section is to propose some initial criteria which can be applied when allocating configuration enabling tasks either to end-users or other work system entities. The criteria are divided into two sets: those concerned with the end-user and those concerned with other parties acting on the end-users behalf. The lists given below are intended to be illustrative rather than exhaustive, and are being refined and developed through experimental work and simulation that is in progress in the RACE 2 project, ASCOT.

5.3.1. *Candidate Criteria for the Allocation of Configuration Enabling Tasks to the End-User*

GUIDANCE identified a number of generic design principles aimed at improving the ease of use of IBC services for end-users (see chapter 3.2 for a fuller description of these principles). Some of these design principles are particularly relevant to the allocation of configuration enabling tasks and are intended to describe desirable system properties from an end-user's perspective. The candidate criteria are:

- **Controllability**:
  The user should be able to control the actions of the machine.
  For example, when using a video conferencing service end-users should be able to split a call into two distinct calls. This is a desirable system property that the user should have under their own control.

- **Flexibility**:
  The user should be able to choose the modalities of their task performance with respect to the input media available and the experience they have.
  For instance, when using a video conferencing service end-users should be able to allocate and de-allocate service components and media components as required.

- **Compatibility:**
  Users should be able to apply knowledge they have gained from outside the system.
  For example, end-users are familiar with enhancing the quality of television output using sliding controls for contrast, brightness etc. By analogy similar controls might be provided for increasing the quality of videocommunications by increasing bandwidth (but from the end-users point of view by simply increasing quality from low to high, say).

- **Coherence:**
  The user should be able to generalise their experience between system components.
  This criterion implies that users should be able to generalise their experience about improving quality of transmission (say) between different services such as conferencing and distribution.

- **User Characteristics:**
  Designers should consider the users' physical, cognitive and affective requirements such as skill use, learning, control and variety of task.
  This criterion implies that when allocating configuration enabling tasks designers should be aware of problems associated with increasing service complexity with respect to the cognitive and affective capabilities of the end-users.

5.3.2. *Candidate Criteria for the Allocation of Configuration Enabling Tasks to Other Parties*

These criteria can be contrasted to those described above in that they refer to other parties acting on behalf of the end-user. An example of this might be a system manager may who is allocated a number of configuration enabling tasks instead of the end-user. Of course, there will be trade-offs associated with allocating a task to the system manager rather than the end-user. The criteria are intended to inform such trade-off decisions. The candidate criteria are:

- **Engineering Operational Requirements:**
  The allocation decision may be constrained when the function under consideration may need to be carried out with a considerable to degree of accuracy, repeatability, or reliability from an engineering perspective. i.e. some configuration enabling tasks cannot be allocated to the end-user because they are not feasible due to either the network or hardware capabilities, or the service platform e.g. when the network is a token ring LAN (Local area network), an end-user cannot change partners or redirect a call.
- **Mandatory Allocations:**
  Some decisions may already have been taken at a management (or application) level that may constrain the designer's options for allocation of function. For instance, senior management may have already taken the decision to utilise a newly developed technology, or may be satisfying other customer demands.
- **Cost:**
  In many systems monetary cost is the major determining factor for determining allocations. Unfortunately, the operational costs are not always traded off against development or user costs. For example, the network manager or Public Network Operator (PNO) may not allow the end-user to retrieve a video clip which involves the enabling task of increasing the bandwidth during a call as organisational policy requires calls to be kept to a minimum cost (increasing the bandwidth increases the call charge).
- **Performance Optimisation:**
  Designers should aim to meet not only the technical but also the Human Factors requirements of the worksystem. This criterion is essentially a reminder that trade-offs are necessary between organisational performance requirements and those of the end-users within that organisation.
- **User Characteristics:**
  Designers should consider the users' physical, cognitive and affective requirements such as skill use, learning, control and variety of task. This criterion implies that when allocating configuration enabling tasks designers should be aware of problems associated with increasing service complexity with respect to the cognitive and affective capabilities of the end-users.

6. Conclusions

In this chapter we have described the potential improvements to telecommunication services that will become possible through the IBC revolution. We have also tried to show that if these services are not implemented in a methodical way, with attention to the needs and requirements of the users, then the result will be a disaster. Fortunately, the research that has been conducted in the past few years by the RACE programme has provided the basis for a coherent approach to the design of integrated services, and the eight-step method presented in chapter 1.1 ensures that users at all levels of an application are taken into account.

Users of a system include not only the person in front of the camera or at the keyboard, but 'above' them the person or organisation who supports their usage and provides the resources for it, and who may be responsible for ordering the services in the first place, and 'below' them the people or organisations who are affected by their communication. For example, a travel agent using a multimedia database will have been provided with the equipment and connections by their company, and will be using them for the benefit of the agencies customers. The users of a service must also interact with the service providers, for example the PNOs, and their commercial and functional needs must also be taken into account.

The IBC pilot projects reported in this chapter have provided a wealth of information about the usage characteristics of the services for all classes of user, and the eight-step method provides a way of allocating the tasks necessary for the accurate yet flexible configuration of services through this hierarchy. In the chapters that follow in this section, we will consider in more detail the needs of the organisations that commission IBC services, and of the support required by people who must design them. If the purchasers of IBC services are not made aware of the importance of usability information, they will not be prepared to invest in services which provide it, and so may waste their investment; and if the designers of the services do not have adequate access to usability information they will find it increasingly difficult to build integrated and usable services.

Acknowledgements

Much of the material reported in this book has been provided by the URM project, especially the descriptions of the application pilots from the first phase of RACE. The project has produced a number of Deliverables describing their work in more detail, and has also produced a Hypercard stack containing summaries of the RACE pilots. Copies of this information can be obtained from IFC Research, 99 High Street, Guildford, Surrey, GU1 3DP, U.K. The project R2089 ASCOT has the objective of providing a toolset for service configuration. Copies of deliverables can be obtained from the prime contractor: MARI Computer Systems Ltd, Old Town Hall, Gateshead, NE8 1HE, UK.

RACE projects

The following RACE projects have been discussed in this chapter. Further information about each project can be obtained from Race Central Office, European Commission, DG XIII/B, 200 Rue de la Loi, BU-9, 1040, Brussels, Belgium.

APPSN	R1054	Application Pilot for People with Special Needs
APTITUDE	R2030	Advanced Platform Technologies in Teleworking for Underpinning a Decentralised Economy
ASCOT	R2089	Ad-hoc Service Configuration Tools
CAR	R1079	CAD/CAM for the Automotive Industry
DIDAMES	R1060	Distributed Industrial Design and Manufacturing of Electronic Subassemblies
DIMUN	R1039	Distributed Manufacturing
DIMPE	R1061	Distributed Integrated Multimedia Publishing Environment
DIVIDEND	R1059	Dealer Interactive Video
ECHO	R1074	Electronic Case Handling in Offices
GEOTEL	R1073	Library Services for Petrochemical Industries
GUIDANCE	R1067	Generic Usability Information for the Design of Advanced Network Communications in Europe
MARIN	R1062	Ship-to-Shore Communications
MULTIMED	R1042	Functional Service Integration in Support of Professional Groups
MUSNET	R1078	European Museums Network
PPV	R1070	Pay-Per-View in Europe
RESAM	R1058	Remote Expert Support for Aircraft Maintenance
TELEMED	R1086	Telemedicine
TELEPUBS	R1075	Telepublishing
TIM	R2078	Tourism Information and Marketing
URM	R1077	The Usage Reference Model

Computers, Communication and Usability:
Design Issues, Research and Methods for Integrated Services
P.F. Byerley, P.J. Barnard and J. May (Editors)

Chapter 1.3

Usability and the Market for New Services

Brian Champness* and Ian Dennis†

*IFC Research Ltd, Sundial House, 99 High Street, Guildford UK

†Plymouth University, Plymouth, Devon, UK

1. Introduction

Human engineering, human factors and usability research has fed largely into the design and manufacturability of new products, not directly into their marketing, or into the marketplace. Over the years various lists of features have been constructed to illustrate the usability approach—from Shackel to Byerley—but few have incorporated the idea of the *market* for new services. One exception was Hewlett Packard in the '80s, which insisted that Human Factors should assist *marketing* in the identification and investigation of users and their needs, and in the analysis of competitive products. Ergonomic and *market*-based requirements were to be researched hand in hand, and the *marketing* department was to be provided with information about product requirements that would satisfy user needs. *Marketing* was an integral part of the iterative process involving human factors working hand in hand with design and manufacturing.

In retrospect this intimate involvement between Hewlett Packard's usability research and marketing, (despite the fact that it appears to have been somewhat transitory), may prove to provide some lessons for us all. There have been a number of examples of large publicly funded R&D projects designed to produce new market opportunities that have patently failed to do so, (the UK's Alvey programme was one). Although RACE is certainly not one of these, it can be argued that usability research within RACE should extend its remit somewhat in reaching out to marketing as well as human factors departments in large organisations.

There are a number of other reasons why this approach may be appropriate now. In the mid '80s Eason (1978) could say that the ultimate measure of usability is whether a system is used. Although he added that this would partly depend on its cost and benefits, today it could be said that the ultimate measure of usability is whether or not a system *sells*, in which case the attention of usability researchers may increasingly be drawn to the market, and all that the concept entails.

It is also the case that usability research has considerably widened its remit in recent years to encompass not simply products, costs, users and tasks but whole systems and their service environments (Byerley and Bruins 1992). It is no longer possible to make a simple distinction between 'appeal' (marketing) and 'usability'. A product or service that may be functional, but which users find unappealing will achieve low ratings on current conceptions of usability. The skills and findings involved in usability research today may be of increasing relevance to those involved in finding markets.

The purpose of this chapter is to suggest that usability researchers should move closer to and work more closely with their colleagues in market research and marketing departments, and to provide an outline model for how this might be begun. There may be some initial resistance or scepticism on the part of those involved in marketing and sales. After all they are, like Anthony, but unlike usability researchers:

> *"Enthron'd i' the market-place";*

but unlike Anthony they are not:

> *"sitting alone, whistling to the air",*

and may need to be persuaded that usability has something powerful to offer them (Anthony and Cleopatra, Act 2 Scene 2).

2. The benefits of the usability approach

The power of usability research both in influencing designers and in influencing decision making by presenting information on user needs is well established and documented. Elements of these twin benefits were discussed by Connell & Arlandis (1982). They concluded that the usability approach, and long term usability research "minimises the risk of designing out key user requirements and maximises the chances of fast development of demand once networks are rolled out". The latter is important at a time and in an area where market success will surely depend on offering users flexible alternatives to existing products and services, and the ability to react quickly to opportunities provided by fast changing technological developments.

The cost effectiveness of usability research has been discussed, but not yet formally demonstrated in the RACE programme. Outside RACE Nielsen and Phillips (1993) are among the few researchers formally to estimate the costs and benefits of usability engineering in monetary terms. They point out that whereas the *costs* of various usability methods are easy to measure by adding up staff time the *benefits* are much harder to measure. However, comparing the use of different computer interfaces, they were able to show that over 3,000 users a good pop-up interface could save $870,000 in lost time, (based on a user costing $100 per day). It might be possible using similar methods, to assess the notional loss of possible revenue resulting from 'designing out key user requirements', or from failing to develop the new services that are being proposed in RACE.

Formal elucidation of the financial benefits of usability programmes would go a considerable way towards persuading market development departments in corporations to take notice of and perhaps co-operate more closely with usability teams. As Hewlett Packard found, long term usability researchers need to develop iterative relationships with both company and Telecom Operators' market researchers, market policy specialists and market strategy departments. There is a long way to go, however, between identifying the requirements for products and services together with demonstrating the possible financial benefits of using the results of usability research, and the market place.

3. Putting knowledge into action

There is a difference between knowledge that is 'applicable' to the real world, and knowing how to put that knowledge into action—Argyris (in press) called that 'actionable' knowledge. His distinction between knowledge that is 'applicable' to the real world, and knowing how to put that knowledge into action ('actionable knowledge') is relevant here—how to turn the results of long term usability research into knowledge that can be acted upon by not only by designers of new services, but also those who turn these into new markets.

Applicable research must be shown to be relevant to contexts beyond that in which it is produced—the URM project and more recently the URSA project have produced service typologies and proposed new services that go beyond the context of the data from which they were produced. To be actionable in the market context, conceptual and behavioural bridges will need to be built between this work and the market place.

An alternative way of describing the process might be to say, for example, that in the URSA project descriptive research tells us what the universe of applications and services looks like; subsequent normative research then goes on to generate ideas for new services. Prescriptive research would then provide a method of moving from usability to the market for new services. Argyris maintains that research that produces knowledge that is genuinely actionable combines descriptive, normative and prescriptive features.

In order to move closer to the marketplace, long term usability research may need to use the language of marketing, and become more prescriptive.

4. Market research and the use of usability

Market researchers in the general area of telecommunications do consider usability, but mostly in connection with terminals, not telecommunications or networks. Survey respondents are asked what they think of new products and services. In telecoms market research these questions tend to concentrate on the end terminal, rather than telecoms aspects, so features, functions and costs are important.

Market researchers are also out to identify reactions to 'designer' features that might make certain chosen products more attractive to targeted markets. A conspicuous recent example is the cellular phone, where the 'concept' was felt to be as important as the actual service. This concept incorporated both aesthetic factors (beauty) and status, together with the fact that it was new, different and special.

It is not difficult to find examples in the hardware field of usability being featured in market plans, promotion and advertising; but examples are rare in the marketing and selling of telecommunications services. In addition, usability is easier for market researchers to conceptualise in terms of 'kit' (hardware), than 'service', particularly telecoms services.

So marketing telecommunications services has always been more difficult than other hardware or software products, and will become more so in the IBC context—when flexibility of use and tariffs, together with the users' opportunity to be unpredictable in their behaviour will be the major benefits to new IBC users (Connell & Arlandis 1992).

5. Components of a bridge between usability research and the market

Usability research can lead to better design of products and services, and, using the broad definition espoused in chapter 1.1, can help to ensure that the chosen new products and services will be cost effective. In turn, better product design may lead to increased sales, and better choices may result in more carefully targeted sales campaigns. Figure 1 illustrates ways in which usability research could extend its usefulness towards the marketplace.

The figure suggests that there is a considerable distance between usability research and the market, although the way to strengthen the lines from applicability to actionability is clear—more contact and collaboration with those closest to the market.

The top left and bottom right corners represent the ultimate question implicit in the title of this chapter: can usability research help to increase sales and enhance profit ratios? Although no direct answer is currently available, it is possible to point to intermediate links in the chain, and to suggest specific ways in which usability may extend its benign and possibly profitable influence.

Extension A suggests a closer relationship between usability's traditional contribution to the design of products or services and the market research survey—it is rare for the two teams or departments involved to work together. Extension B brings usability even closer to the decision making processes involved in the choice of products to market and sell. Market research may for example be able to show that a potential service would be extremely attractive to potential customers, but usability researchers, with their knowledge of current and, more important, likely future tariffs for the sophisticated switched high-band links required may be able to demonstrate that the service is unlikely to be cost effective for x years, but could be so after that time. More dramatically, the kind of research undertaken within URSA—able to estimate and predict demand for new kinds of product such as complex integrated services—could also help to modify and sharpen market strategies.

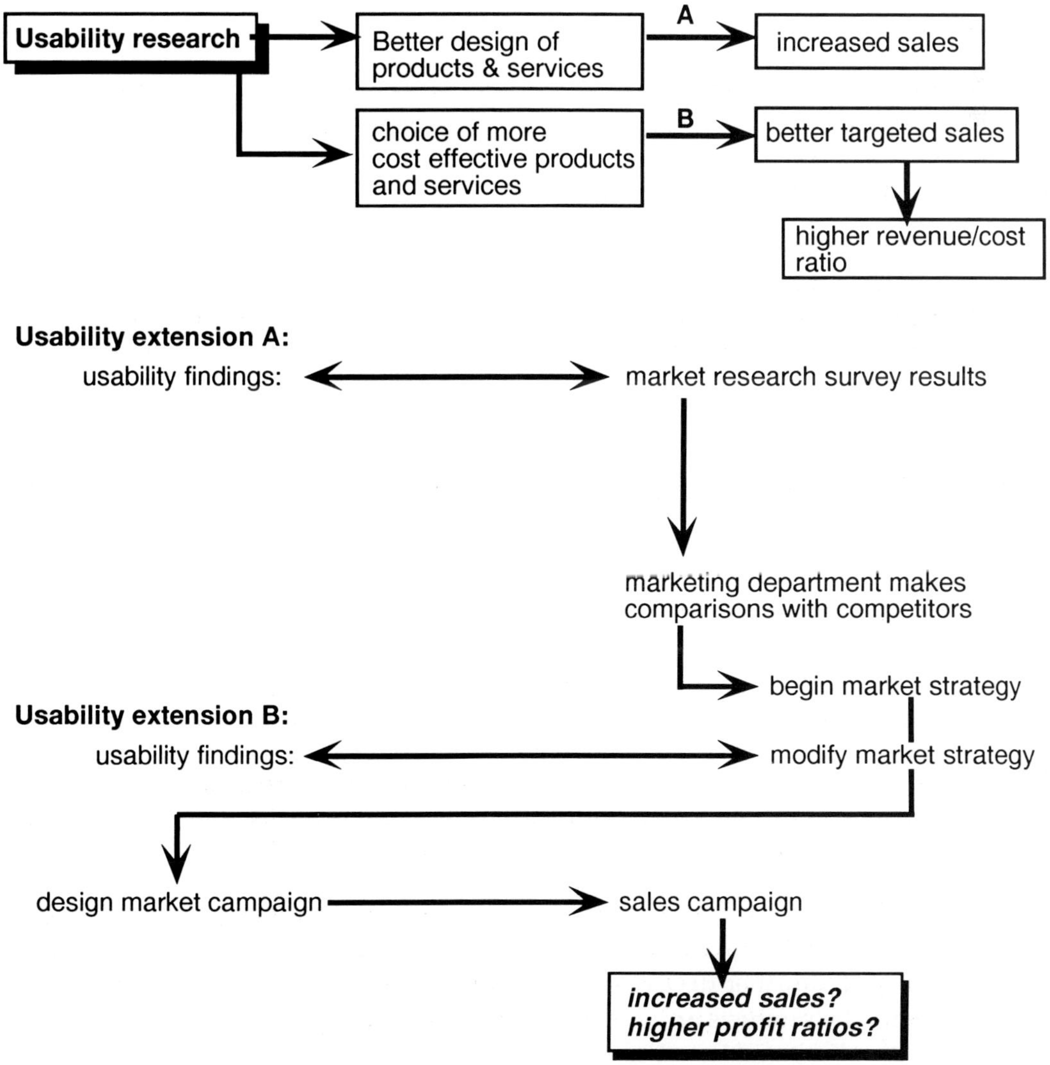

Figure 1: From usability to market research

Surveys or assessments of user needs do not provide the same functions as market research. What the user says is needed to perform a given task, or satisfy an application may not be the same as the arguments that persuade the customer to buy—although there may well be overlaps that can be exploited.

In the figure *market research* is aimed at finding out what will sell, and how best to sell it. Marketing people take the results from market research, previous sales figures from related products, make comparisons with competitors, talk to promotion and sales people, then design a marketing campaign. Usability researchers could, and should, be involved at several stages in the preparation of such a campaign; at the least the design of their own instruments and experiments should be informed by the needs and activities of the marketing department—and vice versa.

The market itself can be described in a number of ways:

- product/service definition
- gross size in terms of:
  - number of sales
  - revenue/profit
- customer characteristics/profile.

Long term usability research certainly contributes to product service and definition, usually via their design. It could also contribute to the building of customer characteristics and profiles if there were closer collaboration between usability assessment of needs and market research surveys on customer reactions. Finally the kind of work carried out in the URM project and more recently in the URSA project could provide useful inputs into the design and targeting of marketing and sales campaigns—particularly for new IBC services.

6. Usability and the market in the RACE programme

Although there are few references to the market as such in the RACE programme, the structure of usability work in the programme is more conducive than most to incorporating market factors. A number of models appearing in RACE literature, and in Byerley and Connell (1992) are capable of absorbing additional modules or components that would encourage the collection of market-related data in addition to micro and macro economic data, and provide the motivation for closer relationships with those involved in marketing.

Usability research that is, to use Argyris' term, 'genuinely actionable' can be found in Whitefield et al (1992), further developed in chapters 3.1 and 3.2 of this book. After pointing out that sometimes the products or human factors research has little influence on the design of interactive systems, Whitefield et al concluded that the reason may be that "their usefulness for design is not demonstrated". The GUIDANCE project (RACE 1067) attempted to deal with this by developing a design method providing a designer not simply with descriptive and normative results but prescriptive advice as well. It uses a clearly formulated method with demonstrable practical and (from the designer's point of view) usable results.

Their perspective, which is reviewed in section 3.2 of the previous chapter, focused on the ease of use of systems—how a number of services could be integrated to maximise ease of use (a component of usability). They pointed out that there are problems involved in integrating services via IBC, when computing, networking and telecoms technologies will merge. They go on to say that even when physical aspects of integration will be solved the "mental problems of learning, conceptualising and recalling how to use services will not disappear".

Equally important and difficult mental problems will arise when marketing departments have to describe these new integrated services in their sales campaigns. The next stage might well be to develop a programme using the perspective of the user as customer or purchaser (in RACE terminology this might well be named SELLWELL). This could draw on the methods and models developed in the GUIDANCE project, but translate these into the prescriptive language needed by those closest to new markets.

As for the RACE programme as a whole, 'market' is usually referred to only obliquely. As the running argument in this paper is that usability research, particularly as represented the RACE programme, should build additional bridges to the market, it is finally worth exploring the question: should RACE participants be interested in the market? After all, RACE is a pre-competitive R&D programme.

The four unifying concepts underlying RACE (RACE, 1992) are:

- implementation
- engineering
- verification
- collaboration

The implementation concept refers largely to the need for Telecoms Operators (TOs) to make 'major infrastructure-led investment decisions', in connection with establishing trunk networks via broadband islands. This must be synchronised with major business and industry activity centres, and the strategy followed should be timely and flexible enough "to follow market demand".

Although commercial services are not thought to be implemented until 1995/6 if the design of and results from the communications experiments are to lead to genuinely commercial services (as they must), then any implementation models in RACE must include ways of making usability research *actionable*, which means that new relationships with marketing departments in the companies involved in user research do indeed need to be forged. This is particularly relevant in RACE, which is designing, describing, and prescribing novel services specifically for IBC. These will involve complex integration in ways that should provide the user/*customer* with increased flexibility in hardware, software and tariffs; and perhaps most important will allow the customer to behave spontaneously and unpredictably. The challenge of describing, marketing and selling these services will require a combination of usability and market research skills.

References

Argyris, C. (in press). *Actionable Knowledge*. San Francisco: Jossey-Bass.

Byerley, P. & Bruins, R. W. (1992). Conceptual framework for the usage of telecommunication services. In *Integrated Broadband Communications: views from Race- Usage Aspects*. North Holland Series in Telecommunication, vol 18.

Connell, S. & Arlandis, J. (1992). Commercial issues in the Definition and Marketing of Broadband Services. In P Byerley and S Connell (eds), *Integrated Broadband Communications: views from RACE—Usage Aspects*. North Holland Series in Telecommunication, vol. 18.

Eason, K. D. (1978). Towards the experimental study of usability. *Behaviour and Information Technology*, 3 (2), 133-143.

RACE (1992) *RACE '92:Research and technology development in advanced communications technologies in Europe*. Commission of the European Communities, Directorate General XIII.

Nielsen, J. & Phillips, V. L. (1993). Estimating the relative usability of two interfaces: heuristic, formal and empirical methods compared. To appear in *Proceedings of InterCHI'93*. ACM: New York.

Whitefield, A, Byerley, P F, Denley, I, Esgate A, & May, J (1992). Integration of Services for Human End-Users (1): Design Principles, Enabling States Analysis, and a Design Method. In: In P Byerley and S Connell (eds), *Integrated Broadband Communications: views from RACE—Usage Aspects*. North Holland Series in Telecommunication, vol. 18.

Computers, Communication and Usability:
Design Issues, Research and Methods for Integrated Services
P.F. Byerley, P.J. Barnard and J. May (Editors)
1993 Elsevier Science Publishers B.V.

Chapter 1.4

Supporting the Designers: Reaching the Users

Deirdre Terrins-Rudge* and Anker H. Jørgensen†

*Department of Applied Computing, Faculty of Informatics, Magee College, University of Ulster.

†Department of Psychology, Copenhagen University, Njalsgade 88, DK—2300 Copenhagen S.

1. Introduction

A lot of effort in the HCI research community is directed towards developing formal methods and principles for design of user interfaces to be used by designers in the software industry—be it in designing IBC systems or less sophisticated systems. However valuable these methods and principles may be, if they are not taken up by designers then they will never be able to affect service engineering. This chapter covers the requirements of the designers themselves by focusing on the designers' working contexts, the support they require, and the difficulty of forcing the design process to fit a formal, structured model.

This chapter is concerned the activities of designers in the software industry. These activities can be referred to as systems design, system development, system engineering or software engineering. The issue of providing support for designers' requirements is relevant to all manifestations of design of computer artefacts. This chapter aims to provide a description of the product design context for usability in current research and practice. The issues of product design focus in particular on the opportunities in the advanced communication services sector within the next 20 year period.

2. Designers designing and designerly thinking.

The design of IBC services is a focus which sharply delineates specific instances of generic requirements of computer artefact designers. As noted by Cameron and Terrins-Rudge (1993) the perceived requirements of end users and the perceived modes of representation for delivery are often at variance:

> "Producers of multimedia applications (for example) have typically assumed a basic community of end-users who find little or no difficulty in interpreting the framework within which the information is represented. The given media utilise modes of representation which are presumed to be lacking in features which may present the consumer with any cause for doubt or uncertainty."

IBC artefact design provides an exemplar instance of designers' requirements by virtue of the fact that it brings together elements which interact to produce an intensified need for timely and adequate support.

IBC service design and delivery is a skill in short supply in a sector of burgeoning requirement. The providers and maintainers of telecommunications networks have an urgent business case to answer. Can they harness and exploit the potential of niche markets for value added services, (e.g. premium rate and global corporate services), or will they watch the growing trend of having their networks used by the plethora of independent premium rate service providers currently operating downstream? This is the straightforward business problem of 'termites in the woodwork', where the traditional core competence of the organisation (a large scale network operation) is at odds with the skills required to meet a market need (bespoke customer services). It is all the more urgent when considered in the light of trends toward corporate downsizing, enterprise-wide integrated communications and computing, and the promise of multimedia and computer supported co-operative working in the delivery of information. The trend towards converging standards and customer perceptions of usability further reinforce the need to understand and deliver timely and acceptable services.

The characteristics which define IBC artefact design as an exemplar are:

- Time as the most critical factor in the design process.
  The design cycle for IBC artefacts must be short to allow speedy time to market and timely response to transient niche markets.
- Design decisions which deal with technical and subjective issues.
  IBC artefact design needs organisation and theory which deals with both technical control aspects and practical issues of intersubjective communication.
- Problem scope
  IBC design activity consists largely of problem setting in order to render the problem amenable to available solutions.

The designers' role in service design is to frame the questions relating to design decisions within the framework of deliverable services.

These characteristics are shared with many design questions. Professional practice in design, as in many professions, is perceived to be a process of problem solving. Problems are addressed by the selection of available means in line with established principles in order to find a solution. This view of professional practice accords with the assumption of technical rationality in professional practice (Schön 1983). Three major components of professional knowledge are generally agreed upon (Schein 1973):

'1. An underlying discipline or basic science component upon which the practice rests or from which it is developed.
2. An applied science or 'engineering' component from which many of the day to day diagnostic procedures and problem- solutions are derived.
3. A skills and attitudinal component that concerns the actual performance of services to the client, using the underlying basic and applied knowledge.'

(from Schein 1973, p.39)

Thus professional practice is seen to consist of the application of basic science to achieve applied science, while applied science in its turn provides diagnostic and problem-solving techniques which can be applied to the delivery of professional services. The delivery of professional services is dependent on the use of applied knowledge which in its turn is dependent on the existence of an understanding of scientific foundations. This rationalistic tradition has become part of the fabric of our assumptions in prescribing methods and principles for design activity, especially in relation to computer related artefacts.

A review of the background to design requirements and support in computer artefact support leads directly to Herbert Simon and his 'program for a rationalistic science of design' (Simon 1969). This rationalistic tradition was based on a critique of the engineering approach to design activity. This critique found no real scientific base for design and viewed the engineering approach as a craft based activity. Simon (1969) stated that our knowledge about design was 'intellectually soft, intuitive, informal and cookbooky'. He suggested we develop a science of design applying the methods of logic and mathematics to solve the problem. Thus the problem became one of defining the functionalities of artefacts and their uses and users and evaluating the outcome across possible alternatives. Unfortunately this loses most of the essential elements of design activity. The creativity of professional designers and the creative input of users which by definition cannot be formalised in a decision theoretic sense. Design is essentially a practical activity involving judgement and experiential learning.

This type of understanding of design requires the inclusion of the subjective elements of design activity along with the objectivity provided by the rationalistic tradition. There is the subjectivity involved in the true inclusion of the users, end users and procurement decision makers. The formalisations of systems thinking are not disputed as being necessary for traditional action within systems design. They are tried, tested and proven. What is required is a supplementary reinterpretation of design methods in the light of what we already know about 'designerly' ways of thinking (Archer 1982). Creative designerly ways of thinking and doing design are inherently co-operative. They involve the co-operative effort of users and designers. Design methods need to reflect the design process to provide mechanisms which allow the uptake of judgements and creative assessment on the part of designers and users. This sort of provision is complementary to the decompositional rational approaches which we now have and are refining in general human computer interaction and usability engineering disciplines. Some recent work in provision of designer support has begun to go further in recognising the need for complementary representation and mechanisms. The issue here is to design usable design methods.

Having considered designers designing and designerly thinking, the following sections begin with a description of some real design behaviour, and then look at the application of principles and working methods for achieving design success. The importance of negotiation in the design process is outlined and the emerging research agenda in understanding and supporting design activity are reviewed.

3. An Empirical study of designers designing

The section centres around one research package in the Esprit project AMODEUS called 'Empirical Designer Studies'. Its purpose was to 'characterise and document the properties and heuristics underlying design decision making'. In this package we adopted two approaches to empirical studies of design. The first approach was to study ongoing projects in commercial settings. We looked at ten systems from a variety of domains: an air traffic control system, a system for animating formal software specifications, a database system for administering soil samples, and a liquidity system in a bank. We employed a number of data collection methods, ranging from participant observation, structured interviews, questionnaires and video/audio recordings. Our main findings were that a common strategy in user interface design is muddling through (Lindblom, 1959). This strategy is a way of coping with uncertainty, turbulence and ambiguity—which was the order of the day in several of the projects being studied. The strategy has many benefits as it is easy to use, it is energy and time saving, and it is politically feasible. Thus it is a suitable way of making decisions in changing contexts. Our finding is supported by the extensive study of design practice by Krasner, Curtis and Iscoe (1988). They found that the three most salient problems in software design were: the thin spread of application domain knowledge, fluctuating and conflicting requirements, and communication and co-ordination breakdowns.

The second approach included workshop studies of design processes in our laboratories based on the concept of design scenarios developed within the AMODEUS project (Young & Barnard, 1987). A design scenario is 'an abstraction of designer deliberations or behaviour which concerns a design issue relevant to usability'. The design scenarios were lifted from real design projects and typically included a specific design issue related to usability and a number of design options. The task of the designers in the workshop was to give advice on the issue of usability. In the workshop two designers worked their way through a design scenario for about one hour while being video recorded. In all fourteen workshop studies have been conducted.

Computer systems designers are often accused of being computer-centric; indeed, a lot of systems witness that. However, in these workshops, we saw little or none of this. The predominant trend in the designers' focus was on user experience, predispositions and cognition, on users' tasks and application context. We must, however, add that the design scenarios carried little information about the technicalities of the systems to be employed, and that in many cases the designers have not been familiar with the technicalities of the particular system.

The most striking feature about the designers' work is that they use themselves and their prior experience as working models. In all the workshops there were references to systems they had designed, had used or had heard about. These served the purpose of building blocks in the discussion and they generated a generic frame of reference.

Intuition played a strong role in some of the workshops with experienced designers. After the designers had read the scenarios silently and started discussing, it only took 3 to 5 minutes before a tentative option cutting across the given design options was suggested. In both cases the designers ended up with that solution—but they spent lots of time on the other options underway.

Another important cognitive capability employed was imagination. As the main focus was on users, tasks, and application context—and this information was only provided sparsely—the designers sought recourse to their imagination. They used 'simulated runs' of their imagined design sketches in order to assess their advantages and disadvantages. They also make use of extreme examples and imagined the impossible in order to try out their design sketches.

An emergent feature is the role of assumptions throughout the process as well as the implicit nature of many of the activities of the designers. Some of the design options are made explicit and discussed—as are the criteria for deciding among the options. But most the elements constituting the grounds for the final design are addressed only implicitly.

The designers also in general avoid problem solving—at least deeply and thoroughly; instead, they go for partial solutions—as opposed to elaborating and analysing the different options under consideration. There is also very little explicit evaluation where options are related explicitly to criteria or goals. The main 'tool' is simulated runs that illustrate features of the proposed design options. These then feed into the process in an implicit manner. Thus, we will claim that we see no decisions being made as such—or at least very rarely. Rather we will characterise the 'decision making' as gradually evolving commitments that build up implicitly as mutual understanding between the designers.

3.1 Information Requirements

As mentioned previously, the designers focused on users, tasks and application context. As the design scenarios provided sparse information on these issues, the designers explicitly and implicitly asked for this kind of information. Thus, they were very aware of their information needs in this area. As a contrast, they never requested user interface expertise—in other words they considered their own skills as being sufficient.

In one of the workshops this point was illustrated clearly. While discussing the information needs in the end of the workshop, we asked the two designers if they could use an 'oracle' in user interface design, either during their daily work or in the workshop? The answer was a prompt 'No!' following by a reflected '...no'. But after a pause one of the designers added 'Yes, when we developed a user interface standard in our company some time ago!' Thus, in their daily work when usability is one factor between many other, they see no need for usability expertise. However, when usability is the only item on the agenda, they realise the shortcomings of their experience and skills.

3.2 Relevance from a Design Perspective

We asked the designers at several occasions whether the workshops seemed familiar and resembled their daily design work. The predominant answer were unqualified 'Yes'es. It surprised us a little—but we were of course pleased with the answer. We all know that the workshops do differ from daily design work, but the workshops seem to have caught some essential aspects of design work.

From a design perspective, a striking question is whether the design scenario approach is representative of design? In spite of the closing comment in the section above, the answer is no. They have been lifted out of real design projects but they carry only a small bit of the design context: history, dynamism, ambiguity, contradictions, roles, constraints, opportunities—all essential features of real design. Although this may seem contradictory, this does not invalidate the above statement that the design scenarios have captured some essential aspects of design work. The design scenarios undoubtedly illustrate some essential aspects of design work, but the question is to what extent and in what respect the results the workshop designers arrived at differ from the results arrived at by the real designers?

In spite of these misgivings, the design scenarios have been an excellent vehicle in the project towards integration between the modelling side and the design side. In the first half of the project when the pivotal vehicle was the behaviour scenarios, we (the design people) often shook our heads in sheer frustration with the narrow perspective of the modelling people (whose work is reported in section 2 of this book). Where was the task? Where was the user? Where was the application domain? In the second half of the project where we used the design scenarios, the modelling people started to behave as designers and cried out for additional information about the users, tasks, and domain knowledge.

A striking feature of the modelling people's application of their modelling techniques was the informality of the application. The modelling people are highly informed HCI specialists with a lot of experience in using and evaluating sophisticated user interface designs. A question arose in the minds of the design people whether they 'really' employed their modelling techniques or they behaved as informed HCI specialists? The answer is that they indeed drew on concepts and structures from the modelling techniques—but behaved as other skilled experts in a field!

This point leads to a brief discussion of the role of formalism in systems development. We would like to cite a study carried by Bansler and Bodker on the use of formal methods in systems development. They looked into the use of the widely known and used methods Structured Analysis and Design as marketed by Yourdon and colleagues (e.g., Yourdon & Constantine, 1979) They claim

> "There is a wide gap between the way systems development is portrayed in the normative scientific literature and the way it is carried out in real life"

and

> "... experienced designers pick and chose among the various formalisms given in the method, adapt them for their own purposes, and integrate them into their own design processes"

These quotes capture some essential features of the nature of design and the potential role of formal methods.

4. Applying the principles, virtual worlds and muddling through.

The context in which professional designers practice is as for many practitioners and professionals one of an going dynamic change. The changes are caused by emerging demands and changes. There are however some relatively stable references which each profession can call upon. These constants are described by Schön (1983). A review of the design and professional problem solving processes across a variety of design professions led to the delineation of a set of constants that various professions bring to the 'reflection–in–action' of their professional activity. These are :

- '• the media, languages, and repertoires that practitioners use to describe reality and conduct experiments
- the appreciative systems they bring to problem setting, to the evaluation of inquiry, and to reflective conversation
- the overarching theories by which they make sense of phenomena
- the role frames within which they set their tasks and through which they bound their institutional settings.'

(from Schön 1983)

Within this relatively solid context a practitioner is free to use test reflect upon, revise and repattern their understanding and practice. The medium in which a professional operates in order to reflect on actions taken can be a sketchpad, client-patient relationship, experimental laboratory models or interactive dialogues. Skill in the manipulation of the relevant conversational media is therefore crucial to the practitioner. No amount of understanding of the underlying rules, principles and theories can substitute for the construction of a virtual world in which to perform generative and imaginative rehearsals of actions, reflect upon their effects and test candidate solutions. These virtual worlds can only be constructed where a constant appreciative system is in place. The framing of a problem is considered to be complete when the problem can be stated in terms of the theoretical and appreciative structure of the practitioner. The constants of media, language, repertoire appreciative system and overarching theory will change slowly overtime as a profession and a professional develop—they are however stable enough to provide the framework within professional problem formation, solution formulation and testing are carried out.

This understanding of professional practice in general can help us to form a check list of the sorts of requirements which designers of ICB and other complex computer related products will require. The core list—media, language, repertoire, appreciative system, overarching theory, role frame—provides a first pass in our assessment of how we are doing in providing support for designers of IBC and other complex computer related artefacts. A surface review will serve to outline the need for a concentrated research agenda in this arena:

- Media–i.e. the languages and repertoires that designers use to describe reality and to conduct experiments.

  Into this category falls a plethora of methods, tools, representational devices (e.g. diagrams, pseudo-code, storyboards etc) which exist for designers of HCI and systems generally. These methods are often rarefied, over simplistic, focused on particular elements of the design process to the exclusion of others, not yet industrially scaled, not yet supported by tools, not yet standardised or not yet self articulate and self communicating.
- Designers appreciative overviews-(i.e. the system of thinking designers bring to the problem setting process, to the evaluation of inquiry and to the process of reflective interchange).

  Into this category fall the educational experiences available for designers operating in a multi-disciplinary arena, the exemplars of good practice available, referents from related domains, awareness of dynamic changes in techniques and approaches, breadth of conceptualisations and ability to envision potential solutions.
- Underpinning/Overarching theory- (i.e. the substrate of agreed potential and proven explanations within which design phenomena make sense.)

  Into this category we can place technical, rational, social, psychological, ergonomical, economical and other related theoretical views which will form the mental furniture of a designer's context and background.
- Role Frames–(i.e. the view of design professionalism held by the designer).

  This represents the accepted and relatively stable view the designer holds about the set of tasks they undertake and through which their professionalism is defined within a societal and institutional/public setting.

Clearly there are a number of unanswered questions here. How does the medium used in representation relate to the quality of explanation provided by a particular solution? What is required to educate a class of interdisciplinary designers with the repertoire and referents necessary to meet the challenge of complex artefact design? For our purposes it seems clear that the quality of our understanding of the design process is weak. There has been a neglect due to the assumption that such soft issues were not amenable to this type of understanding. We can begin to look in more detail at the design process in action and assess what the reality of the design process is.

5. Designing the designer; Infinite regress or successive approximation?

The rationalistic concept of design is that design is a process of problem solving by individual decision makers among sets of possible worlds. The reality of design of computer related artefacts is in the answering of questions about designing for human use. This is not to assert that computer science should not involve analytic and formal theoretic aspects. The generative substructure of computing is mathematical but its superstructure and applications may not be so. The human usage of computers and computer related artefacts and products requires thinking about organisations and environments of use. Designing computer related products (including the special case of IBC services) involves deciding how these products are to be constructed, delivered and used. These decisions can only be taken within a dialogue which involves those who participate in the design and usage.

In our introduction to this chapter we made the point that however valuable design methods and principles may be, if they are not taken up by designers, they will never be able to affect service engineering. We need a closer understanding of how to foster participative dialogue in design. It is the process of dialogue and participation in negotiation which is most lacking. This is central to any discussion of designers requirements for IBC artefact design. The usability of methods and principles for design is primary. A main finding of the AMODEUS 'Empirical Designer Studies' was that a common strategy in design is 'muddling through'. This is and will continue to be the prevailing strategy for as long as 'muddling through' is easy to use, time saving and politically feasible'.

It would seem that the reality is one of experienced designers using such aspects of formal methods as seem most useful on an opportunistic basis and in a highly creative way. Methods are required and adapted to help deal with technical aspects of design, the evolution of commitments and development of design understanding within design teams.

Mechanisms which rethink the representation we use in describing design and which facilitate the development of future scenarios by users and designers. The work of the Race project GUIDANCE, in implementing a software package to deliver the method described in chapters 1.1 and 3.1, has pointed the way in this regard (Terrins-Rudge et al 1992). The inclusion of alternative design representations and the provision of creative envisioning techniques (e.g. storyboarding support in the design of IBC services) begin to make inroads on meeting designers requirements for support (Whitefield et al 1992).

This sort of trend is apparent in the increased need to provide similar sorts of support for network users in telecoms environments. One level of need is the perception of a requirement which will allow corporate network customers and even end-users to design and configure network services themselves (see chapter 1.2). This need for customisation of network facilities and usage levels is consistent with the network architecture but requires careful design in terms of control consistency and billing arrangements for network management. Another level of need is the requirement for service providers to be able to design (and redesign) services in real time. In the premium rate service sector provider organisations serve niche markets which are transient and based on the value of timely and current information to the customer (e.g. weather information services). This type of product needs to be able to update the design of the service without time delay in order to ensure the accuracy of the information. This is currently theoretically possible but would require a facility to allow provider organisations to redesign their usage of the network and the loading of services (and thus service scripts) onto the network. These problems are crucial in fully exploiting the potential of IBC services, they are technologically solvable but they contain a large element of context and end user input to solution design.

Rationalistic design approaches are not enough to provide for the needs of designers working in this area. What is required is a process oriented view. The design process and the disciplines of thought and action required for creative activity are the keys to provision of usable design methods. The focus of thinking and activity thus becomes centred on the human element in design and usage of artefacts. The design process concentrates on the adequacy of the designed artefact for intended usage, and its relevance in terms of its context of use and its intended users. This type of thinking is being fed into the use of usability statementing in usability engineering standards. There is some progress yet to be made in this regard but signs of understanding the value of user centred disciplines are emerging.

6. Negotiation and negotiators

In order to achieve the development of complementary design support approaches we can call upon the rich body of design theoretical work exemplified by industrial designers, architects (Alexander, 1964) and other design professionals. The design methodology approach in other fields has been well documented (Cross, 1974). Developments of this approach include the work of Archer (1982) on the development and support of designerly thinking and an understanding of designerly activity (Broadbent, 1984). Practical design and design knowledge are characterised as reflection-in-action (Schön, 1983) and the education of design skill via master apprentice relations makes suggestions for the development of the art and science of product and artefact design. One recurrent theme in these endeavours is the primary importance of negotiation based interaction both within the design team and between the designers and users. Part of the complexity here is that the user of a computer related artefact is not necessarily the same person as the client in the professional relationship between designer and client. In addition, chapter 1.1 presents a distinction between external and internal design perspectives. The results from these two perspectives are typically from different design teams and also need not be compatible. The encapsulation of both perpsectives in one eight-step method should assist in ensuring that at least negotiation between designers takes place.

According to Schön (1983) the reflective practitioner is one who acknowledges that while they are presumed to know they do not necessarily have all the relevant knowledge. More importantly the reflective practitioner views their uncertainty as a source of learning for themselves and their clients. This philosophy leads to importance being allocated to the involvement of the end user and an understanding of the importance of the information provided by the end user. The client/end user becomes a participant in the process of design and thus causes a challenge to the expert or exclusive status of the professional.

The refinement of practical examples of participative design are increasingly available. For example the PEN&PAD project which undertook participative design in the development of a medical workstation for patient care (Rector et al 1992). Methods for user involvement include adaptations from work by Monk (1984), Mumford (1983) and Norman (1986).

Negotiation in the design process requires methods to support complex negotiation. There are a large number of informal interactions which we use to create and maintain relationships. There are a broad range of mechanisms for social interaction to facilitate discussion, negotiation and planning. Design negotiation involves creative collaborative processes involving people in sharing skills, resources, time and information to achieve design goals. This in itself provides a research agenda on the provision of designer support.

7. Conclusions

Thus it would seem that we have at present begun the process of realisation that both rationalistic and subjective ways of knowing are inherent in the design process. The provision of design support is fast becoming a bottle neck in the design of timely and appropriate artefacts, including IBC services. There is a developing research agenda which requires the synthesis of these contributory supports to put in place the constants designers of these artefacts require. There is also a requirement for curricular development to foster to development of the competencies of designers in the use of these constants.

References

Alexander, C. (1964) *Notes on the synthesis of form* Harvard University Press Cambridge Mass.

Archer, B. (1982)*Designerly Ways of Knowing*. Research Position Paper. Department of Design Research. Royal College of Art

Bansler, J.P. & Bodker, K. (in press) A Reappraisal of Structured Analysis: Design in an Organizational Context. To appear in *ACM Transactions on Office Information Systems*

Broadbent, G. (1984) 'The development of design methods' in N. Cross (ed) *Developments in Design Methodology*. Wiley & Sons: Bath.

Cameron, E. S & Terrins-Rudge, D., (1993) 'An Epistemological Approach to the Interpretation and Representation of Objects in a Distributed Multimedia Environment.' In *The Challenge of Multimedia. Trinity College Dublin 25-26 March 1993* Joint Trinity College Dublin and Massachusetts Institute of Technology Conference Publication

Cross, N. (1974) *Human and Machine Roles in Computer Aided Design* Open University Press: Milton Keynes

Curtis, B., Krasner, H & N. Iscoe (1988) A Field Study of Software Design Process for Large Systems. *Communications of the ACM, 31,* pp.1268-1287.

Lindblom, C.E. (1959) The Science of Muddling Through. *Public Administration Review, 19,* 79-88.

Monk, A. (1984) *Fundamentals of HCI* London: Academic Press

Mumford, E, (1983) *Designing Human Systems for New Technology: The ETHICS Method.* Manchester Business School Press

Norman, D.A. (1986). 'Cognitive engineering'. In D. A. Norman & S. W. Draper (Eds.), *User centered system design: New perspectives on human-computer interaction* (pp. 31-61). Hillsdale, NJ: Lawrence Erlbaum Associates.

Recotor, A.L., Horan, B., Fitter, M., Kay, S., Newton, P.D., Nowlan, W.A., Robinson, D., Wilson, A. (1992) 'User Centered development of a general practice medical workstation: the PEN&PAD experience. In *Proceedings of CHI '92,* pp. 447-453, New York: ACM.

Schein, E. (1973)*Professional Education* McGraw-Hill

Schön, D.A. (1983)*The Reflective Practitioner: How Professionals Think in Action.* Basic Books

Simon, H. (1969) *The sciences of the artificial.* MIT Press: Cambridge MA

Terrins-Rudge, D., Adamson, S.J., Ionescu, R., Peterson, M. & Hedman, L. (1992) 'Designing for Designers' Needs'. In P. Byerley & S. Connell (eds) *Integrated Broadband Communications: Views from RACE—Usage Aspects, (North-Holland Studies in Telecommunication, 18),* Elsevier: Amsterdam.

Whitefield, A., Byerley, P., Denley, I., Esgate, A., & May, J., (1992). 'Integration of services for human end-users (1): Design Principles, Enabling States Analysis and a Design Method.' In P. Byerley & S. Connell (eds) *Integrated Broadband Communications: Views from RACE—Usage Aspects, (North-Holland Studies in Telecommunication, 18),* Elsevier: Amsterdam.

Young, R.M., & Barnard, P.J. (1987). 'The use of scenarios in human-computer interaction research: turbocharging the tortoise of cumulative science.' In J. M. Carroll & P. P. Tanner (Eds.), *Proceedings of CHI + GI '87: Human Factors in Computing Systems and Graphics Interface (Toronto, April 5-9)* (pp. 291-296). New York: ACM.

Yourdon, E. & Constantine, L.L. (1979) *Structured Design* Prentice Hall, New Jersey.

Section 2

Basic Research in Service Usability

Philip J. Barnard and Jon May

MRC—APU, 15 Chaucer Road, Cambridge, UK—CB2 2EF

The development of effective information technology requires design teams to consider a huge range of system, user, and workplace concerns. Obviously, as each new advance in technology occurs, it will typically raise a whole range of new design issues. While structured methodologies provide ways of ensuring that the design process goes through relevant stages, they typically provide little to help designers reason about substantive inter-relationships among system, user and domain concerns. As a part of the ongoing programme within Esprit Basic Research, the AMODEUS project (AMODEUS 1, 3066; AMODEUS 2, 7040) has brought together expertise in formal system modelling, psychological user modelling, and design analysis. The overall goal is to consider how interdisciplinary inputs might usefully be brought to bear, in an integrated way, on the resolution of new design issues and their realisation in software architectures.

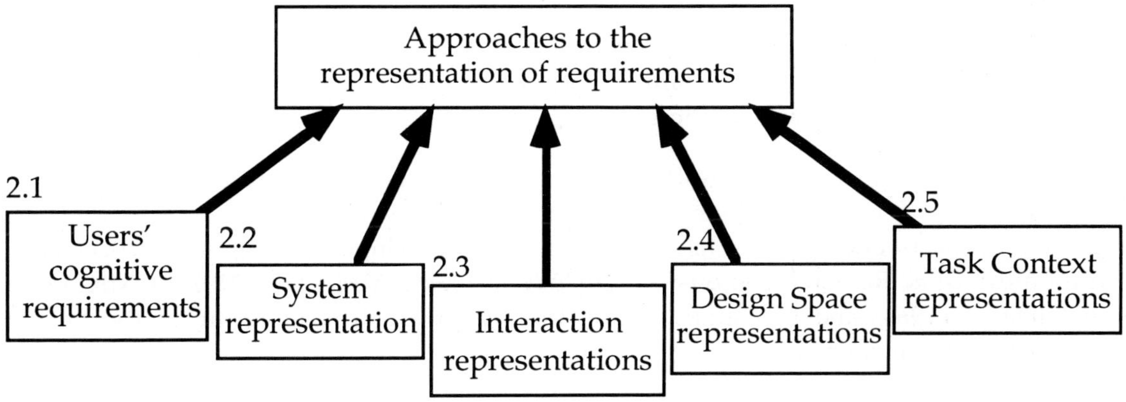

The context for the work described in this section is that of basic research. We are concerned primarily with developing the kinds of capabilities practical methodologies will be looking towards on a 10-15 year time scale. All of the chapters presented in this section are, in one way or another, concerned with the development of ways of representing requirements. The work presented in chapter 2.1 concentrates on the development of a language for representing and specifying human cognitive requirements. This is based on a particular form of human information processing theory. In chapter 2.2, the emphasis moves on to representing the behaviour of the computer system. The concern here is not with the system as a whole, it lies with aspects of the system that specifically impact its interactive behaviour. Interactive behaviour is again the focus of chapter 2.3. However, in this chapter an attempt is made to forge a key link between users and systems. The representation here is directed at capturing, the conjoint behaviour of both users and computer systems. The chapter itself concentrates on how a concern with constraints in the interaction may influence the design of the system. Chapters 2.2. and 2.3 both rely upon formal modelling techniques which imply some form of commitment within the design process to the use of formal techniques. That commitment need not, however, be complete. A good deal can be learned by understanding and applying the principles that are formally expressed without necessarily using the notations within the software engineering process.

In the remaining two chapters, the emphasis shifts from the problems of representing user and system behaviour, to the problems of bringing relevant information together within the design process. In these chapters the representational techniques that are adopted become less technical. Chapter 2.4 illustrates how a particular notation—questions, options and criteria (QOC)—supports the analysis of design spaces. This chapter goes beyond earlier work on the general topic of design rationale by describing a model of the design process which helps guide its use. Chapter 2.5 describes the most recently developed representation. This 'COSITUE' approach was specifically developed to represent in brief form, of an evolving set of design commitments. The approach is oriented towards the support of practical usability engineering. Of the approaches considered here, it takes the broadest view and has within its scope not only user and the system aspects of an interaction but also contextual aspects concerned with the collaborative and organisational properties within which the interaction occurs.

Although each of the chapters deals with a particular form of representation, in each case concrete examples are laid out to illustrate how the analyses can be applied to particular settings. Several of these examples are actually drawn from problems within the scope of broadband telecommunications applications.

The particular five forms of representational technique included here are, of course, not intended to be a comprehensive review of what is available. Other techniques for representing users and systems have and are being developed within AMODEUS itself, within other Basic Research projects—MOHAWC (3105); KAUDYTE (3219); COMIC (6225) and GRACE (6296)—and of course, within the wider scientific literature of the field of human-computer interaction.

Computers, Communication and Usability:
Design Issues, Research and Methods for Integrated Services
P.F. Byerley, P.J. Barnard and J. May (Editors)

Chapter 2.1

Cognitive Modelling for User Requirements

Philip J. Barnard and Jon May

MRC—APU, 15 Chaucer Road, Cambridge, UK—CB2 2EF

1. Introduction

The general objective of this chapter is to explore the potential benefits that explicit cognitive modelling techniques offer for the definition and understanding of user requirements. In doing so, we shall not be arguing for the replacement of usability testing or more heuristic forms of evaluation. Rather, we shall be arguing that more adequate theories than we have developed to date have a very real part to play commercial design contexts.

In the first section, we briefly contextualise this argument by exploring a particular view of applied science. Here it is generally assumed that the 'science base' is developed via empirical methods and theoretical synthesis. It is also assumed that knowledge in the science base is not just 'applied', but that we need specific representations that support the application of science in particular contexts. In developing this view we briefly examine the kinds of representations developed in the last 20 years in the specific context of human-computer interaction. Our argument is that although existing theories have substantial drawbacks, the effort should not be abandoned.

The next section summarises a particular form of general theory, Interacting Cognitive Subsystems (Barnard, 1985; 1987; Barnard & Teasdale, 1991; Teasdale & Barnard, 1993). This theory is intended to support the development of applied techniques of broad scope. To achieve this, the basic theory is itself translated into an approximate form that describes properties of cognitive activity. It is at this level that we seek to formulate general theoretical claims that are potentially applicable. Indeed, the essential argument of this chapter is that valid principles at this level are within reach; that their application in novel contexts can be supported without the expenditure of massive resources or time; and that they can provide an input into the design process that is of greater value than simply determining or predicting which of a number of design options would be best.

In the context of this book, we can develop this argument one stage further. The language that is used to describe approximate models of cognitive activity can be used *to articulate and hence specify a significant range of user requirements*. Although lacking the kind of formal rigour possessed by some methods for modelling properties of computer systems (see chapter 2.2) and interactions (see chapter 2.3), the process outlined here acknowledges the indeterministic nature of much human behaviour, and derives authority for its principles from the cumulative outcomes of behavioural experimentation.

We conclude by addressing some of the wider issues that relate the long-term basic developments presented in this section of the book to the shorter-term applied methods encapsulated in the chapters of Section 3.

2. A framework for the development of applied science

In discussing the support for interaction development practice that can be offered by applied science, Long (1989) defines it as 'principles which, when applied by practice, ensure that specifications of computer behaviour and user behaviour, when implemented, interact to produce a desired performance of work'. Principles should address both the design and the design process, giving support in the form of *what* decisions should be made and *how* they should be made. Where successive experience of design can lead to the development of craft skills that can only be communicated by general 'rules of thumb' or guidelines, cognitive ergonomics attempts to develop principles that are applicable in specific design contexts, and which make specific design prescriptions. The application of engineering principles should produce a design that can be guaranteed to meet its specifications, which craft skills alone cannot do.

The contribution of cognitive science to HCI research is to provide the conceptual background against which the engineering principles can be understood. Engineering practice (showing that the principles work) is one form of validation; scientific explanation (showing why the principles work) is another. Together engineering and science produce a synthesis to support design.

For the purpose of this chapter, an important aspect of Long's description of the support that engineering and science knowledge can offer to design is that knowledge is a *representation* of the real world (figure 1—although note that Long's terminology is not always followed here). The real world must be *analysed* to produce a *discovery representation* which allows the observable phenomena to be described and compared. This is a bridging representation that allows *assimilation* from the real world to the theoretical concepts that form the scientific and engineering representations of the world. The knowledge embodied in this representation is similarly applied via a second bridging representation. It must first be *contextualised* to a specific design context via an *application representation* that then allows scientific and engineering knowledge to be *synthesised* with the real world through design decisions.

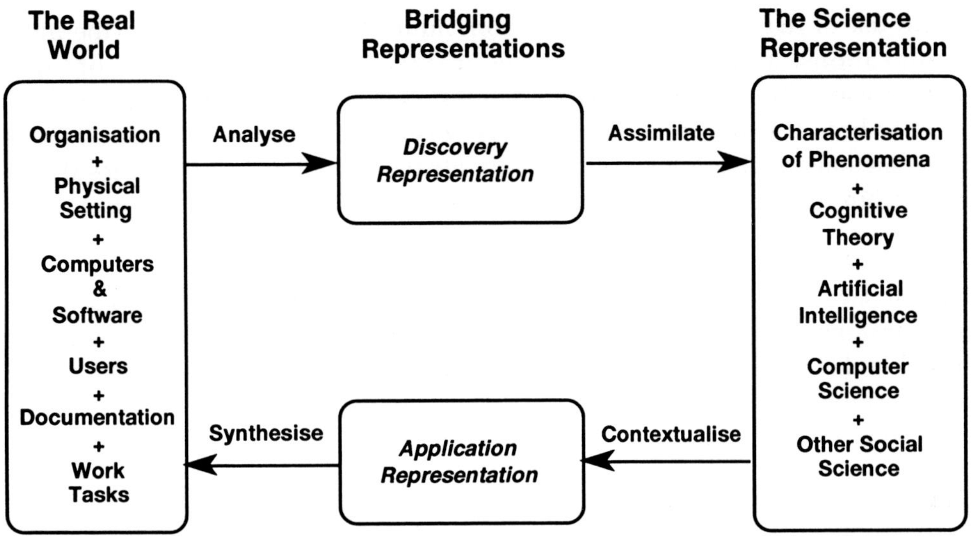

Figure 1: The bridging representations between the real world and its scientific representation (figure from Barnard, 1991)

Each part of the cycle is dependent upon its precursor. The knowledge embodied in the science representation will depend upon the adequacy of the discovery representations and the material they deliver for theoretical assimilation. Obviously, the ability to construct useful applications representations will, in turn, depend upon the existence of a broad and valid body of knowledge within the science representation. It is not only the content that is contextualised into an application representation that is important. The precise form of an application representation is crucial. The application representation must be appropriately situated with respect to the design process. Both the form and content of an application representation must be meaningful to the people who are to carry out the final synthesis in the real world design context. Failing this, it will at best be misapplied, and at worst not applied at all.

In contextualising the science knowledge to form an application representation some information will inevitably be lost or modified with the aim of meeting these constraints. If the application representation is too close to the science representation, it will not be usable in the design situation; and if it is too far from the science representation it may no longer be valid. In the following sections we will examine the application representations that are used by several different approaches to the application of HCI knowledge.

3. Discovery and application representations

In the early 1970s computers went well beyond being number crunching devices—they were increasing used to support cognitive work. For this class of application, there was little relevant empirical evidence and virtually no applicable theory. Of course, much was known about human use of hardware, particularly from human factors research for the military and telecommunications authorities. In these areas, the standard form of application representation was a 'traditional' set of guidelines. However, the new computer applications posed an ever broadening range of questions (Grudin, 1990). Providing answers to these questions seemed an ideal 'applied' opportunity for those methods and theories that had been developed within the emerging paradigms of cognitive psychology, linguistics and artificial intelligence.

Over the last two decades HCI research has called upon a broad range of discovery representations, including standard taxonomic, observational, and experimental techniques. From these a substantial body of empirical evidence has been accumulated and brought together in handbooks such as Hellander (1991). Although much of the research was itself directed at providing immediate support for application developers, a significant part was primarily directed at the development of new forms of theory-based application representation. Indeed, a seminal text by Card, Moran & Newell (1983) argued for the development of approximate, engineering-style theory. This form of theory was intended to be used early in the design process to provide predictive evaluation of user learning or performance with an envisaged task or interface.

In developing these early forms of applied cognitive theory, the researchers relied very heavily on the laboratory-based theoretical products of their parent disciplines. Grammatical techniques originating in linguistics were used to characterise the complexity of interactive dialogues; artificial intelligence-style models were used to represent and simulate the knowledge requirements of learning. Finally, of course, derivatives of human information processing models were used to calculate how long it would take users to do things. Although these approaches all relied upon some form of task analysis, their apparatus was directed towards some specific function.

Some of the models were principally directed at capturing knowledge requirements for dialogues for the purposes of representing complexity, such as BNF grammars (Reisner, 1982) and the Task Action Grammar developed by Payne & Green (1986). Others focused on interrelationships between task specifications and knowledge requirements, such as GOMS analyses and cognitive complexity theory (Card et al., 1983; Kieras & Polson, 1985). Yet other apparatus, such as the model human information processor and the keystroke level model of Card et al., (1983) were primarily aimed at time prediction for the execution of routine cognitive skill. Few models incorporated apparatus for integrating over the requirements of knowledge acquisition and human information processing constraints (e.g. see Barnard, 1987).

There can be no doubt that some of these theoretical ideas have been successfully used within the research community. GOMS, of course, is the most notable example. Yet, as Carroll and others have frequently pointed out (e.g., Carroll, Kellog & Rosson, 1991), there are few examples where cognitive theory has had a major and direct impact on design within the commercial community. On this criterion, cognitive science can more easily provide illustrations of its potential value through the application of empirical methodologies, through the data they provide and through the direct application of psychological reasoning in the invention and demonstration of design concepts (Anderson & Skwarecki, 1986; Card & Henderson, 1987; Carroll, Kellog & Rosson, 1991; Hammond & Allinson, 1988; Landauer, 1987).

As the field of HCI moves into its third decade, fundamental questions are now being asked about whether theory-based application representations have any value at all in design (Carroll, Kellog & Rosson, 1991). Behaviour in human-computer interactions may simply be too complex for basic theory to have anything other than a minor impact (Landauer, 1987). The early vision of a strong theoretical input to design runs the very real risk of being overtaken by heuristic methods. Unless developed more profitably, the kind of deep understanding potentially offered by cognitive theory may once again become neglected in design practice.

Many of the difficulties associated with the development of truly applicable theory are well known, so we need do little more than summarise them here. One problem is that the life cycle of theoretical enquiry and synthesis typically post-dates the life cycle of the products that it seeks to deal with; in addition, the theories are too low level; they are of restricted scope; as abstractions from behaviour they fail to deal with the real context of work; and they fail to accommodate fine details of implementations and interactions that may crucially influence the use of a system (Carroll & Campbell, 1986; Newell & Card, 1985; Whiteside & Wixon, 1987). Similarly, although theory may predict significant effects and receive empirical support, those effects may be of marginal practical consequence in the context of a broader interaction, or less important than effects not specifically addressed (Landauer, 1987).

Added to these there are profound practical difficulties. Developing appropriate theory in an applied context is a very demanding process. Techniques like cognitive complexity theory (Kieras & Polson, 1985), programmable user models (Young, Green & Simon, 1989) or the various grammars (e.g. Moran, 1981) require detailed specifications to be generated for each application modelled. Often each application can require the specification of many rules, the construction of which requires a modelling expert. All of this work has to be redone for each problem considered. Therefore application of such techniques requires a large commitment of resources. In many design contexts, this is difficult to justify. In addition, analyses that appear straightforward with small–scale examples often prove hard to scale up to full systems. Given the overall theoretical rhetoric concerning scientific adequacy and replicability, it is perhaps paradoxical that the actual application of theory can involve what Long & Dowell (1989) have called 'craft skill'—not only in terms of the modellers requiring knowledge of theoretical technique, but also in their ability to recognise what has to be modelled. The availability of this particular form of craft skill imposes real practical barriers to the meaningful application of advanced theory.

There are three general reactions to these problems. One reaction is to abandon theory altogether and get on with design, supported where necessary by empirical evaluation. There is now ample evidence for the overall cost effectiveness of both empirical and heuristic evaluation (Nielsen & Phillips, 1993). A second alternative is to abandon traditional forms of *cognitive* modelling and adopt a different set of basic concepts (eg Whiteside & Wixon, 1987; Bannon & Bødker, 1991). An aim here is to develop new forms of more relevant 'theory' that meet the requirements of the design process (Carroll, Kellog & Rosson, 1991, see also Draper,

1992). Extensions of this general strategy include the development of semi-formal notations through which theoretical assumptions (however derived) can at least be represented. Two variants of this are represented in subsequent chapters—design space analysis (chapter 2.4) and a frame-based approach for representing the state of a design in a way that is tailored to the requirements of usability engineering (chapter 2.5).

The third alternative, and the one adopted here, is to do some iterative design on cognitive theory itself and on the discovery and application representations involved in its development and use. Such redesign should, of course, be based upon the lessons learned from the relevant research to date. From a methodological point of view, Barnard (1991) has pointed out that HCI research would benefit from the further development of discovery techniques and applications representations. In the case of discovery representations, three examples were used. First, experimental techniques could be further developed to concentrate rather less upon the isolated effects of individual variables and rather more on complex task performance and the interactions of variables therein. This would potentially do much to improve our understanding of what make up the important sources of variability in practical contexts. Second, development of scenario methodologies could do much to help improve the scope of modelling techniques (see also Young & Barnard, 1987). Third, knowledge elicitation techniques could be enhanced to inform our theoretical understanding of error causation in human-computer interactions.

Barnard (1991) also argued for the development of novel forms of application representation. He pointed out that those seeking to apply cognitive theory in HCI have, in fact, behaved just like early system designers. Like glass teletypes, basic psychological technologies were under-developed for their new purpose and, like the early design of command languages, the interfaces (application representations and engineering models) were heuristically constructed by the theorists around what they could provide rather than by any deep analysis of requirements, extensive study of their target users (designers), or of the actual contexts in which design occurs.

A number of techniques are being developed that could be regarded as iterative improvements upon earlier forms of modelling technique. Many earlier theoretical ideas assumed an idealised user. Developments of programmable user models - PUMS , for example, involve a high degree of operationalisation of task knowledge (Young, Green & Simon, 1989). Additionally, this approach provides a more adequate understanding of less than ideal human learning and performance because the knowledge is processed in the context of a constrained cognitive architecture. In principle, this technique relies for its effectiveness on skills that designers already have - programming ability. It is, however, rather time consuming in its application. The particular feature of requiring little expertise in the technicalities of cognitive modelling is shared by another novel theory-based technique, the 'Cognitive Walkthrough' (e.g., Polson, Lewis, Rieman & Wharton, 1992). However, in contrast to PUMS, the Cognitive Walkthrough was designed to require few resources in terms of time, effort and prototype building.

Barnard (1987) has also proposed an overall strategy for the cognitive modelling of less than ideal user learning and performance. The approach takes as its point of departure a particular model of the human information processing mechanism, Interacting Cognitive Subsystems. This provides a basic decomposition of the architecture underlying human perception, cognition and action. It is then assumed that we can describe the activity within this mechanism in a principled but approximate way. It is further assumed that systematicity in the description of underlying cognitive activity will be reflected in regularities in overt behaviour. Principles governing cognitive activity and its relationship to behaviour can then be inferred from experimental evidence which maps out the empirical phenomena of human-computer interactions.

The final step in the strategy is produce an expert system which actually automates the process of building approximate models of cognitive activity for the purposes of supporting design reasoning. The ultimate intention is, of course, to provide a validated tool which requires relatively little in the way of detailed knowledge of the underlying theory and which can be used on time scales compatible with design and development schedules.

Many aspects of this strategy have now been documented in one form or another. Barnard (1985) laid out the underlying architecture based upon an analysis of human memory tasks. Subsequent papers analysed a range of experimental evidence to produce principles governing the operation of the architecture specifically for the complex tasks of HCI (eg see Barnard, 1987; Barnard, Grudin & MacLean, 1989; May et al., 1993). Illustrative sets of principles have also been implemented in a series of demonstrator expert systems (Barnard, Wilson & MacLean, 1987; 1988; Barnard, Blandford & May, 1992).

In the remainder of this chapter, we shall concentrate on one specific aspect of the strategy. This concerns the specification of cognitive activity itself. If we are to represent users' cognitive requirements within this approach, we need to specify exactly what we mean by cognitive activity and how it is constrained by both knowledge-based and process-based limitations. The next two sections outline the architecture and illustrate its potential functioning in an HCI context.

4. Interacting Cognitive Subsystems (ICS)

Interacting Cognitive Subsystems (ICS) is a part of a wider theoretical movement that considers the human information processing mechanism as a highly parallel architecture, with no 'central processor' or 'limited capacity working memory'. The ICS architecture is represented as a set of functionally distinct subsystems, each with specialised yet equivalent capabilities. The core assumption is that we are dealing with a *system* of distributed cognitive resources in which behaviour arises out of the co-ordinated operation of the constituent parts.

As a fundamentally *systemic* approach, ICS seeks to be comprehensive, in the sense that its constituent parts are seen as capable of supporting all aspects of perception, cognition, emotion, as well as the control of action and internal bodily reactions.

4.1 The basic architecture of ICS

Specifically, cognitive organisation is described at two levels: the systemic level and the subsystem level. At the systemic level, ICS comprises a set of nine functionally independent subsystems, each specialised for handling some aspect of peripheral or central processing. At the level of individual subsystems, although each one operates on a different form of information, or mental code, all are assumed to have the same simple internal structure. The basic architecture of a cognitive subsystem is illustrated in figure 2.

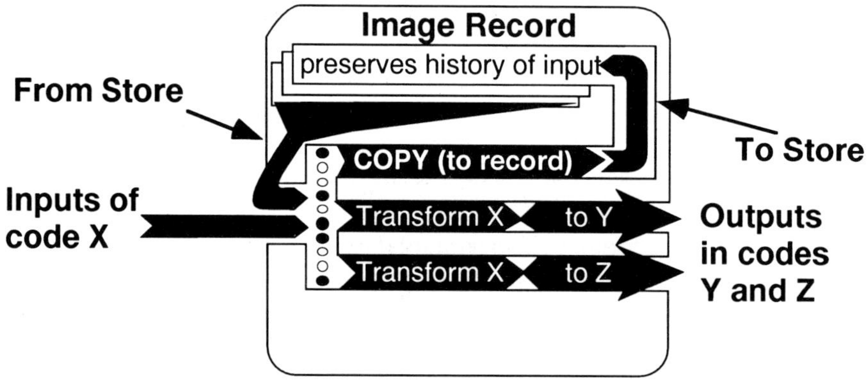

Figure 2: The basic architecture of a cognitive subsystem

Subsystems have the capability to store information in an image record. All incoming information in the appropriate mental code is continually transferred into that image record. This process does not change the way information is encoded and hence is called a COPY process. It not only stores information; this continuing transfer is also the basis of our subjective experience. Other processes within subsystems serve the function of transforming information from one mental code into other mental codes. These processes are organised in parallel both with the transfer of information into the memory record and with each other. They form the basis of information flow from one subsystem to another.

The systemic level of description is given in figure 3. Each of the nine subsystems (see table 1) is essentially specialised for handling a particular type of information, or mental code. On the sensory side, there are subsystems specialised for dealing with visual and auditory information as well as one specialised for handling proprioceptive information, here referred to as body state information. On the effector side, there are subsystems for handling bodily movements (Limb subsystem) and the output of vocal/speech-based information (articulatory subsystem). The central subsystems are specialised for handling higher order representations, dealing specifically with structural descriptions of information in sound space (Morphonolexical subsystem), visual space (Object subsystem), semantic descriptions of referents, their properties and inter-relationships (Propositional subsystem), and a more wholist level of knowledge integration (Implicational subsystem) which deals with schematic models of general regularities in both the internal world of thought and bodily feeling as well as the 'external' world with which the individual interacts.

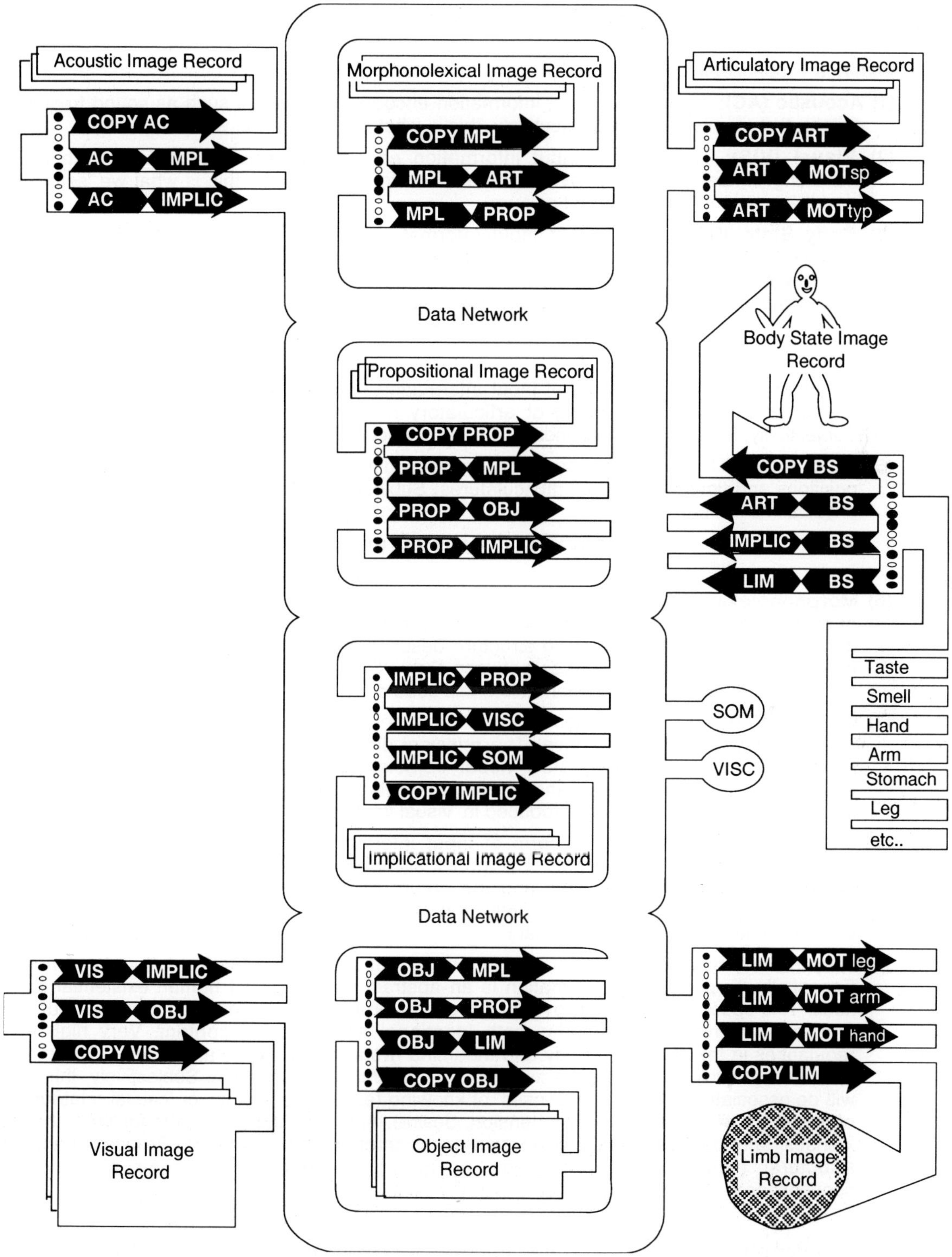

Figure 3: The overall architecture of Interacting Cognitive Subsystems.

PERIPHERAL SUBSYSTEMS

a) Sensory & Proprioceptive

(1) Acoustic (AC): The form of input information encodes dimensions such as sound frequency (pitch), timbre, intensity etc. Subjectively, this is what we 'hear in the world'.

(2) Visual (VIS): The form of input information encodes dimensions of light such as its wavelength (hue), brightness over visual space etc. Subjectively, this is what we 'see in the world' as patterns of shapes and colours.

(3) Body State (BS): The form of input information encodes dimensions such as type of stimulation (e.g., cutaneous pressure, temperature, olfactory, muscle tension), its location, intensity etc. Subjectively, this code corresponds to bodily sensations of pressure or pain, positions of parts of the body, as well as tastes and smells etc.

b) Effector

(4) Articulatory (ART): The form of input information encodes dimensions such as the force, the target positions and timing of articulatory musculatures (e.g., place of articulation). Subjectively, this code corresponds to our experience of subvocal speech output.

(5) Limb (LIM): The form of information encodes dimensions such as the force, the target positions and timing of skeletal musculatures. Subjectively, this code corresponds to 'mental' physical movement.

CENTRAL SUBSYSTEMS

(6) Morphonolexical (MPL): The form of input information is an abstract structural description of entities and relationships in sound space. Dominated by speech forms, where the principal dimensions encompass a surface structure description of the identity of words, their status, order and the form of boundaries between them. Subjectively, and when uncorrelated with either acoustic or articulatory information, this code corresponds to what we 'hear in the head.'

(7) Object (OBJ): The form of input information is an abstract structural description of entities and relationships in visual space. The principal dimensions encompass the attributes and identity of structurally integrated visual objects, their relative positions and dynamic characteristics. Subjectively, and when uncorrelated with real world events, this code corresponds to what is actively produced in 'visual imagery.'

(8) Propositional (PROP): The form of input information is a description of entities and relationships in semantic space. The principal dimensions encompass the attributes and identities of underlying referents and the nature of relationships among them of either a concrete or abstract type. Subjectively, this code corresponds to being aware of specific semantic relationships ('knowing that').

(9) Implicational (IMPLIC): This is the most difficult form of information to describe. Essentially, 'implicational' information is an abstract description of human existential space itself. This space is abstracted over both sensory/proprioceptive and propositional input. The principal dimensions encompass both ideational and affective content. Very high level regularities in the world, the body and the 'mind' would be captured at this level. The mental entities constructed can be seen as *schematic models* of experience. Subjectively, this code will be associated with holistic 'senses' of knowing (e.g., of 'familiarity' or 'causal relatedness' of ideas), or of affect (e.g., apprehension, desire). Actual content and form for these holistic 'senses' only becomes available to awareness through their semantic (Propositional), and structural (Object/MPL) refinement; or through feedback from their physiological consequences (Body State) and effects in the world.

Table 1. The Peripheral (sensory, proprioceptive & effector) and Central subsystems within ICS and the type of information with which they deal (based on Barnard & Teasdale, 1991).

Because each subsystem operates upon a different class of information, information is repeatedly transformed, interpreted and re-represented throughout the overall cognitive system. It is this dynamic processing of the representations of information that gives rise to the phenomena that can be used to make predictions about the patterns of cognitive activity. These patterns are 'approximate' in that they do not describe precisely what is happening to a given token of information, nor do they directly produce estimates of latency or values for memory load. Instead of such quantitative values, the model generates qualitative assessments of cognition, which vary according to the dynamics of the flow of information between the subsystems.

From the point of view of user modelling for interface design, the salient features of an interface are those which affect the course of the processing of information, and hence the patterns of cognitive activity. These aspects relate to the matching by each subsystem of incoming representations in their appropriate code to representations that they have operated upon before (the 'image record'), and to the development and function of skilled performance (which in ICS terminology is called 'procedural knowledge'). The image record contents of a subsystem represent a complete record of all inputs that it has received – in other words, all representations that the cognitive system has processed that are in the appropriate code for that subsystem to operate upon. Over time, and with the repeated processing of similar information, these *experiential* records give rise to generalised *abstracted* records which can be used to elaborate and interpret incoming representations, which may be novel, contain errors or be incomplete.

Procedural knowledge is likewise acquired through the repeated processing of similar information and the production of similar outputs, and enables one subsystem to produce an output representation for another subsystem directly, rather than by going through an intermediary set of transformations. Image record contents and procedural knowledge allow the subsystems to build representations of information, and to elaborate them or to transform them into other codes. A major element of the ICS modelling approach concerns the degree to which an interface allows or demands the recruitment of existing image record contents and procedural knowledge, the support that it gives for their acquisition and development, and the extent of any conflicts between the requirements of the interface and existing knowledge.

4.2 An example of information flow through the system

We shall illustrate information flows through the system as a whole by reference to a specifically HCI example. The example concerns a situation in which an author is reviewing some text that has just been entered. Figure 4 traces the sequence of processing activity.

The text itself must obviously be processed through the visual subsystem (1) where the primary copy process creates an image record of all the information in the visual field. The full display thus appears as the representation in the proximal end of the image record of the VIS subsystem. In parallel the VIS→OBJ process recodes the raw visual data into a higher order representation which reflects a more abstract structural description of visual form. This process of recoding involves information reduction. For example, the higher level object code would not represent gradations of brightness (cf. Table 1) but would distinguish the form of upper and lower case letter shapes. Since these processes can only recode a single data stream at a time, the actual recoding would also be selective. With the recoding of a visual representation, selectivity would operate spatially and only part of the visual field would be undergoing recoding (the 'just-read' text). The object level description of the selected area of the visual field is passed over the data network where it is copied into the image record of the Object subsystem.

A process (OBJ→MPL) in the object subsystem (2) then recodes the orthography of words into a speech based code, the morphonolexical (MPL) code, which represents the surface structure of language. This latter code specifies lexical items in phonological form and marks the higher order constituents (e.g., phrases and clauses) within the linguistic string. In the normal course of understanding spoken speech, this surface structure representation would be computed from an initial acoustic representation (AC→MPL). However, since it originates from a visual input, reading demands cross modal recoding. The speed of normal visual processing is incompatible with the normal, slower rate at which auditory speech is heard and hence processed. Accordingly, the relevant transformation process in the MPL subsystem must operate in a buffered configuration to interpret the incoming representation (3). For this to happen, the MPL representations originating in the object subsystem are copied into the image record of the MPL subsystem, from where they can be accessed at the appropriate rate by MPL transformation processes.

It is important to note that it would not have been possible for the OBJ→PROP process to represent the linguistic meaning of visually presented sentences. The linguistic knowledge required to construct propositional representations of sentences is viewed as confined to the MPL subsystem where the MPL→PROP process constructs a representation of the meaning of the text in propositional format (3). In MPL→PROP recoding, the details of surface structure form are discarded, such as specific lexical items and their order. The resulting representation is transmitted over the data network and copied into the image record of the propositional subsystem. Hence, the propositional image record contains abstract semantic symbols to represent the meaning of recently interpreted text (4).

The propositional subsystem incorporates a process which interprets relations among propositions. In this recoding (PROP→IMPLIC) details of the individual propositions are lost and the highest level cognitive representation of the text is constructed, the 'implicational' code. This code specifies relations among interlinked knowledge structures at a level of description similar to Minsky's (1975) 'Frames' or Schank & Abelson's (1979) 'Scripts'. In order to avoid confusion with the specific properties of other formulations, the implicational level of representation will be described as involving schematic models. The PROP→IMPLIC process involves interrelating propositions in the text both among themselves and in relation to prior experience as represented in the propositional image record. This process is therefore inferential in nature. The resulting schematic representation is transmitted over the data network and copied into the implicational image record and information encoded within it reflects the overall conceptual structure of the text (5).

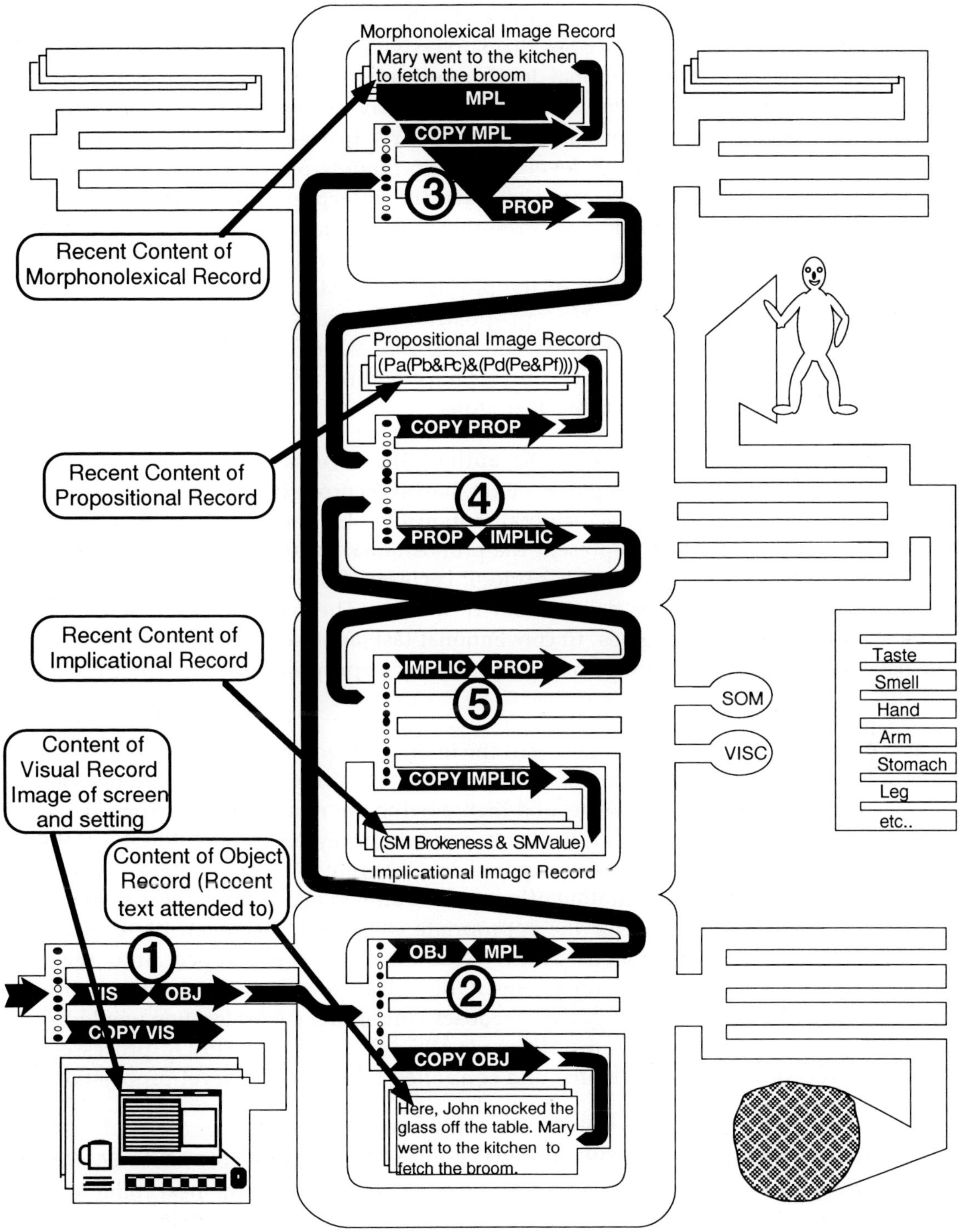

Figure 4: The sequence of processing activity for text comprehension during on-screen editing

The cognitive activities associated with understanding text on a VDU are thus viewed as involving a number of discrete processes that successively recode the visual image of the text through a number of representational levels until an implicational representation of its content is formulated. The interpretation of the text may nevertheless develop further. The implicational subsystem incorporates a secondary process that computes detailed propositional structures from high level schematic models of world and self knowledge (IMPLIC→PROP). This process can elaborate the interpretation of the text and any propositions actually output would be transmitted over the data network and copied into the image record of the propositional subsystem. Thus, propositional content would come together both as a result of the interpretation of the physical text (MPL→PROP) and via reciprocal inferential mapping between the propositional and implicational subsystems (PROP→IMPLIC & IMPLIC→PROP). When these representations are consistent, they can blend at the input to the propositional subsystem, providing a more elaborate representation than that available from the displayed text alone.

As the sentence is being re-read by the author, the form and content of the derived propositions and propositional complexes might, however, prove inconsistent with an accessed representation of the author's original intention. Under such circumstances the IMPLIC→PROP process might represent the propositional content of the discrepancy and a subsequent PROP→IMPLIC & IMPLIC→PROP cycle might then construct a representation specifying that the sentence be deleted and re-written. The propositional contents of this representation correspond to 'goals' in conventional task analyses. The representation would then have to be mapped through output processes appropriate for the interface to that particular word processor.

For the purposes of illustration, suppose that the interface involves a linear menu of command words at the top of the display and a mouse-based dialogue. The sentence would be deleted by pointing at the command word 'delete' and then moving the mouse to select appropriate start and end points. For an inexpert user, further computational cycles through PROP→IMPLIC & IMPLIC→PROP processes might be required to elaborate a representation of the action sequence. In addition, the configuration of resources outlined above for reading would also be involved in scanning the menu for the appropriate command. However, the propositional representation of the intention 'to delete' would have to be mapped onto the object (PROP→OBJ) and then the limb subsystems (OBJ→LIM) to control the motor act of mouse selection (LIM→MOT). During this process feedback concerning cursor location would be made available to the representational subsystems via VIS→OBJ and OBJ→PROP processes. Each step in the process would be evaluated and controlled by the higher level propositional and implicational representations.

As soon as the evaluation by PROP→IMPLIC and IMPLIC→PROP processes confirm that physical deletion has occurred, these processes would establish that the next thing to do was to re-write the text. Possibly by a combination of re-reading prior sentences and/or accessing prior episodic representations stored in their image records, these same processes would elaborate a fresh propositional representation for the new sentence (IMPLIC→PROP). A configuration of processes would then be required to map this representation onto direct effector action.

Each of the representational subsystems, outlined more fully in figure 3, incorporate parallel processes for recoding both 'incoming' and 'outgoing' representations. The propositional representation computed in the implicational subsystem can therefore be transmitted over the data network, where the propositional subsystem would assign it a surface structure form (PROP→MPL). This in turn would be transmitted and recoded in the MPL subsystem into an extended articulatory representation (MPL→ART). The actual act of typing characters would then be controlled by the articulatory subsystem's ART→MOT_{typ} process, operating in a buffered configuration with its associated image record. It might also reasonably be assumed that some spatial aspects of motor control requires concurrent processing via the PROP→OBJ, OBJ→LIM and LIM→MOT processes. In addition, auditory feedback from keystroke activity would be automatically copied into the acoustic image record, but further processing of this information would presumably be suppressed by the dominant flow of information.

4.3 Describing the properties of cognitive activity

Within the ICS framework, processing activity is not seen as under the control of a 'central executive'. Rather, individual processes, such as those discussed in the above example, are seen as mutually self-controlling via their interactions. In order to frame a theoretical account of the cognitive underpinnings of any form of behaviour, it is therefore necessary to consider system-wide properties of processing activity. Clearly, highly detailed accounts of what is going on within any one part of the system would lead to models of enormous and unmanageable complexity. However, for the purposes of attaining a molar level description of the essential properties of cognitive activities, it is possible to consider the functioning of a parallel architecture in terms of the major considerations which constrain the pattern of information processing. From the technical properties of the architecture outlined in figure 3, four classes of consideration naturally emerge.

First, the behaviour of the complete mechanism will obviously depend on the identity of the particular processes that are required to support a particular kind of mental activity. The set of processes or recoding operations required to support a given activity will be referred to as a *configuration* of processes.

Second, the action of each and every process in a given configuration will obviously be constrained by what recodings it can perform more or less automatically, and this in turn will depend on what that process has learned. In line with many other theoretical perspectives, this is referred to as the *procedural knowledge* embodied in the process. Obviously, if any given process does not embody the procedural capability to transform a particular pattern of incoming information, then this will affect the overall operation of the complete system.

Third, it is widely held that not all knowledge is represented in a proceduralised form, but rather is held in the form of declarative representations stored in some form of episodic format. Within ICS, such declarative representations are stored in the image records of the different subsystems. So, for example, in order to determine the name of a particular computer file which contains a given document, a memory record may need to be accessed. The performance of the architecture on any given task may therefore also depend upon the nature and properties of the particular kinds of memory records that may need to be accessed. This third consideration governing the operation of the complete system relates to the *record contents* accessed or used in the task setting.

The considerations relating to *procedural knowledge* and *record contents* deal specifically with the capabilities of individual subsystems. Since the dynamic course of mental activity depends upon interactions between subsystems within some defined *configuration*, we additionally need to define a fourth consideration which deals with properties relating to the overall dynamic co-ordination and control of the complete mechanism. This construct will be referred to as *dynamic control*, and is intended to capture the status of particular representational resources, the extent of their use, as well as how information flow is governed and used in the internal monitoring and evaluation of configural activity.

As a framework, the ICS perspective holds that a rich theoretical understanding of the cognitive underpinnings of behaviour in complex tasks can be achieved by specifying more detailed properties of *configurations, procedural knowledge, record contents* and *dynamic control* and by defining how specific interdependencies between these factors influence overt behaviour or in the case of HCI, interactions with computer systems.

The essential idea pursued here is that ICS can be used to construct relatively precise descriptions of cognitive activity, and that these descriptions can be directly related to behaviour as observed in user-system interactions. The basic properties of the architecture outlined in the previous section suggested that overt behaviour would be a complex function of four factors in a manner that can be summarised as follows:

Properties of Behaviour ← Fn (Configuration of processes
Procedural Knowledge
Record Contents
Dynamic Control)

5. Specifying approximate models of cognitive activity

The complexities of human cognition cannot be expected to yield to a very simple minded analysis of this form. The function outlined above relates primarily to approximation over activity that occurs within the very short term dynamics of human cognition. The time frame over which principles at that level might apply can be expected to be very short. Interactions with a computer require the cognitive mechanism to be co-ordinated to control extended sequences of behaviour. Any unified sequence of behaviour is likely to involve a number of types or phases of cognitive activity (cf. Norman, 1986). So, the correction of material during text editing may require reading existing sentences, pointing to and selecting a 'delete' command using a mouse, and the keystroking of new text. Each phase will recruit a different configuration of processing resources. These phases will require different but inter-related descriptions.

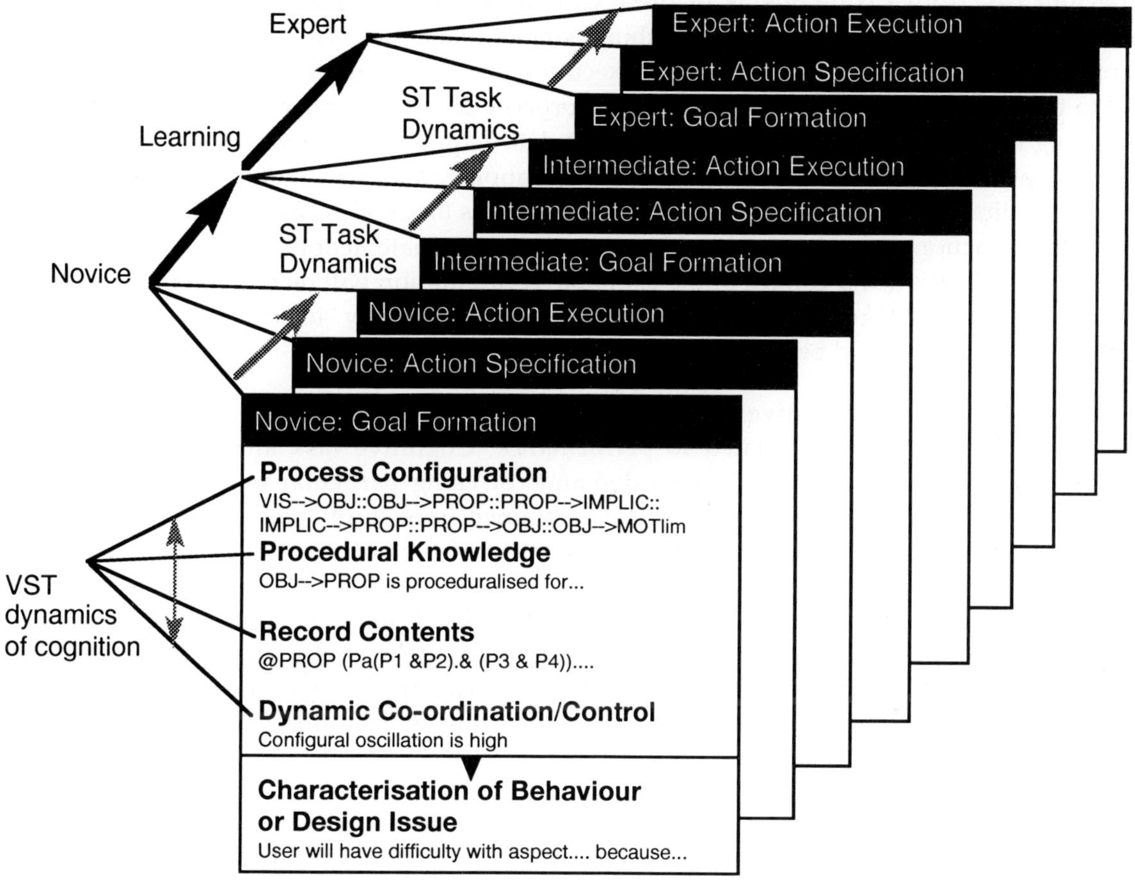

Figure 5: A family of cognitive task models

Alternative configurations may also be recruited to support the same overt actions. Novices engaged in creative composition may use one configuration to control keystroking and expert copy typists another. Similarly, for a novice, the descriptions of procedural knowledge and of record contents will be different from those for an experienced user. As learning progresses, users will not only have potential access to a larger range of relevant memory records, but their response patterns will increasingly acquire the characteristics of fully proceduralised knowledge. Thus, in order for the modelling approach to have significant scope, it becomes necessary to approximate over the short term changes involved in the co-ordinated control of sequences of action and over the longer term changes that occur as a result of learning and experience. It is therefore appropriate to consider families of inter-connected models for the different phases of activities and for different levels of user expertise (see figure 5). This general idea is obviously extensible to include both alternative strategies for coping with a task, and individual differences within a particular class of user population. Variation in the descriptive models should nevertheless be related systematically to varying attributes of user behaviour.

The four descriptive components of these Cognitive Task Models (CTMs) should also be systematically inter-related, not only within a single 'plane' but also between the planes that represent the short term dynamics of task execution and the longer term dynamics of learning. So, to take but a simple example, if any given process does not contain a fully proceduralised mapping from the incoming data to the output data required by the task, then obviously the overall pattern of data flow through the configuration cannot be smooth. Extra transactions will be required among processes in different subsystems to access memory records that may be pertinent to action execution, or to infer what an appropriate action might be. In effect, the overall co-ordination and control of the complete configuration will depend upon the status of procedural knowledge, and any record contents that may need to be used. These interdependencies should be systematic.

This forms a basis for cognitive task analysis through which a theoretical decomposition of cognitive resources can be related to performance. Cognitive task analysis examines the representational and mental processing demands of using particular interface features and relates these to patterns of user performance obtained in experimental research. The aim of the analysis is to infer rules interrelating properties of the complete cognitive mechanism. However, if we are systematically to infer rules that govern activity, we must have a more detailed specification of what we mean by configurations of processes, procedural knowledge, record contents and dynamic control. Furthermore, what we mean by each of these constructs should be clearly related to the underlying theory drawn from the science base itself. Using graph structures as an illustrative medium, this section will outline fuller specifications of these concepts.

5.1 The core constructs

The concept of 'phases' of cognitive activity helps us to think about how to approximate over the short term dynamics of cognitive performance, but we also obviously need to address issues of expertise and strategic variation. In figure 5, the stack contains a number of descriptions for novice, intermediate and expert users. In this case, moving deeper into the stack represents long term change in cognitive structures that come through learning and experience. In practical terms, if we are to build families of inter-related models, we require a specification of the concepts needed to constrain the set to be built at any one time, to organise the parameters used in modelling, and to determine how to actually gather the relevant information in the course of an enquiry. Several of the points already made through the 'stack' representation are reformulated in figure 6 as a graph structure.

Figure 6 captures the notion that the family of cognitive task models has *content* (the attribute spaces of configurations, procedural knowledge, record contents, and dynamic control). Second, it captures the notion that the particular models that will actually get built should have a specific *scope* with respect to user properties; the types of cognitive activity (goal formation, action specification, action execution), and the required stage or strategy breakdown. Scope is also determined here according to features like the type of system, the task set and the domain of application over which the modelling will be done.

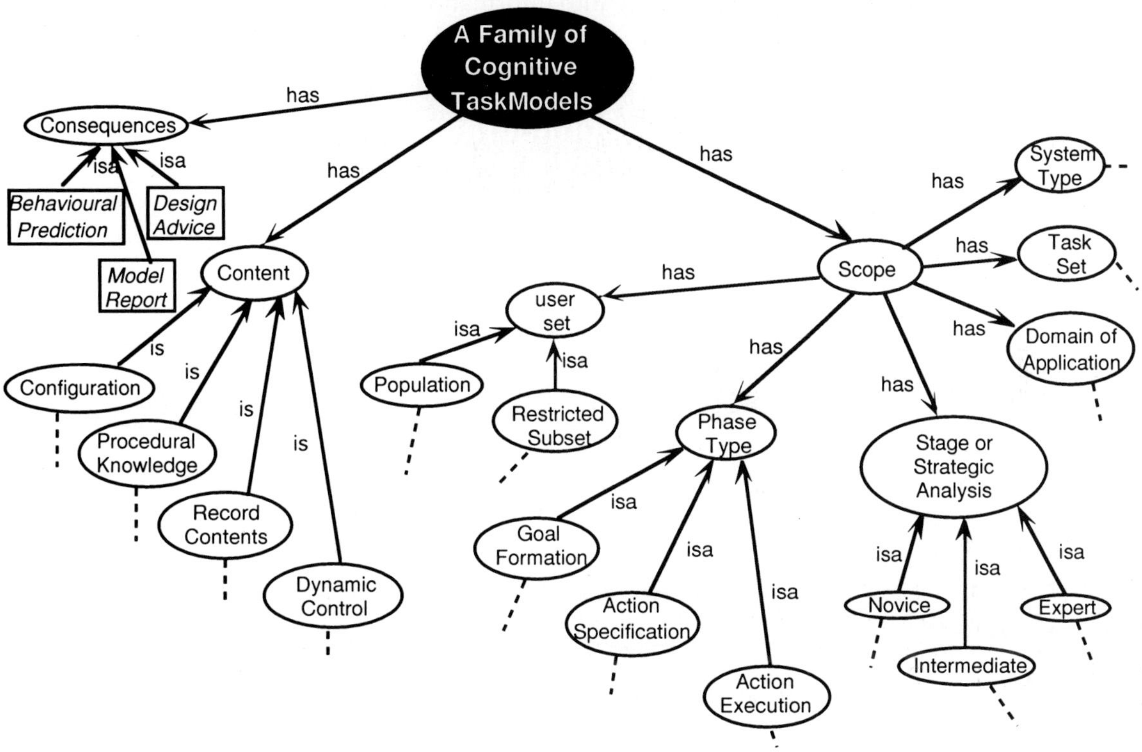

Figure 6. A graph structure representing core constructs required to determine a family of cognitive task models.

The third major conceptual organiser is concerned with the required *consequences* of the model. The four component attribute space of cognitive task models might be used to produce classic predictions about properties of user behaviour in terms of times and errors. Obviously, we could also potentially go beyond simple predictions and model reports to use the attribute space to help the designer understand the underlying psychological issues through a *model report* or to help produce direct *design advice*.

5.2 A Configuration of processes

A *configuration of processes* is simply a definition of those parts of the mental mechanism that are required to carry out a particular type of mental task. Configurations are built from a small number of standard constituents. They make up a set of interacting information processing resources (figure 7). According to the basic ICS theory (Barnard, 1985), the human mental mechanism consists of nine subsystems each of which is built out of only three basic resources (see figure 2): a *primary 'COPY' process* which represents information in a particular mental code and transfers that information into the second basic resource, *an image record*. Each subsystem has its own image record. The third type of resource, *secondary processes*, transform information from one mental code into information in another mental code.

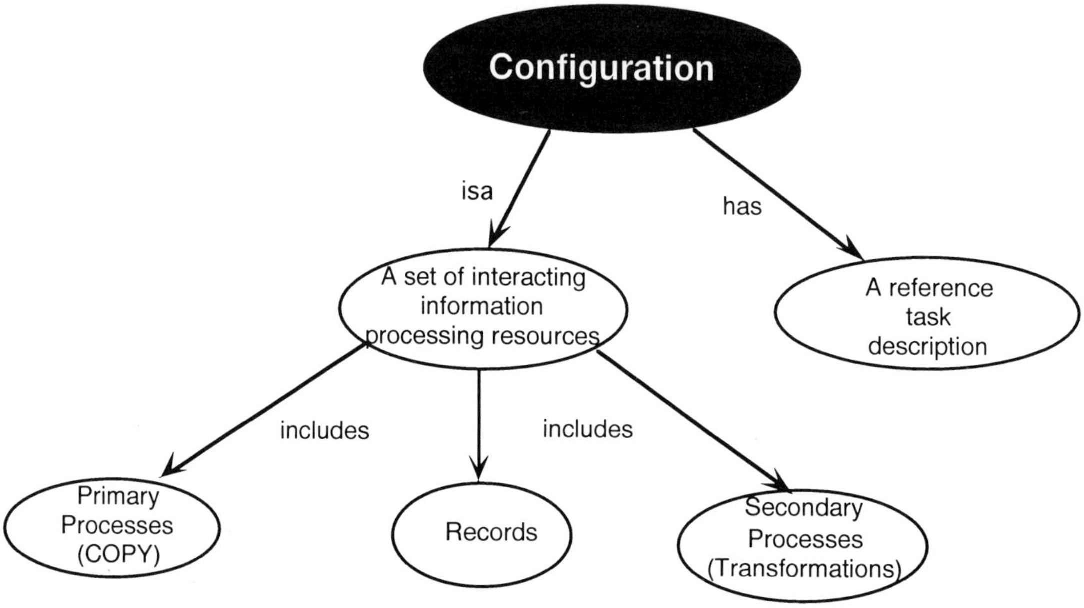

Figure 7. Configurations of resources.

Given that the resources of ICS are distributed over many subsystems, a typical mental 'task' will draw up several individual processes and optionally, image records. Table 2 illustrates two configurations of processes required respectively for the comprehension of words on a visual display unit and command entry. The command entry example incorporates a reciprocal loop between propositional and implicational processes governing the output of keystroke commands through their surface structure specification (MPL) to their articulation (ART) and motor output (Mot). In an earlier section we outlined graphically (figure 4) how information would flow through the cognitive architecture for interpreting information on a VDU, deciding to delete something, and then entering new text. The processes within subsystems provides a direct means of specifying, for any given task, a set of cognitive processes necessary to carry out that task. Hence, a *reference task description* is associated with particular configurations of resources in figure 7.

Obviously these configurations would differ if commands were entered or selected using a manipulative device, and if 'commands' were non-lexical icons or buttons. In effect, the theoretical framework can be used to describe just those processes that form the active organisation of resources underlying cognitive activity of different types. (in table 2, :: signifies transmission over the data network, while * represents an exchange of data with the world).

(i) Comprehension:	(ii) Command Entry
*VIS → OBJ::	::PROP → IMPLIC::
::OBJ → MPL::	::IMPLIC → PROP::
::(copy MPL →MPL→PROP)::	::PROP → MPL::
::PROP→IMPLIC::	::MPL → ART::
::IMPLIC→PROP::	::ART → MOT*

Table 2 : Process Configurations for Task Comprehension and Lexical Command Entry.

5.3 Procedural Knowledge and approximations over it.

Whereas the concept of a *configuration* of resources captures the idea that only specific parts of a complex architecture are required by particular mental tasks, the concept of *procedural knowledge* deals with the particular capabilities of individual processes that transform information from one mental code to another. This notion of representational change is a central feature of the ICS approach. In terms of its actual action, a particular mental process can be usefully thought of as either a collection of production system rules or as a connectionist network. The essential feature of both is that they effect a *mapping* from a particular input in one mental code to an output in another mental code. Clearly, any given process effects a substantial number of mappings. In the case of a transformation from Morphonolexical code (MPL) to a Propositional code, we are talking about a process that can map from any sentential surface structure representation in a language to the meaning of that structure. Thus, while the basic concept is one of mapping from a particular instance in one code to another (*A* mapping), the complete process effects many such mappings (*A set* of mappings).

For the purposes of building approximate models of the functional capabilities of a process, we need to consider the properties of sets of mappings. Figure 8 shows our current organisation of concepts for procedural knowledge. This diagram is divided in the centre by a vertical dotted line. On the left hand side are the basic theoretical constructs, that can be usefully thought of as applying to a single mapping, whilst those on the right hand side are intended to capture salient approximations over subsets and sets of mappings. The ultimate success of approximate modelling of cognitive activity will depend on our ability to use such set-based constructs effectively in reasoning about overt behaviour.

The left hand side of the diagram shown in figure 8 captures several basic concepts. By definition mapping has *an input* and *an output*. The inputs and outputs are *data structures* in particular *mental codes* whose identity and basic properties are specified by the ICS theory. The particular *dimensions of encoding* for each type of mental representation were summarised earlier in table 1.

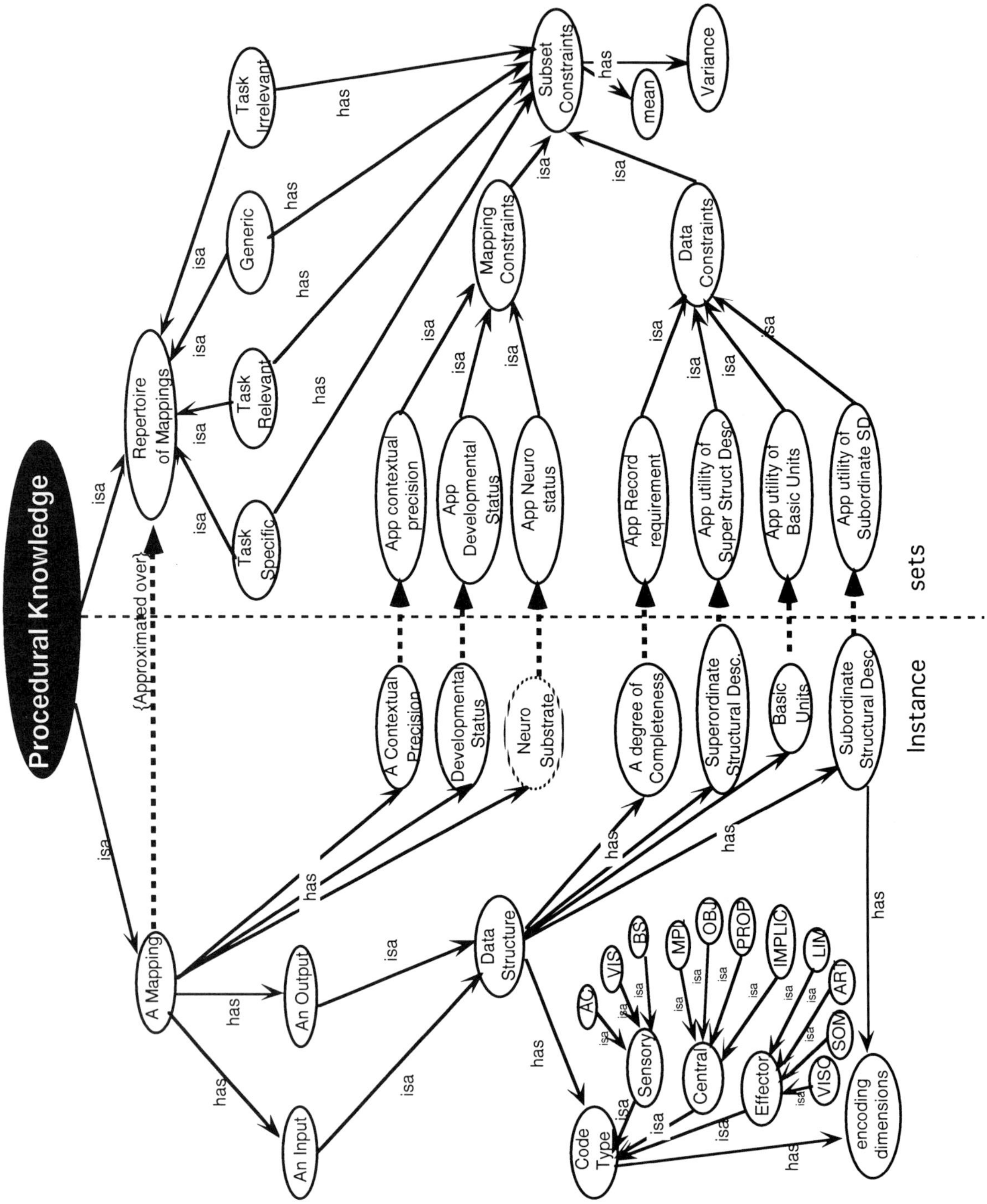

Figure 8. Graph structure representing properties of Procedural Knowledge and approximations over it.

Several other basic theoretical properties of procedural mappings are also shown in figure 8. First of all, the data structures have *a degree of completeness*. They may be only partial specifications of what is required for the input to generate an output. Secondly, those data structures themselves require specification in terms of a detailed *structural description* (SD) of what is in them. A data structure in any mental code has *basic units*, which have *a constituent structure*, and are themselves organised into *superordinate structures*. These definitions are in effect characterisations of the general properties of the representations that are input to a process and output from it.

Properties of the mappings themselves must obviously include *an identity*. Given a particular input the mapping will usually produce a particular output. However, this mapping actually effected may be subject to additional constraints. For example, it may be more or less practised (*developmental status*) and its operation may be influenced by the status of the underlying mental device—as perhaps degraded by brain damage or alcohol abuse (*neuro substrate*). A mapping may give rise to an error because the underlying procedural mapping is of low *contextual precision*.

So, for example, with an input data structure at a propositional level specifying 'put information on the VDU', the underlying mapping from that propositional to an MPL representation may be equally likely generating the command words 'display' or 'show'. The precision of the mapping would be relatively low. The mapping may have similar information processing consequences in either case—but one would be an error in the context of the total set of operations available.

The right hand side of figure 8 depicts the concepts required to approximate over subsets of individual mappings. Naturally any given process within ICS can effect many mappings and we can view them as a repertoire whose potential influence on cognitive performance can only be adequately gauged relative to the setting. In the case of human-computer interaction that would be constrained by interface properties and the tasks that can be accomplished with the application software. Accordingly, for the purposes of analysis, we subdivide a repertoire of mappings into those that are *task specific*; those that are *task relevant*; those that are *generic*; and those that are *task irrelevant*.

We can operationally distinguish between mappings essential to the tasks and interface, those that would be closely related to those essential mappings, and highly generic mappings. Each of these three classes could potentially influence performance. To take an obvious example, a desk top metaphor provides a range of established 'procedural mappings' that can be drawn upon in the understanding and control of an interface. Such mappings would be task relevant. Other well-established mappings, like general problem solving approaches, would be generic mappings that could be called into play. Naturally, the vast bulk of mappings in any process would be largely irrelevant to any well focused analysis of cognitive activity based upon a well-delineated context. It has to be acknowledged at the outset that firm boundaries between these classes would, in all probability, prove hard to sustain. Nonetheless, if we are to build effective approximations, we must establish a tractable number of categories rather than an exhaustive number of distinctions. If the overall approach proves to be of value, then the number of distinctions and the boundary definitions can be extended and refined.

The right hand side of figure 8 distinguishes between two general classes of property—a class that applies primarily to approximations over properties of individual mappings (*mapping constraints*), and those that are derived from the data input to and output from the actual process (*data constraints*). Since both are sets of mappings we can think of properties of the individual cases having something like the character of a statistical distribution—with means, variance, skew and so on. In the case of mapping constraints, we can assign a simple interval scale to a property like contextual precision (set size n, precision 0%; 25%; 50%, 75%, 100%).

We can make similar assignments with specific types of data constraints. In preliminary analyses of alternative task structures (Barnard, MacLean & Hammond, 1984; Barnard, Wilson & MacLean et al, 1988), we could do useful reasoning about cognitive activity by assuming that constituents of a task could be assigned 'item and order uncertainty' at each level of a structure. This reflects the degree to which the accessibility or positioning of a given constituent is constrained by knowledge already embodied in the mapping. By attempting to combine such concepts, we can develop simple approximations about the utility of superordinate structures, basic units, and subordinate structures.

A good grasp of this problem comes by considering what happens when an average British tourist is trying to converse in German—when their vocabulary size is low (functional set of basic units @MPL, size = 100); their syntax impoverished (utility of superordinate structure @ PROP→MPL 25%); and their pronunciation laughable (utility of subordinate structure @ MPL→ART <25%). Obviously, the data *output* at a syntactic level of this example comes to form an *input* data constraint to mappings that are subsequently executed (e.g., articulation). We can reason from such properties to the likelihood that a massive amount of inferential processing and memory access would be going on across most of our hypothetical tourist's mental mechanisms. Under such circumstances, it would be difficult smoothly to co-ordinate and dynamically control the inter-operation of all the processes in the mental configuration for speech production. The overt behaviour would be typified by hesitant performance and not necessarily satisfactory results. In terms of the potential generality of the underlying theoretical characterisation, we would expect the same kind of analysis to apply not only to the occasional, infrequent and untutored user of the German language, but also to the occasional, infrequent and untutored user of the Unix command language.

5.4 The nature of Record Content and approximations over it.

There is substantial overlap between the organisation for procedural knowledge (figure 8) and that for record contents (figure 9). Within ICS, record content refers to representations stored in the image record of a particular mental code. There is a record resource associated with each and every subsystem.

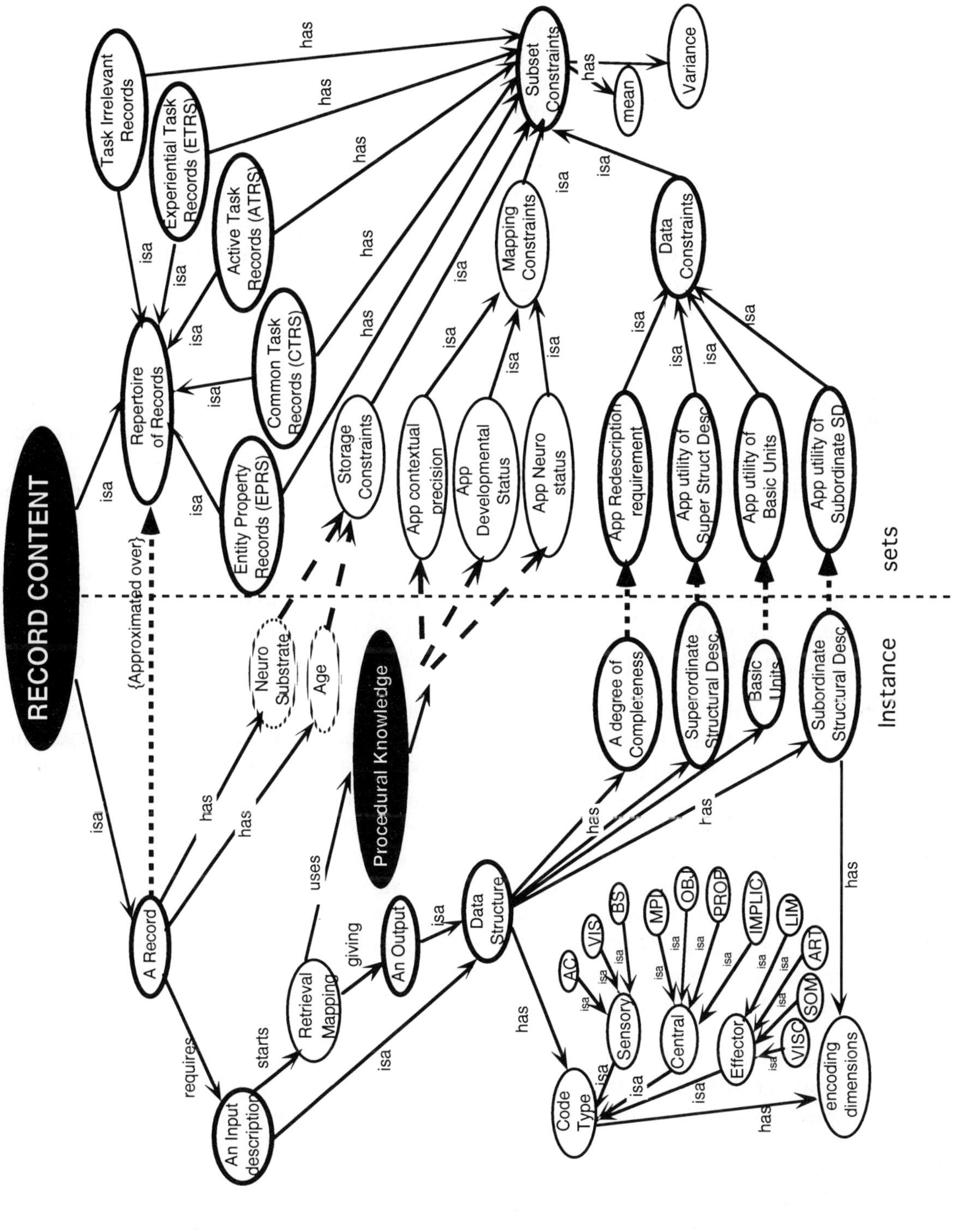

Figure 9. The organisation of concepts for *Record Content* and approximations over it.

The actual representations stored are themselves, of course, data structures in just the same way as 'inputs' or 'outputs' to procedural knowledge. Indeed, one basic principle of ICS that all incoming data structures are *copied* into the image record in parallel to any transformations carried out within the subsystem. Furthermore, the output of one subsystem become the inputs and hence record content of another.

The basic characterisation of data constraints for both individual records and approximations over sets is essentially the same as that for procedural knowledge. Data constraints are not, however, the only theoretical communality between our characterisations of procedural knowledge and record content. Within the ICS theoretical framework, a target description of information in a particular code is used to access the record structures. If a match is found, then any additional information contained within the stored record is effectively made available. The choice of words here is crucial. In fact, there is no special purpose apparatus for the 'retrieval' of information in the ICS theoretical vocabulary. What happens is that a particular process re-configures from processing incoming data to access data structures preserved in the image record. Any information located is then receded to generate an output from that process. Strictly speaking, information in the preserved code is never 'recovered' from a record. The contents of a record are simply 're-used' as the basis for the generation of an output.

In theoretical terms, a input (description) starts a retrieval mapping to generate an output (see figure 9). The retrieval mapping, of course, uses precisely the same procedural knowledge as would be used in the transformation of any incoming data pattern. Hence, precisely the same concepts apply to a retrieval mapping as apply to a simple input-output transformation.

A basic principle of the operation of the ICS architecture is that any one process can only recode one coherent stream of data at a time. Hence, a given process within a configuration must switch from a direct input to access the image record associated with its subsystem. It cannot do both. Furthermore, the theory holds that only one process within a subsystem can gain access to its image record at any one time (see Barnard, 1985). Thus, any access to an image record has implications for the dynamic control of an integrated configuration of processes. Likewise, if an input 'description' is insufficient to locate a target record through similarity or under-specification, or if the record access fails to meet the target match, then there is a requirement to generate a fresh description with significant implications for the dynamic control of the complete system. There must be extra processing activity and configural shifts to cope with both re-description and access.

The types of record content that need to be considered in building approximate cognitive task models do however differ from that for procedural knowledge. The general view of learning and skill acquisition adopted within the ICS framework is one in which record content plays a major role in the development of procedural knowledge. Indeed, the development of procedural knowledge depends upon, and cannot occur without, record content. Each subsystem can represent data and copy it to its image record before the relevant transformations are developed. This follows from the basic principle of where the record content comes from and from the 'primacy' of the COPY process. Repeated co-occurrences are then abstracted from the record structures and become embodied within the transformational processes (hence their labelling as secondary processes in figure 5). Whilst many forms of perceptual motor (playing tennis) and cognitive skill (e.g., language production) may develop into fully automated mappings after massive practise, other complex tasks may continue to rely on stored representations rather than automatic mappings. Indeed, with many of the tasks involved in human-computer interaction, we might expect some elements and constituents to become fully proceduralised (e.g., hitting the return key) whilst others rely for their execution on access to memory.

In figure 9 the particular distinctions again have theoretical motivation. There is a major distinction of relevance that links a given analysis into the specific context to which it is addressed. The complete repertoire of memory records is divided into those that are task relevant records and those that are not task relevant. For any given analysis, the vast bulk of memory records would, of course, be irrelevant. Within the relevant structures we distinguish four basic subsets:

Active Task Records (ATRs). In the course of processing information in any task, a whole variety of records will be generated. At any given moment, the information that has just been acted upon by a subsystem forms its ATR. ATRs constructed in the immediate past are thus readily available for recovery—across sensory, central and effector subsystems.

Experiential Task Records (ETRs). Records laid down through past experience accumulate over time. We can, for example, remember what we did last time we used a particular computer system. We have a repertoire of such experiences and those laid down within the image records of the central subsystems can be accessed via descriptions originating in some other subsystem capable of generating the description in the required code (the direct activation principle). The revival principle of recovery will not usually come into play with these records.

Common Task Records (CTRs). As individuals repeatedly perform a particular task, we assume that they abstract a representation of what they have to do in that task. Individual details (like the exact command) are dropped from the complete information contained within an ETR. In terms of our structural descriptions of representations, the constituent structure of basic units (i.e., subordinate structures) may become proceduralised and effectively discarded from the representation, as may some basic units themselves. The operative representation may contain only a superordinate organisation for a given task plus, for example, some information on what basic units can go where. In the specific context of human-computer interaction, CTRs represent what is often called how-to-do-it knowledge.

Entity Property Records (EPRs). Like CTRs, EPRs are also abstractions. However, this time the abstractions are intended to refer to how-it-works knowledge or indeed any other property of tasks, interface objects, or other operational constructs in the domain of use. An understanding of how the desk top metaphor works with a particular application would, for example, be characterised as a set of EPRs, as would knowledge of what could be achieved with a particular common task record (e.g., the knowledge that I can access my database in three ways).

Again, although the particular contrasts may prove hard to sustain, they are simply intended to be adequate for the purposes of modelling a significant range of important phenomena in the domain of human-computer interaction. We would nevertheless hope that their utility would extend to other domains without too much modification.

5.5 The nature of Dynamic Control and approximations over it.

In providing a conceptual structure for the dynamic control of a multi-processor architecture, we move from the static concerns of characterising mappings and sets of records to describe the behaviour of a complete system *over time*. As has been stated, a key feature of the ICS architecture is that there are no general purpose resources such as a working memory or central executive. Rather, the whole system is seen as self-controlling via the passage of data and representations among individual processes. In order to describe dynamic control, we need to characterise configural patterns of cognitive activity across many different resources and how those patterns change as that activity evolves. Indeed, a number of the main ideas that we need have already been foreshadowed earlier when we discussed the consequences of missing procedural knowledge or dealing with the re-description requirements of record access.

A pattern of activity is something that applies over a configuration of resources. However, given the potential interconnectivity of the whole mechanism, it seems useful to distinguish a *core* configuration for a particular task from *subsidiary* configurations that may nonetheless play a significant role. Given the wider architecture, it is possible to perform certain tasks concurrently. We can, for example, readily drive a car whilst holding a conversation with a passenger. This can happen because the configuration of processes for driving need not necessarily draw on any of the processes required for language production. However, if the driver wishes to assess their passenger's reaction to a particular point of discussion, processing resources in the visual and object subsystem must be diverted from the driving task and incorporated into the configuration for interpreting the conversation. There must be a shift in the configuration of processing activity, however momentary.

Similarly, if we are in a phase of cognitive activity involving action specification for issuing an interactive command, there may be a similar configural shift whereby the user changes from simply searching for and accessing a memory record, to evaluating menu options in which the uptake of visual information becomes directly relevant. Likewise, the user may notice a peculiar pattern of pressure at a finger tip that signals that they have typed a pair of letters where only one was intended. For the brief moment in which they are 'paying attention to the finger tip' information, a central process may switch from handling information originating from a visual data stream to one involving a data stream within a proprioceptive process, and then back to the visual data stream. These classes of consideration are captured in the left hand side of figure 10.

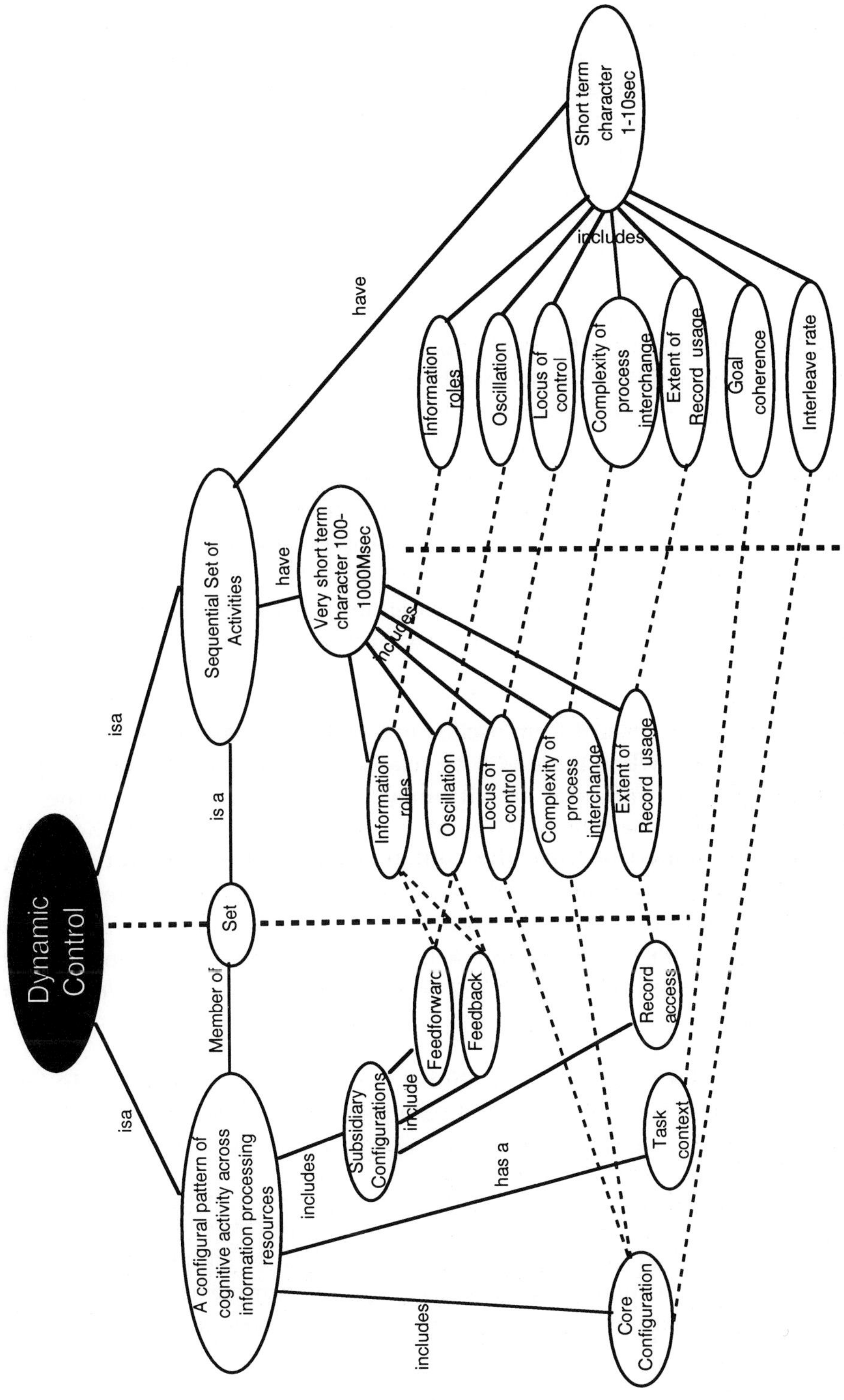

Figure 10. The organisation of concepts for characterising *dynamic control* and its approximation over time.

For the purposes of approximation, the operation of a subsidiary process configuration can fulfil a number of *information roles*. So for example, in a conversation, the arms and hands may be gesturing in a manner semantically related to the ongoing conversation. In this case, the subsidiary configuration would be fulfilling the function of providing task relevant feedforward information. Alternatively the hand might be holding onto a small child, in which case the operation of the same process would fulfil no informative function in relation to the conversational activity in the core configuration that is controlling speech production.

Unlike figures 8 and 9, there are two vertical dotted lines in figure 10. These distinguish approximations over different durations of the dynamics of cognitive activity. The first partition concerns dynamics in the very short term. It is topicalised upon the approximate description of activity within a single member of a family of models (see figure 5). In any given time slice of (say) 50 msec of activity, it is unlikely that there can be much in the way of significant shift from one coherent activity to another. In the range between say 100-1000 msecs, configural patterns may be relatively stable or subject to significant shifts. The kinds of shifts outlined above where there are momentary alterations is captured in the very short term by the concept of a configural *oscillation*. A process in a core configuration may change from one originating data resource, be it access to a record, or a switch from data coming from one process to data coming from another. The defining characteristic of an *oscillation* is that it is subservient to the immediate requirements and purposes of activity within the core configuration.

Although there may be very short term oscillations within and among specific configurations, there will also be periods when there will be relatively stable patterns of configural activity. Even so, the patterns of configural activity may differ in terms of both the *complexity of process interchange* and the *extent of record access*. In the earlier section dealing with configurations we illustrated how processes within the central subsystems could form a simple 'loop' of reciprocal processing activity—as for example when the Propositional to Implicational transformation and the Implicational to Propositional transformation enter into an extended period of reciprocal interaction for planning an utterance. Activity within this loop may be more or less stable and more or less extended in time. However, more complex patterns may develop.

Within the simple propositional to implicational loop, there is no necessity to 'verbalise' the thoughts involved. However, the configural pattern might involve:

::IMPLIC→PROP::PROP→MPL::MPL→PROP::PROP→IMPLIC::

in which case the thoughts would be verbalised into a surface structure code and then fed back to the propositional subsystem and then back to the implicational subsystem. The pattern would be more complex than the simple reciprocal loop. Adding significant record access at one level would also serve to increase both the configural complexity and extent of activity. Adding further oscillations or interdependence on subsidiary configurations would make matters worse.

Within any core or subsidiary configuration of processes, the ICS framework assumes that one process will be 'buffered'. In buffered processing, a particular process is configured to recode information just copied into the image record of its subsystem. In effect the COPY process and the process of transformation are operating in series rather than in parallel. This means that, rather than dealing with a direct input, the process is recoding an extended representation. This is viewed as a more robust mode of processing than when the relevant process is configured to take input directly from the data network of Figure 1. Buffered processing is described in detail in Barnard (1985). Although we will not pursue the details here, buffered processing is associated with information being the focus of conscious awareness.

The ICS theory assumes that, within a given configuration, only one process can be configured for buffered processing. The act of buffering controls the rate of data flow throughout the configuration. Thus, one representation is effectively dominant within a pattern of configural activity. This is captured by the concept of *locus of dynamic control*. So, for example, when we are concentrating upon the sound of a speakers voice, the locus of dynamic control will be in the acoustic subsystem; when we are concentrating on how someone is saying something it may be at the MPL subsystem; when we are concentrating on what they mean, it will be at the propositional subsystem; and when we are concentrating on the emotional feelings generated by what it means the locus will be at the implicational subsystem.

Extending the analysis of dynamic control over longer periods is a relatively straightforward matter. We can consider the properties of information roles, oscillation, locus of control, complexity of process interchange and extent of record usage in some sense 'averaged over' the co-ordination of activity within the short term dynamics of task performance (1-10 secs and up). Here though, we need to draw in two further considerations. First, cognitive activity obviously operates in a *task context* and subserves task goals. We therefore need to consider the overall *coherence* of that activity with respect to the task *goals*.

The activity may subserve the goals in a coherent way or, for example, extended activities may be less than coherent, as when an individual switches from one set of actions to another in an apparently aimless fashion. However, the switching between phases of cognitive activity may be very coherent indeed. When phases of cognitive activity switch from one to another, this is captured in the concept of *interleaving*. Over the short term, users characteristically do not plan their actions, obtain all the visual information they require and then execute the actions. While entering a computer command sequence, they may type a bit, look, think, look again and type some more. On a larger scale they may interact at the terminal, answer the phone, service an interruption from a visitor and then resume activities at the terminal. Taken together, the concepts of configural oscillation, interleaving and goal coherence enable us to restrict our phase analysis to goal formation, action specification and action execution. Rather than finessing the problem of cognitive activities in a task being not strictly seriated (e.g., see Norman, 1986), we seek to explicitly capture properties of non-seriality.

6. Concepts and principles to support reasoning about cognitive activity

The whole approach is predicated on the assumption that particular configural patterns of activity within the human cognitive mechanism will be associated more or less directly with behavioural consequences. The detailed concepts introduced in the previous section expanded what is meant within the ICS framework by configurations of processes, procedural knowledge, record contents and dynamic control. These were formulated to enable us to reason about cognitive activity and its applied consequences. In the very short term dynamics of cognition we would expect overt behaviour to reflect systematic descriptions in these terms.

Concepts such as these enable us to make explicit and discuss cognitive requirements such as the overall 'load' on the wider system. Obviously, immediate cognitive load is not thought of here in terms of a limited capacity working memory. Nevertheless we can specify conditions within a multiprocessor architecture that would have equivalent effects. So, for example, we would expect time on task and errors to be globally related to the complexity of dynamic control, and features such as oscillation. We would also expect slip errors to arise when crucial data structures remain unevaluated because the locus of dynamic control is at a different subsystem, and so the information that they contain is not available.

Similarly, in the short term, we can think of overt performance as being constrained by properties of several features of the cognitive system. These include knowledge of tasks (common task records); relevant entity property records or experiential task records; static features such as representations of the current state of the interface (active task records); and dynamic features such as interleaving of activities subserving different goals.

As we move to consider the longer term dynamics of learning, we can talk in terms of the initial acquisition of experiential task records and the gradual build up of sets of task specific common task records and entity property records. Even at this level, properties of members of the set can be mapped back down to the original theoretically motivated concepts—once again the detailed properties associated with record content and the procedural knowledge required to recode and use them.

Having a set of constructs with which to reason about cognitive activity is one thing: having a set of validated principles that govern that activity is quite another. An important component of our wider research strategy is to ensure that principles interrelating procedural knowledge, record content and dynamic control are firmly grounded in experimental evidence. There are now a number of examples of the motivation of such principles from experiments in the HCI domain.

On the basis of a substantial number of experiments on the learning of different sorts of command names we have inferred a number of principles concerning the effects of pre-existing procedural knowledge upon performance. So, we have argued elsewhere (e.g., Barnard, 1987; Barnard, Grudin & MacLean 1989) that the number of wholly proceduralised process in a configuration can be directly related to the complexity of process interchange; that substructures within a representation constrain the effectiveness of a mapping; and that ambiguity of meaning among command terms within a set influences conceptual precision within that set. Ineffective substructures and low conceptual precision increase complexity of process interchange by requiring additional record access and inferential processing activity.

Likewise, the practical effects of detailed properties of record content can be inferred from relevant experimental evidence. Figures 11a and b show two ways of accomplishing the same electronic mail task. The two versions differ in the ordering of the individual operations. Barnard (1987) shows that there is greater uncertainty associated with the (a) structure than the (b) structure.

For example, users can be expected to have pre-existing knowledge (entity property records) that enable them to infer that establish operations that get information used in another command are a pre-requisite for invoking the latter command. This means that the order individual commands within the pairs of structure (b) is constrained whereas the order of the operations within the groups of four in structure (a) is not.

In this way, we can move towards principles specifying that the complexity of dynamic control will depend upon the extent of item and order uncertainty within the superordinate structure of common task records. In turn, the particular properties of uncertainty that apply in a given instance of a common task record will depend upon the existence of entity property records or experiential tasks records that may function to reduce or increase the particular levels of uncertainty (Duff & Barnard, 1990).

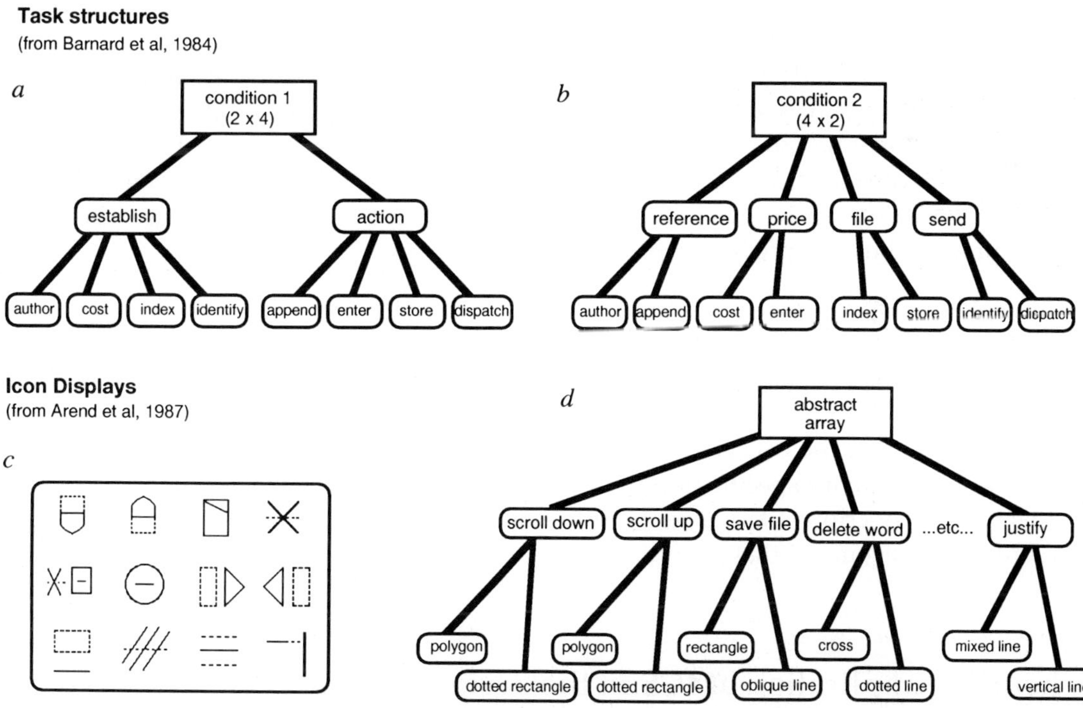

Figure 11: structures from the email task of Barnard et al (1984) and the icon search task of Arend et al (1987)

The kinds of representation that support memory for the operations in a task are most likely to involve representations that support propositional and morphonolexical aspects of language production. Similar principles can be inferred concerning the processing of specifically visual aspects of interfaces. So in the case of the icon array shown in figure 11 c, a property applying to the superordinate structuring of the array of icons concerns positional uncertainty. Obviously, if the position of icons vary from one user encounter to another, a greater degree of visual scanning will be required than if the items are in fixed positions (Green & Barnard, 1990). Similarly, the structure of the icons as visual objects (the lower levels of decomposition in figure 11 d) can be expected to determine the amount of processing activity. Indeed, May, Barnard & Blandford (1993) use evidence of this sort to argue that the complexity of dynamic control in visual scanning will depend, among other things, upon the depth of predicate evaluation that must be achieved in order to distinguish a particular target from others in the operational set. In this case the array actually present upon the screen constrains the form of active task records and bottom up aspects of mental processing activity. The characteristics of knowledge about the form of icons representing particular operations (entity property records) constrain top down aspects of mental processing activity.

This treatment of the derivation of principles based upon experimental evidence has necessarily been brief. Our illustrations are not intended to provide detailed justification for the principles. In the present context we merely wish to indicate that such principles are not conjured out of thin air. They may be systematically derived by interrelating the theory-based analysis with the available evidence. Armed with systematic principles that are specified in terms of such properties of human information processing, it becomes possible to build approximate models that describe cognitive activity in terms of abstract properties. User requirements can be interpreted, at a cognitive level, with respect to these properties. With the addition of rules that relate these models to characteristics of user performance, the underlying principles can be generalised to novel task and interface settings. The whole modelling process thus has predictive properties.

The use of systematic experimental evidence to support the derivation of principle is an important step for the production of a modeller that would lay claim to any sort of deep validity. Clearly, much experimental evidence is obtained in rather tightly constrained task environments. If this sort of modelling is to find a role in the design of interfaces, it is equally important to show that the concepts used to describe cognitive activity have face validity in helping us to reason about specific design options. The next section addresses the problem of using this type of descriptive language both to characterise cognitive issues in a design space and to direct the outcome of that analysis.

7. A heuristic application of ICS in the design context

The general idea of 'characterising the psychological aspects of the design problem' has, of course, numerous familiar rings to it. It is a core aspect of the use of traditional behavioural science 'guidelines'; it forms a central plank of Design Rationale (MacLean et al. 1990, and see

Chapter 2.4); as well as the Task-Artifact Cycle of Carroll and his colleagues (Carroll, Kellog & Rosson, 1991). Last, but not least, the psychological aspects are presumed to be implicitly derivable by the designer using a programmable user model (PUM—Young, Green & Simon, 1989). However, our current train of thought does appear to have a novel twist by comparison with these familiar rings.

Traditional guidelines have been general formulated as rules of thumb that link some feature of tasks, users or interfaces more or less directly with a design solution or class of design choice. Design Rationale links questions through options to criteria. Except perhaps through the possibility of a library of previous cases, it does not, in and of itself, suggest what the psychologically motivated questions should actually be. Similarly, the Task-Artifact cycle and Carroll's associated Claims Analysis is a description of a methodology, not a means of inspiring the content of specific claims about usability—that requires a 'theory of tasks'. In the case of PUMs, the psychologically relevant concerns are not posed—they implicitly emerge through the activities involved in specifying the knowledge analysis required to use the instruction language and running it through the PUM architecture. In none of these instances is the nature of the deeper psychological issues directly suggested by the analytic representation.

This then, is the twist. A cognitive task model representation describes properties of cognitive activity at some level of approximation. It incorporates components that describe attributes both of mental processing (configurations and dynamic control) and of knowledge used in it (record contents and procedural knowledge). It therefore provides an abstract, but explicit, representation of what is going on in the head of the user. From this, it should be possible for an analyst—assumed here to be knowledgeable in cognitive psychology—to gain a direct understanding of the psychological issues associated with a design in a particular domain.

Again, at his stage we would not want to make any strong claims about the actual utility of the representation, merely that it is there and could plausibly be used as a basis for formulating the design problems that must be addressed within the design space from the psychological perspective. The information provided would structured to feed the more creative craft skills that come into play when actually producing a design to overcome the problem. In certain respects, the cognitive task model representation should be directly suggestive of a substantive design question and possible means of answering it.

As an example of the way in which we have been using (and developing) this approach to cognitive modelling, the next section presents an 'HCI scenario'. This scenario was used by the Esprit Basic Research Action AMODEUS in the manner recommended by Young & Barnard (1987), to provide an integrative focus for a variety of different modelling techniques—system modellers and modellers of the design process as well as user modellers (Hammond & Myers, 1990). The scenarios that the project used were chosen from a number of 'real world' design problems that had been encountered by project members. In modelling this scenario, our aim was to show that the 'identifier space' represented by attributes of the Cognitive Task Models could be used not just to map onto predictions of expected user behaviour, and hence the acceptability or otherwise of the suggested design options, but also as a basis for generating new design options with a theoretically grounded rationale.

7.1 The 'small screen map' design scenario

In this scenario a system is being developed which will allow a wide range of users from the general public to use a local information map to find their way around, get an overview of the general area, and to identify key detailed points. The problem is that these tasks require different levels of detail, and the screen area available is not sufficient to allow the whole map to be shown at the highest required magnification. A method is sought that will allow users to see both low magnification overviews of the whole map without details and high magnification views with sufficient detail but lacking relationships to the overall map.

7.1.1. Analysis of split screen option

The first alternative splits the screen into two areas. Approximately two thirds of the screen will be taken up with the low magnification overview map, giving enough information to relate places to each other across the whole city, and to find routes, but not to allow the users to identify places, or to find out their names. The bottom third of the screen will be taken up with a high magnification view of a subsection of this map, at a high enough magnification to show place names and so on. The area of the overview that is covered by this close-up would be shown by changing the background colour of the relevant part of the overview. A trackerball would be used to move the contents of the high-magnification area 'around' the map.

The configuration of actions required to use this design involves the users alternately attending to information in the two windows, so necessitating a transition between elements of the visual image record. On attending to the overview section of the window, the section with the inverted background would clearly form the focus of the visual representation, and so no further transitions would be required to manipulate it (via a *VIS→OBJ:: OBJ→PROP:: PROP→MOT* feedback sequence that could easily become proceduralised to *VIS→MOT* without extensive requirements for dynamic control).

Following a transition to the close-up window, however, users would need a further transition to its constituent substructure to operate upon the elements within it, and then transitions between those elements. Moving the close up while attending to this view could potentially cause problems. If the focus of the representation is an element of the view, moving the trackerball left results in the element moving to the right. If the view as a whole were the subject, however, moving the trackerball to the left would result in an interpretation of leftwards motion from the rightward flow of the texture within the view. The user would therefore have inconsistent ETRs (Experiential Task Records) for motion while attending to this view, and this would result in ambiguity. Users would probably have to make a transition back to the view before trying to move the window. Such inconsistent ETRs would retard the acquisition of task-action mappings for action execution.

7.1.2. Analysis of bifocal screen option

In this alternative, the screen is divided into nine rectangles (a 3x3 grid). The central region of the screen is devoted to a close up view, and can be moved around the background low-magnification views. The background views do not remain constant, however, but are dynamically compressed and stretched as the close-up box moves, to ensure that no information is ever hidden.

Unlike the split screen version, here we have to deal with systematic, but unfamiliar distortions of a basic visual image. Both VIS→OBJ recoding and OBJ→PROP recoding would be largely unproceduralised. Given the ingrained nature of our mental processing of normal visual structures this would have the effect of substantially increasing the complexity of dynamic control of the whole cognitive mechanism. In addition, the continuously variable distortion effects due to motion of the central area would give rise to highly variable ATRs (active task records) in the visual domain and any resulting ETRs (experiential task records). Similarly, the discontinuities that must exist at the boundaries between the central and compressed areas would also give rise to difficulties in analysing the presented visual inter-relationships. Indeed, it is not at all clear how the processing mechanism could deal effectively with such distortions and discontinuities given that the intended user class is of the walk-up-and use variety.

Apart from the geometrical problems of what happens at the corners of the screen and what happens where the eight different compression zones meet, this alternative destroys the visual structure of the low magnification overview. The distortions introduced make it impossible for the user to interpret location or distance of any of the points on the map, since their representation upon the screen is related to the position of the close up window. This introduces not just ambiguity in the VIS→OBJ::OBJ→PROP mappings. It also means that only coarse relationships (such as above/below and left/right) can be determined. The user would therefore not use the low magnification area of the screen for relative comparison of elements at all, and would instead attend just to elements shown within the high magnification area. The only use for the low magnification area would be to identify new elements to examine with the close up view.

Route information that included geographical information such as direction could only be obtained reliably through the close up, and if the route extended outside the close-up box, the box would have to be moved towards the approximate location of the destination. The focus of the visual representation here would have to alternate between the target in the low magnification view and the contents of the close up box (to register the route information ie 'am I still on a road?'), hence requiring complex transitions between the elements in the substructure of the close up, this substructure and the close-up box itself, and the close-up box and other (unproceduralised) analysis of elements of the overall view. The latter transition would require processing to correct for the motion of the target (since it would accelerate in the opposite direction to one of the components of motion of the box as the box approached it, and if it crossed the boundary of two compression zones, it would actually move in a different direction), and this VIS→OBJ::OBJ→PROP::(PROP→IMPLIC→PROP)n processing would interfere with the processing required to move and recall the motion of the box, resulting in ambiguity in the image record of the MOT and VIS subsystems, and so hindering an accurate assessment of the geographical dimensions of the route.

The general layout of the whole area would be very difficult to determine from this sort of system, because the low magnification areas of the screen could not be used to obtain it—it would have to be built up laboriously through moving the close up over the whole map and remembering what was in it, with the problems described above.

7.1.3. Analysis of fish-eye screen option

In this option a central area of the screen is again used to present a close-up, being moved around a background view which distorts to present a gradient of magnification, but this time the close up and the magnification zones are round, resulting in a fish-eye lens effect. This alternative has all of the disadvantages of the previous version with respect to the relative proceduralisation of VIS→OBJ and OBJ→PROP processing. The more continuous nature of the motion of elements within the low-magnification view would make the motion towards a target easier because the target would always be moving towards the close up, instead of at an angle to it, but with the added problem that the circular zones remove the discontinuities that would occur at the edges of the square zones. This could potentially allow extensively proceduralised VIS→IMPLIC::IMPLIC→BS mappings to trigger bodily effects which, when fed back to the centre (BS→ IMPLIC), could conceivable give rise to a feeling of disorientation and a sensation of motion at odds with other sensory inputs to the BS subsystem.

7.1.4. Suggestions for design alternatives

All three of these alternatives attempt to allow the user to make use of the low level and high level magnifications simultaneously. Two of them, however, introduce distortions into the low level view that make it impossible for the user to extract the relevant information form it. On these grounds alone the split-screen view can be identified as the preferred option. The other two alternatives can be seen from the analyses of cognition required to use them to actually hinder or prevent the user carrying out the tasks that the alternatives are supposed to support.

The analyses have identified in these designs a recurring problem for the user in relating the motion of the close-up to the overview, and of making the transitions between the content of the overview and the content of the close-up window. Both of these aspects of the problem involve transitions between different elements of the visual structure of the screen, and transitions into the substructure of these elements. Even the preferred alternative is not ideal in this regard.

Having identified this as the key to the usability of the system, it is possible to ask how the nature of the transitions around the visual structure could be made use of in the design. An assumption in the design requirements is that it is necessary and advantageous for the user to have both the close-up view and the overview visible simultaneously, yet the cognitive models point out that the user must make transitions between the two views, and will not be able to extract information from them both at the same time. The designer could therefore question this earlier design decision, and think about designs which do not attempt to show differently scaled maps within one screen display.

One such possibility is to replace the transitions that the user would need to make between different areas of the screen by changes in the screen display, so instead of the user having to attend to a different part of the screen, the whole screen would be taken up by the new scale view. This could by managed by, for example, using the trackerball to move a set of cross hairs around the screen to locate a point of interest within the map, and then on an event like a mouse click or a button press, 'zooming in' so that the point under the cross hairs becomes the centre of a new screen display, but at a higher magnification. A different button would similarly allow 'zooming out'. Scrolling of an image could be allowed if the crosshairs are pushed against an edge of the map.

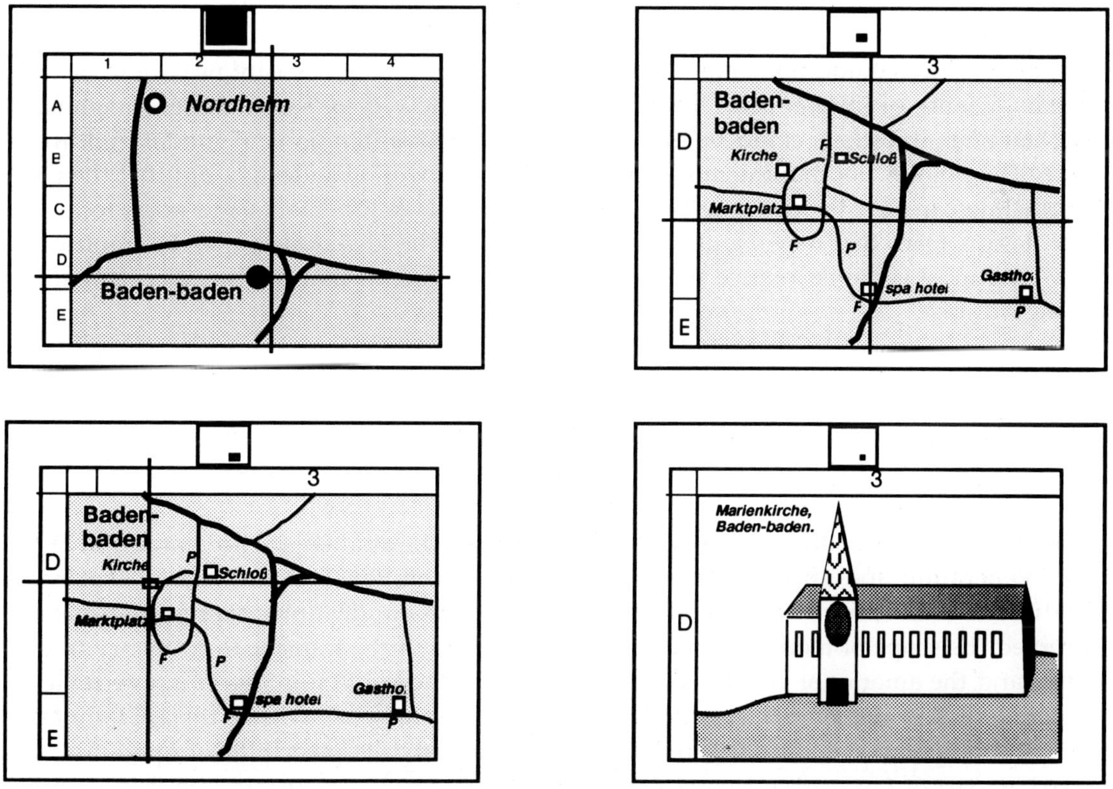

Figure 12: a redesigned interface, showing four successive screens as the user clicks to focus in on Baden-Baden, moves the trackerball to the Kirche, and clicks to focus in again.

This design would not require dynamic transformations of an underlying map to be made, but could be based upon a number of static maps. In fact the design would allow for any number of levels of magnification, each suited to a particular aspect of the users' tasks: a low magnification view for general orientation and major routes but with little other information; a medium magnification showing more minor roads and place names; a detailed high magnification view showing individual places of interest; and a set of close-ups of the actual churches, museums, colleges, etc (figure 12). This might prove compatible with normal cognitive processing transitions, and would certainly be familiar in terms of what is traditionally done in classic 'town guides'.

In this case the modelling was able to reveal a major usability problem with two of the proposed designs which would lead to their being rejected in favour of the third. However, it was also able to identify some problems with this design, and by drawing the designers' attention to the aspect of cognition that was causing the problems helped them come up with a new design that only shows a single scale at a time. Because it requires additional user interactions (to zoom in and out), the new design would be more complex than the three proposed designs, but by matching these actions and changes in the screen display to the transitions in the user's visual structures (in this case, between superstructures and substructures) it could actually be easier to use, since it removes superfluous visual elements that could distract the user.

There are additional questions raised by this design, such as how easily users will be able to maintain a sense of their general location if they are allowed to 'scroll' a high magnification view, but such questions could be answered by successive iterations of model building and design. The point is not that the model itself generates a workable design, but that the process of modelling can alert the designer to weaknesses in their requirements and point to the areas of the design space that they should concentrate upon. The idea is also open to empirical test, and would most certainly reduce the computing requirements both with respect to hardware and algorithms.

8. Conclusion

In the first section we outlined a way of thinking about the applied science paradigm in terms of the generation and application of scientific representations. At the end of that section, three issues were raised for the improvement of our applications representations. They concerned the scope of modelling techniques, the nature of the transfer from theory to design context and the amount of effort involved. In this concluding section, we can now review the development and deployment of the ICS theoretical framework within the HCI context. In figure 13, the phenomena of HCI, the ICS architecture and its principles of operation are all viewed as part of the science base. Its initial development was supported by using empirical evidence from laboratory studies of short term memory phenomena (Barnard, 1985).

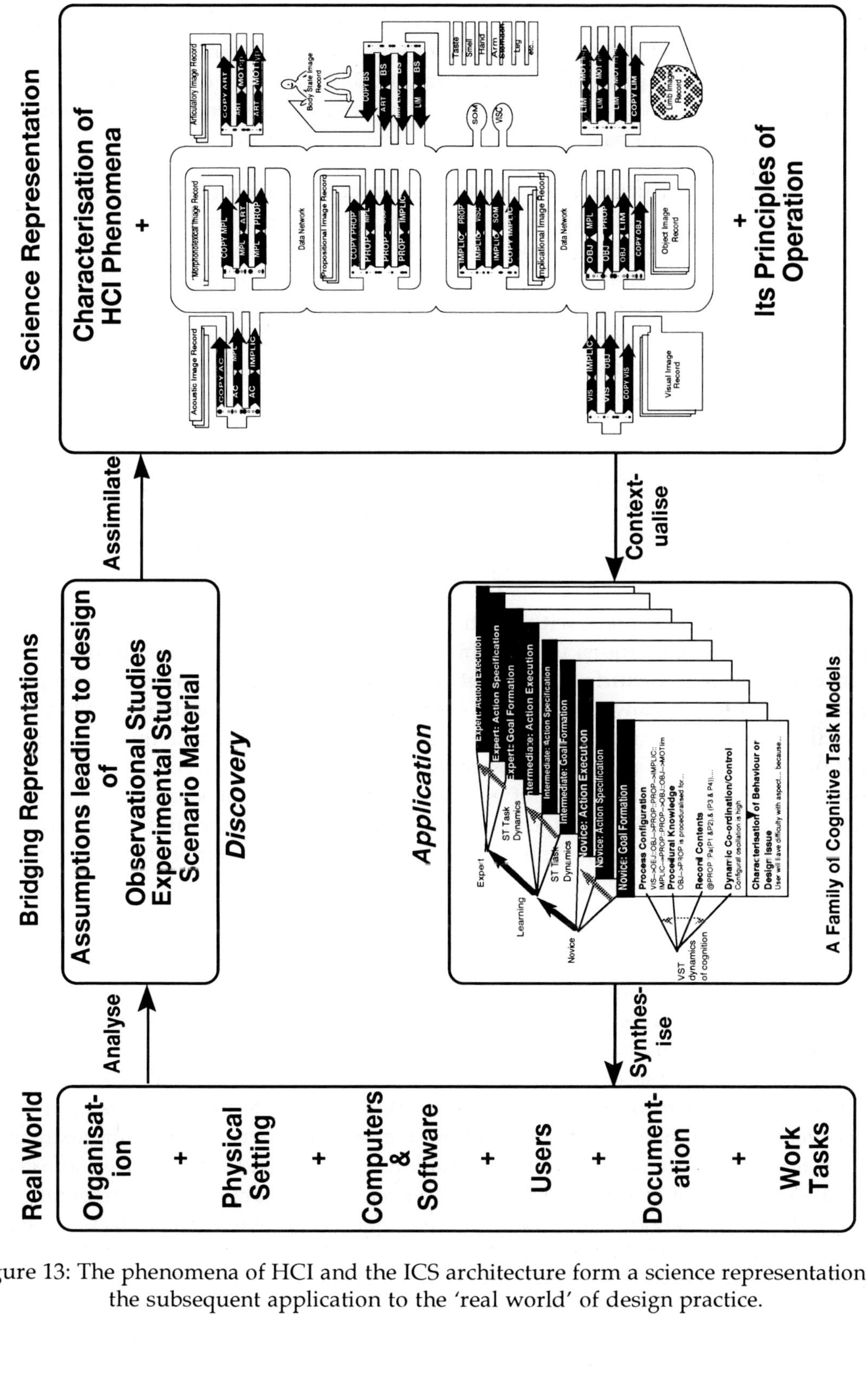

Figure 13: The phenomena of HCI and the ICS architecture form a science representation for the subsequent application to the 'real world' of design practice.

By concentrating on the different types of *mental representation and processes that transform them*, rather than task and paradigm specific concepts, the model can be applied across a very broad range of settings (e.g. Teasdale & Barnard, 1993). In the case of HCI specific discovery representations are used to generate and characterise the relevant phenomena for assimilation to that science base.

For the purposes of constructing a representation to bridge between theory and application the approach makes use of explicit, yet approximate, characterisations of cognitive activity. These were systematically derived and specified. The way in which the overall architecture will *behave* is dependent upon four classes of factor: the precise configuration of processes and memory resources required to support a given phase of cognitive activity; the relevant procedural knowledge embodied in each mental process within the configuration; the form, content and accessibility of any memory records that are needed in that phase of activity; and the activity will depend on the overall way in which the complete configuration is dynamically co-ordinated and controlled.

This form of approximate model *describes* the nature of cognitive activity underlying the performance of complex tasks. The process of building them is, given the principles, relatively automatic. To demonstrate this, we have developed an expert system that embodies the relevant theoretical knowledge, and which can construct cognitive task models for a range of HCI problems (May, Barnard & Blandford, 1993). The system 'knows' what kinds of configurations are associated with particular phases of cognitive activity; it 'knows' something about the conditions under which knowledge becomes proceduralised, and the properties of memory records that might support recall and inference in complex task environments. It also 'knows' something about the theoretical interdependencies between these factors in determining the overall patterning, complexity and qualities of the co-ordination and dynamic control of cognitive activity. Abstract descriptions of cognitive activity are constructed in terms of a four component model specifying attributes of configurations, procedural knowledge, record contents and dynamic control. Finally, in order to produce an output, the system 'knows' something about the relationships between these abstract models of cognitive activity and the attributes of user behaviour.

Obviously, no single model of this type can capture everything that goes on in a complex task sequence. Nor can a single model capture different stages of user development or other individual differences within the user population. It is therefore necessary to build a set of inter-related models representing different phases of cognitive activity, different levels and forms of user expertise and so on (figure 5). The basic modelling unit uses the four component description to characterise cognitive activity for a particular phase, such as establishing a goal, determining the action sequence, and executing it. Each of these models approximates over the very short term dynamics of cognition. Transitions between phases approximate over the short term dynamics of tasks, while transitions between levels of expertise approximate over different stages of learning. In figure 13, the envisaged application representation once again consists of a family of inter-related models depicted graphically as a stack of cards.

As an applications representation, the expert system concept is very different from cognitive walkthroughs (Polson et al. , 1992) or programmable user models (Young, Green & Simon, 1989) discussed earlier. Like PUMs, the actual tool embodies explicit theory drawn from the science base. Likewise, the underlying architectural concept enables a relatively broad range of issues to be addressed. Unlike PUMs, it more directly addresses a fuller range of resources across perceptual, cognitive and effector concerns. It also applies a different trade-off in when and by whom the modelling knowledge is specified. At the point of creation, the expert system must contain a complete set of rules for mapping between the world and the model. In this respect, the means of accomplishing and expressing the characterisations of cognition and behaviour must be fully and comprehensively encoded. This does not mean that the expert system must necessarily 'know' each and every detail. Rather, within some defined scope, the complete chain of assumptions from artifact to theory and from theory to behaviour must be made explicit at an appropriate level of approximation. Equally, the input and output rules must obviously be grounded in the language of interface description and user-system interaction. Although some of the assumptions may be heuristic, and many of them may need crafting, both theoretical and craft components are there. The how-to-do-it modelling knowledge is laid out for inspection.

However, at the point of use, the expert system has some properties more closely akin to those associated with the cognitive walkthrough. Both require considerably less precision than PUMs in the specification and operationalisation of the knowledge required to use the application being considered. As with walkthroughs, the expert system can build a family of models very quickly and without its user necessarily acquiring any great level of expertise in the underlying cognitive theory. In this way, it is possible for that user to explore models for alternative system designs over the course of something like one afternoon. Since the system is modular, and the models are specified in abstract terms, it is possible in principle to tailor the systems input and output rules without modifying the core theoretical reasoning. The development of the tool could then respond to requirements that might emerge from empirical studies of the real needs of design teams or of particular application domains.

References

Anderson, J.R. & Skwarecki, E. (1986) The automated tutoring of introductory computer programming. *Communications of the ACM, 29,* 842-849.

Arend, U., Muthig, K., & Wandmacher, J. (1987) Evidence for global feature superiority in menu selection by icons. *Behaviour and Information Technology, Vol. 6, No.4,* 411-426

Bannon, L.J & Bødker, S. (1991) Beyond the Interface: encountering artifacts in use. In J.M. Carroll (ed.)*Designing Interaction,* Cambridge: CUP, pp. 227-253

Barnard, P.J. (1985). Interacting cognitive subsystems: A psycholinguistic approach to short term memory. In A. Ellis, (Ed.), *Progress in the psychology of language* (Vol. 2, chapter 6, pp. 197-258). London: Lawrence Erlbaum Associates.

Barnard, P.J. (1987) Cognitive resources and the learning of dialogues. In J. M. Carroll (Ed) *Interfacing Thought* Cambridge, Mass: MIT

Barnard, P.J. (1991) Basic Theories and the Artifacts of HCI. In J.M. Carroll (ed.)*Designing Interaction,* Cambridge: CUP, pp.103-127

Barnard, P.J., Blandford, A.E & May J. (1992) *Documentation accompanying Amodeus Deliverable 19*, Brussels: CEC DG XIII

Barnard, P.J., Grudin, J., & MacLean, A. (1989). Developing a science base for the naming of computer commands. In J. B. Long & A. Whitefield (Eds.) *Cognitive Ergonomics and Human Computer Interaction*, Cambridge: Cambridge University Press. 95-133.

Barnard, P.J., MacLean, A. & Hammond, N.V. (1984) User representations of ordered sequences of command operations. In B. Shackel (ed.) *Proceedings of Interact '84: First IFIP Conference on Human-Computer Interaction, Volume 1*. London: IEE, pp. 434-438

Barnard, P.J. & Teasdale, J. (1991) Interacting Cognitive Subsystems: A systemic approach to cognitive affective interaction and change. *Cognition and Emotion* **5** (1), 1-39.

Barnard, P.J., Wilson, M. & MacLean, A. (1987) Approximate modelling of cognitive activity with an expert system: a strategy for the development of an interactive design tool. In J.M. Carroll & P.P. Tanner (eds.) *Proceedings of CHI+GI '87*, New York: ACM. pp.21-26

Barnard, P.J., Wilson, M., MacLean, A. (1988). Approximate modelling of cognitive activity with an Expert system: A theory based strategy for developing an interactive design tool. *The Computer Journal, 31*, 445-456.

Card, S.K. & Henderson, D.A. (1987) A multiple virtual workspace interface to support user task switching. In J.M. Carroll & P.P. Tanner (eds.) *Proceedings of CHI+GI '87*, New York: ACM. pp.53-59

Card, S.K., Moran T.P. & Newell, A. (1983) *The psychology of human computer interaction* Hillsdale, NJ: Lawrence Erlbaum

Carroll, J.M. & Campbell, R.L. (1986) Artifacts as psychological theories: The case of human-computer interaction. *Behaviour and Information Technology, 8*, 247-256

Carroll, J.M., Kellog, W.A. & Rosson, M.B. (1991) The Task-Artifact Cycle. In J.M. Carroll (ed.)*Designing Interaction*, Cambridge: CUP, pp.174-102

Draper, S. (1992) Activity theory: the new direction for HCI? *International Journal of Man-Machine Studies, 37*, 812-821.

Duff, S. & Barnard, P.J. (1990) 'Influencing behaviour via device representation: decreasing performance by increasing knowledge'. In D. Diaper, D. Gilmore, G. Cockton and B. Shackel (eds) *Proceedings of Interact '90*, Amsterdam: Elsevier Scientific Publishers B.V., pp 61-68

Green, A.J.K, Barnard, P.J. (1990) 'Icon Interfacing: The role of icon distinctiveness and fixed or variable screen location'. In D. Diaper, D. Gilmore, G. Cockton and B. Shackel (eds) *Proceedings of Interact '90*, Amsterdam: Elsevier Scientific Publishers B.V., pp 457-462

Grudin, J.T. (1990). The computer reaches out: The historical continuity of interface design. In J. Chew & J. Whiteside (Eds.), *Proceedings of CHI '90: Human Factors in Computing Systems* (pp. 261-268). New York: ACM.

Hammond, N. & Allinson, L. (1988 Travels around a learning support environment. In E. Soloway, D Frye & S.B. Sheppard (eds.) *Proceedings of CHI '88*, New York: ACM pp.269-273.

Hammond, N.V. & Myers, K. (1990) *Definition of scenarios for M1 Workshop*. ESPRIT BRA project 3066 (AMODEUS) Internal Report RP8/IR6.

Hellander, M. (1988) *Handbook of Human Computer Interaction*. Elsevier Science Publishers: Amsterdam.

Kieras, D.E. & Polson, P.G.(1985) An approach to the formal analysis of user complexity. *Int. J. of Man-Machine Studies, 22*, 365-394.

Landauer, T.K. (1987) Relations between cognitive psychology and computer systems design. In J. M. Carroll (ed.) *Interfacing Thought* Cambridge, Mass: MIT

Long, J.B. (1989) Cognitive Ergonomics and Human-Computer Interaction. In: J.B. Long & A. Whitefield (eds.): *Cognitive Ergonomics and Human Computer Inter-action*, Cambridge: CUP, pp. 4-34

Long, J.B., Dowell, J. (1989). Conceptions of the discipline of HCI: Craft, applied science and engineering. In A. Sutcliffe & L. Macaulay (Eds.), *People and computers V (pp.* 9-32). Cambridge: Cambridge University Press.

MacLean, A., Bellotti, V. & Young, R. (1990). What rationale is there in design? IIn D. Diaper, D. Gilmore, G. Cockton and B. Shackel (eds) *Proceedings of Interact '90*, Amsterdam: Elsevier Scientific Publishers B.V., pp. 207-212.

May, J., Barnard, P.J., and Blandford, A. (1993) Using Structural Descriptions of Interfaces to Automate the Modelling of User Cognition. User *Modelling and Adaptive User Interfaces, 3(1)*

Minsky, M. (1975) A framework for representing knowledge. In P. Winston (ed.), *Psychology of Computer Vision*, New York: McGraw Hill , pp.211-277.

Moran, T.P. (1981) The command language grammar, *International Journal of Man Machine Studies*, **15**, 3-50.

Nielsen, J. & Phillips, V. L. (1993). Estimating the relative usability of two interfaces: heuristic, formal and empirical methods compared. To appear in *Proceedings of InterCHI'93*. ACM: New York.

Newell, S. & Card. S.K. (1985) The prospects for science in human computer interaction. *Human Computer Interaction, 1*, 209-242

Norman, D. (1986) Cognitive Engineering. In D.A. Norman and J.W. Draper (eds.) *User Centered System Design*, Hillsdale NJ: Erlbaum, pp. 31-62.

Payne, S., & Green, T. (1986). Task action grammars: A model of the mental representation of task languages. *Human-Computer Interaction, 2*, 93-133.

Polson, P.G., Lewis, C., Rieman, J., Wharton, C. (1992) Cognitive walkthroughs: a methodology for theory-based evaluation of user interfaces. *Int. J. of Man-Machine Studies, 36*, 741-773

Reisner, P. (1982). Further developments towards using formal grammar as a design tool. In *Proc. Human Factors in Computer Systems (Gaithersburg)*. New York: ACM, 304-308.

Teasdale, J. & Barnard, P.J. (1993) Affect, Cognition and Change: Re-modelling depressive thought. Hove: Lawrence Erlbaum Associates.

Whiteside, J. & Wixon, D. (1987) Improving Human computer interaction: A quest for cognitive science. In J. M. Carroll (ed.) *Interfacing Thought* Cambridge, Mass: MIT pp. 353-365

Young, R.M & Barnard, P.J. (1987) The use of scenarios in human computer interaction: turbocharging the tortoise of cumulative science. In J.M. Carroll & P.P. Tanner (eds.) *Proceedings of CHI+GI '87*, New York: ACM. pp.291-296

Young, R.M., Green, T.R.G. & Simon, T. (1989) Programmable user models for predictive evaluation of interface designs. In K. Bice & C. Lewis (eds.) *Proceedings of CHI '89*. New York: ACM, pp 15-19.

Computers, Communication and Usability:
Design Issues, Research and Methods for Integrated Services
P.F. Byerley, P.J. Barnard and J. May (Editors)

Chapter 2.2

Analysing Display Oriented Interaction by means of System Models

M. D. Harrison†, G. D. Abowd* and A. J. Dix†

†Human Computer Interaction Group, Department of Computer Science, University of York, Heslington, York, YO1 5DD, UK

* School of Computer Science, Carnegie Mellon University, Pittsburgh, PA 15213.

1. Introduction

It is clear that software engineers should design and implement interactive systems with their users in mind. At the same time system evaluators should have a precise understanding of the system if they are to understand the consequence of their evaluation. In this chapter we consider instances in the use or design of graphics based systems so that we might understand how formal specification techniques can be used to assess the design implications of usability issues more effectively. A commitment to formal specification is presumed here in the sense that we assume, without discussion, that the specification of a system should be expressed as a collection of mathematical objects. This view concurs with that of a particular section within the software engineering community (see, for example Nichols, 1992). The advantages of such an approach are precision and clarity of expression, mathematical tractability and the provision of an established set of rules for refining a specification to a computer program. Given formal specification as our baseline, we would like to express system requirements concerned with usability as properties of such specifications. We would also wish to be able to demonstrate that such properties hold true.

We recognise that there are substantial problems of scale associated with large formal specifications. However the potential advantage of a formal approach are such that we believe it is important to explore the development of formal specifications for a wide domain of application and variety of purpose. It is worth noting that the use of rigorous techniques is commonplace within other engineering disciplines.

Formal specifications of realistically sized systems can be large and as a consequence the proof that the specification satisfies given properties is complicated. Complexity can be reduced by including in the design process more abstract descriptions which emphasise aspects of behaviour such as performance, interactivity and security. Models may be specified as a basis for expressing and capturing these properties in the design of the system. Here we are concerned with the specification of the external behaviour of interactive systems; hence, our models are referred to as *interaction models*. Our aim in this chapter is to show how such descriptions may be used to formulate *user recognisable* structures to provide criteria for designing and evaluating interactive systems. In the next chapter the related problem of analysing user requirements and applying them to system specifications is discussed. The user recognisable structures here are less concerned with functionality and more concerned with the nature of the interactive behaviour of the system. We contend that formal models which are specific to the issues of interaction can clarify those parts of the system specification that affect its interactive properties. Thus we offer an approach to the formalisation and integration of user requirements into system specification.

In this chapter, we shall demonstrate our approach through two *scenarios* that have been chosen because they demonstrate different aspects of the requirements on interaction models. The scenarios arise as a result of the research of the ESPRIT Basic Research Action 3066 (AMODEUS), and were initially posed as problems for different modelling techniques in HCI research (Hammond & Myers, 1990). The scenarios relate to graphical drawing packages implemented on personal computers, specifically SuperPaint for the Macintosh and MicrografX Graph Plus for the IBM PC. Fragments of user behaviour are employed to demonstrate problems in the use of those systems. In both cases, the features of the scenario can be abstracted away from the particular implementation, as they both represent common problems that can arise in a variety of graphical drawing packages. We will provide more abstract descriptions of the scenarios based on these package dependent descriptions to avoid confusion that might arise as a result of unfamiliarity with the packages.

The rest of this chapter is divided into two main sections, one for each scenario. To discuss the first scenario we will derive a description of the relationship between the system state and the display that is factored into task relevant features of both, called *attributes*. This leads to a formal description of *state display conformance*, a principle of the interactive system which guides our analysis. The second scenario involves the relationship between the *history* of interaction and *predictability*. We will suggest ways in which we might analyse how the user predicts the outcome of future interaction with the system. It is worth noting that although we see the primary value of these techniques in the construction of interactive systems, we will concentrate mainly on the analysis of existing systems here.

Though the discussion of a scenario concentrates on a particular interactive property—be it state display conformance or predictability—the impact of these properties can be seen in each. For this reason, it is insufficient to discuss a scenario in terms of only the one property. For purposes of presenting the models, we must forego a desire to be complete, though we will occasionally point out potential overlaps between the scenarios.

2. Relating attribute algebras in a layered graphics package

2.1. Scenario description

Many drawing programs support *layers* for the construction and manipulation of pictures. These layers should not be confused with the levels used by many systems to simulate a 3-dimensional structure. A layer in our sense is essentially an independent canvas upon which pictures can be constructed. The two layer system of concern in this scenario contains an *object* layer, in which text, boxes, circles, etc. can be created and edited, and a *pixel* layer, in which freehand pictures are constructed and manipulated at the lowest level of screen detail (the pixel). The design of the system has a number of interesting features but for our purposes we note the following:

- There are commands that relate distinctively to the two layers. For example, within the pixel layer *paint* and *rub out* commands are used, and their function only makes sense in the pixel domain. Within the object layer, the *select* and *group* commands are used, and their function only makes sense as it relates to objects.
- There are a large number of commands that the two systems appear to have in common, i.e., they create visual images that appear to be identical and the commands are invoked by identical icons.
- The display representations created at the two levels are often ambiguous and therefore it is not clear what operations are appropriate at which level.

We will explore the last feature with the aid of an example from SuperPaint. A circular image can be drawn within either layer. Figure 1 shows a picture of two circles which have been drawn in separate layers using the circle drawing facility in each layer. It is impossible to identify, visually, which circle belongs to which layer.

The icon in the upper left-hand corner is an indication of the active layer, i.e., the one whose construction and manipulation commands are currently indicated in the palette on the left side of the screen. If the user wants to move the circle on the left, she must make sure that she is in the appropriate layer to manipulate that circle. The circle on the left provides no clue about the layer in which it was constructed and as a result the user must either remember how it was constructed or guess. In terms of the predictability, which will be discussed in more depth in the next section, because there are ambiguities in the display, there is not enough information for the user to unambiguously determine how her future interactions will affect the system.

Although it might be argued that in its current context this is a trivial problem, it is not difficult to imagine situations in which ambiguities of this kind might lead to uncertainties that could lead to error in critical situations.

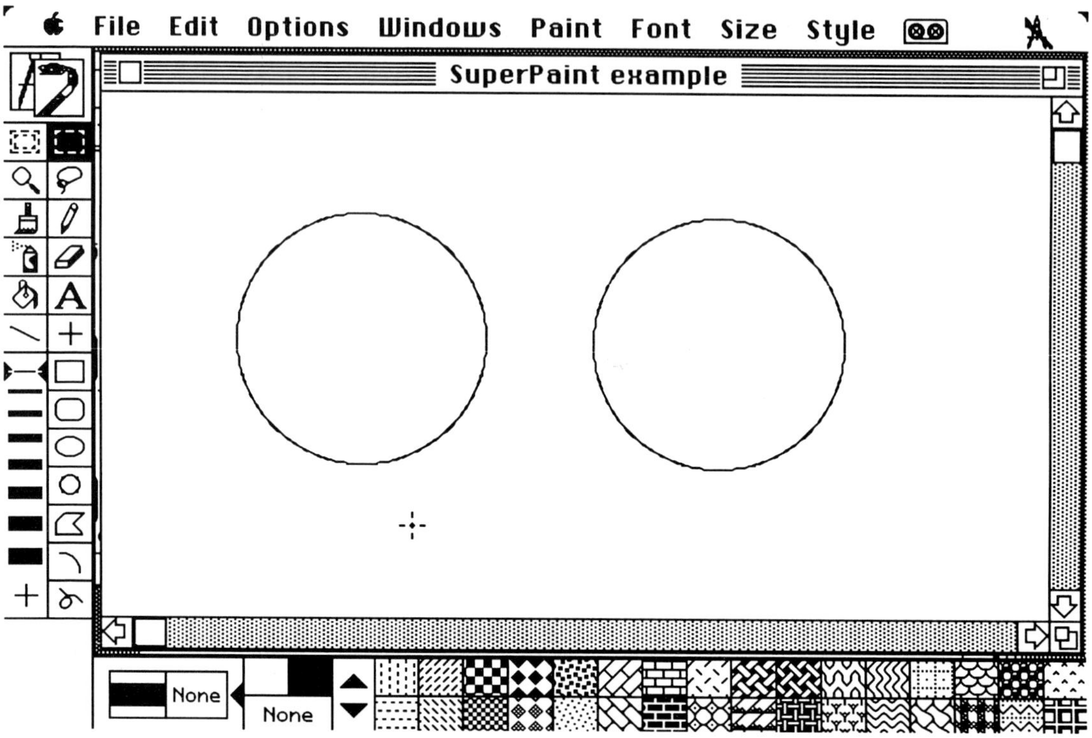

Figure 1: To which layer does each circle belong?

A more complicated situation in which confusion arising from layer ownership affects interaction is as follows. Figures 2 and 3 show how switching between layers can produce interesting variations. The difference between the two canvases displayed in figures 2 and 3 is a result of the difference between system modes associated with object construction and picture construction respectively. What is contained in the underlying state of the picture is the same in both cases, but the displays associated with the object and pixel layers are different because the image created in each layer, if it overlaps something constructed in the other layer, occludes it. In the object layer the 'screen' and 'light bulb' are visible as entities containing a 'white' canvas. In the paint mode we can see that the white backgrounds of the object layer have hidden 'circles' and 'filaments' created in the pixel layer. That these different objects were constructed in the different layers is not clear from the canvases, nor is it easy to construct additional components of the state if the information that the additional objects relate to were constructed in the other layer and are now hidden from view.

The picture created in figure 2 represents the same underlying state as that contained in figure 3. In figure 2 a white background hides the circle and filament depiction that was constructed in the pixel layer. The difference becomes important in use when information that has been created in the other layer, and is only visible in the other layer, is required in order to complete the picture but cannot be seen unless the mode is switched.

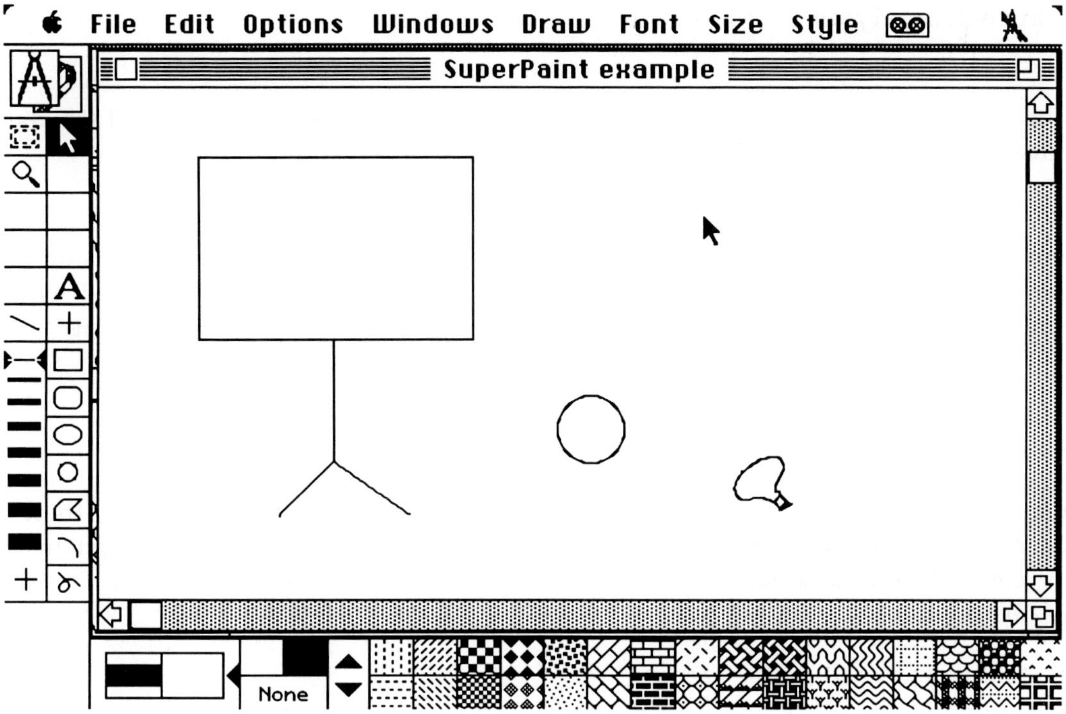

Figure 2: The light is 'off' in the object layer

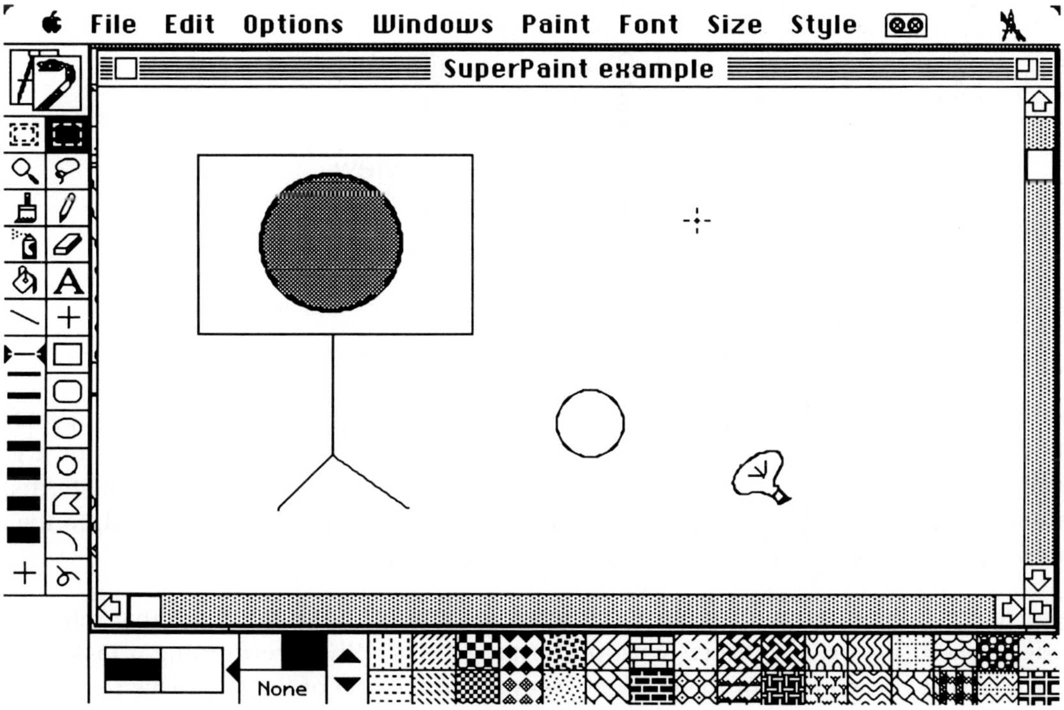

Figure 3: The light is 'on' in the pixel layer

2.2. Formal analysis

At the most general level, the state of the system, represented by the set S, and the state of the display, represented by the set D, are related by a *view* relation.

$$view: S \leftrightarrow D$$

Operations are defined as relations on a state set. System operations are relations on S and display operations are relations on D. Hence the operations that are associated with the state and the display are taken from the power set, P(), of state and display relations.

$$S\_Ops \equiv P(S \leftrightarrow S)$$

$$D\_Ops \equiv P(D \leftrightarrow D)$$

The two state sets are said to *conform* if for each operation on one state set, there is a corresponding operation on the other state set. When an operation on the system state, $c \in S\_Ops$, corresponds in such a way with an operation on the display state, $o \in D\_Ops$, we have the relation

$$conforms: (S \leftrightarrow S) \leftrightarrow (D \leftrightarrow D)$$

$$(c,o) \in conforms \textbf{ iff } \forall d \in dom(o); s \in dom(c) \mid (s,d) \in view \bullet$$
$$\exists d' \in D; s' \in S \mid (d,d') \in o \textbf{ and } (s,s') \in c \text{ such that } (s',d') \in view.$$

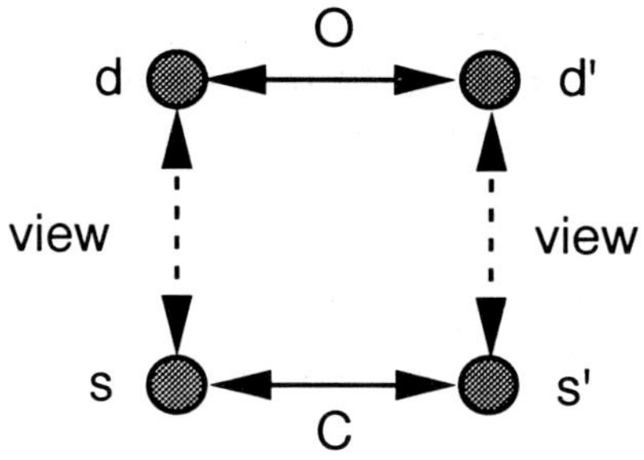

Figure 4: The conformance diagram

This relation between state and display does not take into account how the user's task affects state display conformance. So, assume the set T represents the tasks of interest for a given domain of application. We now proceed with a refinement of the definition of state display conformance that factors the system and display states into more manageable and, to the user or designer, semantically meaningful components, called *attributes*.

In practice, computer users often have a different picture of the system from that intended by the designer. Users are concerned with performing tasks. When performing a task the user recognises that components of the state are relevant to the *goal* of the task. This recognition is reinforced by components of the display. As the components of the state are modified in progress towards the task goal the display is used to emphasise the fact that progress is being made and to help the user formulate the next command whether it carries out a further enabling task or is actually part of their goal task. Progress toward the satisfaction of a goal, therefore, is enhanced by a close correspondence between system and display as described above.

Assume we have a set of all possible attribute names, denoted by A, and a set of all possible attribute values, denoted by V. A state is simply defined as any finite partial mapping from attribute names to values. For a given interactive system, we stipulate that all states are defined over the same set of attributes. In other words, if the set of system states is given as a set of finite partial attribute-value mappings,

$$S \equiv P(A \rightarrow V),$$

then the domain of every state's attribute-value mapping is the same, or

$$\forall s,s' \in S \bullet dom(s) = dom(s').$$

We refer to this domain as the attributes of the system state set S, and we write this as $attr(S)$. Similarly, we can refer to the attributes of the display state set D, and for this we write $attr(D)$.

In recognising display attributes of the system, the user engages in two processes: identification and interpretation. In the case of identification, the display attribute is considered in a way that is similar to the notion of *display template* (Harrison, Roast & Wright, 1989), a notion that is concerned with the component of the display that is relevant to the task. The interpretation captures the significance and meaning of the attribute. For example, when we are talking about a display we can consider two components: a positional component, and an interpretative component. Hence if the attribute happened to be the date we would want to know something about the date and how it fits on the display (for example, the top right hand corner) but we would also need to know more than that it is a grouping of pixels. The display attribute illuminates the state attribute, that is a display attribute may reveal information about a particular attribute of the state. Hence the state attribute for the date will extract the field within the state that contains the internal representation of the date.

We note that another attractive feature of the attribute model is that it resembles notions of semantic features used by user modellers, for example, in the work of Young & Whittington (1990) and Payne & Green (1986).

We can refine our model so that it uses state and display attributes that are relevant to the tasks in T to determine the view relationship. We are interested in how tasks can be used to constrain the relationship between system and display attributes. This can be represented by the mapping *taskview,* which specifies a set of task views that relate individual state attributes to display attributes.

$$taskview: T \rightarrow (attr(S) \leftrightarrow attr(D))$$

Hence, for any task t, the connection between relevant task and display attributes is established. We can use this mapping of attributes to represent a claim about how a typical user understands the effects of a set of operations and thus it can be seen as a component of a designer's view of the user's model of the system.

A given operation transforms the state; a designer specifies an operation in terms of the changes that it causes to the attributes. These changes to attributes are also expressed in terms of the value of certain attributes of the state. We refer to the *signature* of an operation as those attributes which can be changed by the operation and those attributes whose value before the operation determine the effect of the operation. We can choose the signatures for any operation on the system state, representing this by the function *S_sig*.

$$S\_sig : S\_Ops \rightarrow P\ (attr(S))$$

Hence the state signature associates a set of state operations with the attributes of the state that they affect. The function *D_sig*, represents the display attributes that are affected by operations on the display.

$$D\_sig : D\_Ops \rightarrow P\ (attr(D))$$

If we were also to relate attributes and signatures to task, and to user perception, then we could define the signatures that the user might recognise which might be different from the actual signatures. We can use this information to provide a component of a claimed user model if we contend that the user understands the performance of operations on the display state in terms of signatures. Now, if an operation on the system state is to correspond with an operation on the display state, then it will only be recognised by the user if the signature of the state operation is related by the current task via *taskview* to the signature of the display operation. Ultimately, this leads to the user's formulation of the state display correspondence in terms of the task, which we represent as the function

$$user\_conforms : T \rightarrow (S\_Ops \leftrightarrow D\_Ops),$$

We may further define this relationship in terms of the signatures of operations on system and display state. The interactive system has the property of state display conformance for a given task, t, if *user_conforms(t)* is not defined for all state operations, but no pair of state operations maps to the same display operation (in technical terms a partial injective function). We only require a partial function because some operations on the system state may not be relevant to the execution of some tasks.

In the scenario described above, we can split the system up into two separate sub-states, one for each layer. In isolation, the interactive system may well satisfy the constraint that for a given task, t, we have that *user_conforms(t)* is a partial injective function. This is so because in isolation, the attribute of an entity that identifies it as belonging to 'object' or 'pixel' layer is not contained in the signature of the operations on the system state, so it need not be displayed. When the two systems are combined, this attribute becomes part of the signature of many operations on the state, but it is still not displayed. Therefore, the user does not benefit from a strong state display conformance. Using these sorts of techniques the designer develops a claimed model of how the user understands the system.

2.3. Conclusions

One critical issue which we have highlighted in the discussion of this scenario is that the user relies on a strong correspondence between the operations performed on a display and those performed on the underlying system state. This notion has been previously formalised and presented as a state display conformance. The original formulation of state display conformance (Harrison & Dix, 1990) gave us a start in pinpointing the problems that arise in interacting with the overlapping graphical layers. We further refined this definition in terms of attributes which highlight features of a state that can aid the user in understanding how a system works. The attributes are also used by designers in specifying both display and system states, and the user-centred design principle represented by the function *user_conforms,* provides a constraint on the design that can lead to more effective interaction.

The determination of these attributes is one of the key areas where the formal and informal elements of the design process meet. Determining what parts of the state are important to the user will involve some elements of task analysis. Similarly, determining the appropriate display attribute requires some notion of *salience* that requires either the input of an experienced designer or experiment. This can also be related to other work reported in this book. If the attributes of the system can be formally specified then this makes it possible to identify accurately the values of the 'machine' parameters in the usability principles that are presented in chapter 3.2. The prescriptions of these principles allow design expertise and empirical human factors data to be codified and made available at an appropriate point in the design process.

The scenario that has just been discussed is essentially static in the sense that it relates to the user's current understanding of the state of the system as a result of what she sees. The next example deals with properties of the system that enable the user to predict the effect of an operation on the system.

3. Dynamic properties of the interface: determinism

3.1. Scenario description

We describe an object-oriented drawing package with a graphical interface. The objects in a picture have three relevant hierarchies, a structural hierarchy, which is dictated by grouping achieved in the process of constructing the picture, a visual hierarchy, which is dictated by current positioning on the canvas, and a temporal hierarchy dictated by the order in which objects are created. The following sequence of actions in relation to the use of a drawing package demonstrates the relationship between the three hierarchies. It should be noted that in this example grouping is only carried out in the last step of the sequence.

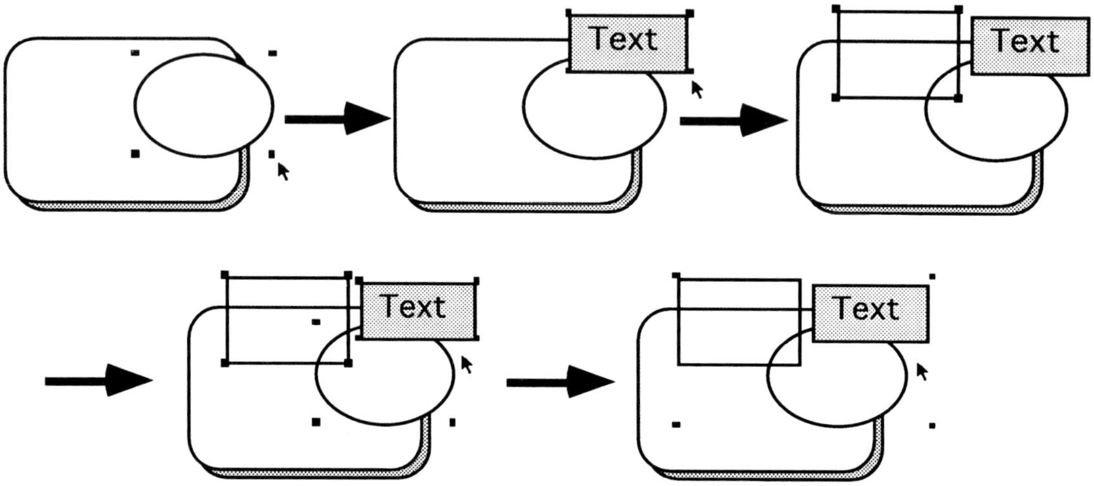

Figure 5: constructing the diagram, the temporal sequence.

In this series of pictures the user has created an oval, then created a text shaded square, then created a rectangle, then selected all three objects and finally grouped them. The operation of selecting an object is complicated by this distinction between structural, visual and temporal hierarchies. The user manual for this system informs the user that a mouse click enables selection. As it happens, the selection cycles amongst the objects that contain the mouse pointer. We may presume, therefore, that the operation of clicking will select an object containing the mouse pointer; each subsequent click will select another object containing the mouse pointer. After each possible object has been selected, the first object will be selected again, and so on. For example, if the mouse were positioned as below and no objects were currently selected, the selection cycle might go square, circle, grouped square/circle/text, shadowed box, and back to the beginning.

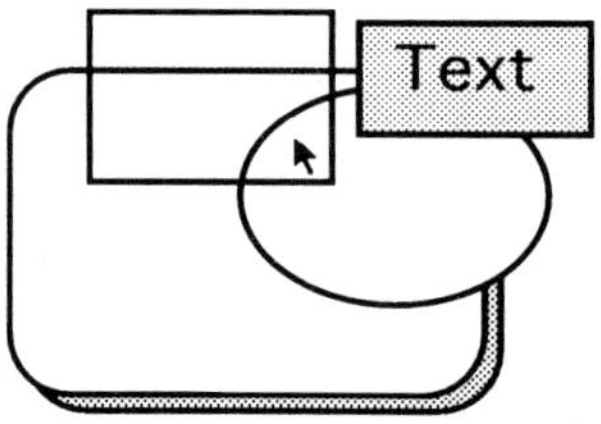

Figure 6: If the mouse is positioned as above what will be selected?

The goal in analysing this scenario is to explain why even though the system's algorithm for selection is deterministic, it is often perceived by the user as non deterministic. The user's response to such perceived non determinism is to adopt the approach 'I will just keep on clicking the mouse until the object I want to select is finally selected.' This 'click until you drop' strategy assumes that the user can always tell when an object is selected. There are likely to be design possibilities in which the user cannot recognise which subset of the diagram has been selected. In this case the lack of visual feedback will mean that this experimental approach is inappropriate.

3.2. Formal analysis

An interactive system, as we have said, is used to achieve a set of task goals. We might informally consider a predictable system to be one which supports these tasks by providing enough information (in sound or vision for example) to indicate what the effect of a new action will be. It is important to recognise that there will often be a difference between the actual behaviour of the system and the user's perception of the behaviour of the system. Predictability of a system can operate at a number of levels: at one level the system may be unpredictable in that it fails to be deterministic as a system (for example executing a command might have an arbitrary outcome); it may be unpredictable because it is impossible to tell from the outward manifestation of the system what effect a particular action will have; it may be entirely predictable in fact but be apparently unpredictable to the user because of the way that relevant information is presented by the system. Hence in the last two cases, the possibility of non determinism is a matter of user interpretation.

3.2.1 System non determinism

A computer system may be unpredictable because an input sequence does not have a unique effect on the state of the system. This notion of system non determinism may be expressed as follows:

We assume the existence of a function *interpret* which maps the input sequence to a set of possible state transformations. The choice between the alternatives is arbitrary.

It is hard to imagine a design that would be non deterministic in this sense unless the hardware upon which it was based was very unreliable. Some systems, for example computer programs designed to select the winner of a lottery might be designed to be deliberately non deterministic. A more substantial class of such systems may be identified if we consider certain kinds of dynamic system in which autonomous activity may complicate the task for the operator. In these cases, something like system non determinism is true of the system from the perspective of the user. In practice we are more interested in other more subtle forms of perceived unpredictability that do not correspond to non determinism of the system. Here we are concerned for example about whether the system provides enough information on the display to allow the user to predict what effect a command will have. These notions of perceivable predictability capture the system's ability to express enough information about its underlying state for the user to make use of the system's determinism.

3.2.2 Observational equivalence, indistinguishability and predictability

A system is observationally predictable if current display and future input fully determine the resulting behaviour. Imagine that two identical versions of the same system are represented by two keyboard/display pairs. At each keyboard a different input sequence is issued by the user, sequence p to system 1 and sequence q to system 2, after which the displays, or current outputs, are identical. In this case, we say that p and q are *observationally equivalent*. The same sequences p and q are *observationally indistinguishable* if any input sequence r subsequently issued to both systems results in the same final display, that is, any further action will betray no difference between the two versions of the system.

A system is *observationally predictable* if observationally equivalent sequences are also observationally indistinguishable.

Hence, for a given display, a user can in principle predict the future behaviour of the system because the state of the system is somehow characterised by what can be perceived of the current display. However it should be noted that, apart from certain types of system, where the full state is represented in the display, the observational predictability property is unrealistic, since the underlying state may be far too complex to display on one screen. In fact we may generalise this notion of predictability by defining it in terms of sets of display attributes. A system will be predictable to the user if all the information is there on the display and the user can appropriately interpret it as a means of deciding how a command will function. Hence the drawing package described earlier could be predictable with respect to the drawing and selection operations if the effect of these operations was entirely governed by the content of the display.

3.2.3 User non determinism

At the third level it is important to determine whether the user actually benefits from the predictability of the system. The user may benefit in two ways:

- Does the user's effective interaction with the system rely upon its predictability?
- Is the information that is provided to inform the user sufficient (given cognitive constraints) to predict the future effect of the system?

Just because a user can predict what output will arise from a given input, does that improve interaction with the system? There is a complementary notion to predictability, which we might call assessment, that deals with reasoning from perceived outputs back to inputs. When a user interacts with a system, achievable goals are continually being assessed within the task domain. In order to accomplish the goals it is necessary both to be able to predict how subsequent inputs will affect the goal, and how a goal was (or was not) affected by prior inputs. In addition, experimental evidence may show that predictability provides no measurable increase in the usability of a system. Here we assume there are situations in which predictability does enhance usability. Conversely the system may appear to be unpredictable because features of the display (attributes) required by the user are not appropriately identified and interpreted by the user at the time that they are required.

In chapter 2.3, a framework will be presented in which these properties of the interaction that cannot be expressed in terms of the system may be represented. Properties are defined in terms of the relationship between collections of information exchanges.

Predictability is defined in terms of:

- a sequence of inputs that is interpreted as a state transformation;
- the initial states;
- a display of the state after the transformation.

The interpretation is denoted by a function $interpret : P \rightarrow S\_Ops$, where we represent non empty sequences of input as coming from the set of programs P and the state transformations are denoted by the set S_Ops defined as above. At a finer level of detail, the input is formed from a set, I, of all possible user inputs—for example, keystrokes, mouse movements, mouse clicks, etc.- so that we can represent P as the set of non empty sequences of elements in I. In chapter 2.3 we consider an event based notion of predictability. The display is again represented by the *view* relation, which for simplicity we assume is a function in the remainder of this account. For simplicity's sake again, we will assume a unique initial state, s_0. We can express observational predictability described above in terms of this model.

A system is *observationally predictable* if, for any input sequences p and q satisfying

$$view(interpret(p)(s_0)) = view(interpret(q)(s_0)),$$

we have for any subsequent experimentation using input sequence r,

$$view(interpret(pr)(s_0)) = view(interpret(qr)(s_0)).$$

An important class of systems that are not observationally predictable have the property that the history of the interaction is required in order to predict what the next command will do. We can capture this notion of history based predictability by defining the input sequence as $p = a_1 a_2 ... a_n$ and a relevant sequence of states as

$$s_i = interpret(a_1 ... a_i)(s_0)$$

where the a_1 are inputs and a sequence of displays as $d_1 ... d_n$ where $(s_i, d_i) \in view$. We will use functional notation to express the same relation $d_i = view(s_i)$. Given any sequence of actions p

we want to model how the user might predict by using knowledge of the history of the interaction. Usually not all the history is required; rather, only some components of the history are important. We can define such *historical predictability*. First, we let

$$history(p) = \langle \langle p_1, d_1 \rangle, \langle p_2, d_2 \rangle .... \langle p_n, d_n \rangle \rangle$$

where $p = \langle a_1, ... a_n \rangle$ and $p_i = \langle a_1, ... a_i \rangle$ and $d_i = view(interpret(p_i)(s_0))$. The history generates a record of all the information that may be recorded about the system in terms of inputs and outputs at the various stages of the interaction.

A system is *historically predictable* if, for any input sequences p and q satisfying

$$view(interpret(p)(s_0)) = view(interpret(q)(s_0))$$

and

$$history(p) = history(q),$$

we have for any subsequent experimentation using input sequence r,

$$view(interpret(pr)(s_0)) = view(interpret(qr)(s_0)).$$

This notion of predictability assumes perfect memory. A user of the system will find it possible to predict with certainty what is happening only if every detail can be recalled. In practice there is a spectrum between requiring perfect memory and requiring no memory (as in the case of observational predictability). We therefore specify a further notion of predictability that depends on the user extracting the requisite information from the history. We assume a function *extract* and define *partial predictability* in a similar manner.

A system is *partially predictable* (based on a function *extract*) if, for any input sequences p and q satisfying

$$view(interpret(p)(s_0)) = view(interpret(q)(s_0))$$

and

$$extract(history(p)) = extract(history(q)),$$

then for any subsequent experimentation using input sequence r,

$$view(interpret(pr)(s_0)) = view(interpret(qr)(s_0)).$$

Here *extract* $\langle\langle p_1, d_1 \rangle, \langle p_2, d_2 \rangle .... \langle p_n, d_n \rangle\rangle$ will pick out attributes from the history of the interaction. For example it may define that part of the history that contains the key incidents in the construction of the diagram.

We can learn more about the nature of *extract* from this scenario. As we have said, the selection method is clearly not observationally predictable in the sense described above. The operations necessary to select a required object cannot be foreseen unless the user is familiar with the history of how the picture was created. For example, when the user clicks in an area containing overlapping objects, it is not possible on the basis of visual structure alone to predict which object will be selected. Thus, two identical drawings may behave differently. We must now ask in what sense this system is predictable. It is possible to predict the effect of selection if it is known how the drawing was constructed. Predictability is dependent on certain features of the preceding interaction history, namely those features which determine how the drawing was constructed. We have a notion of partial predictability in which *extract* identifies those components of the history that relate to the way in which the system was constructed. The function *extract* can be constructed in steps in this scenario.

First, we identify a set C that represents groupings of inputs into commands that are intelligible to the system. We require a function that operates on input/display pairs to determine if the input in the context of the display represents a significant command in the construction of the graphical objects. We represent this function as

$$construct: (P \times D) \rightarrow C.$$

construct(p,d) will yield a nontrivial command only if p produces a change to the temporal or structural hierarchy, noticeable in the display d, and that was not noticeable for any prefix of p. If we map *construct* onto the *history*, we will get a trace of the commands that affect the temporal and structural hierarchy.

It is not necessary to remember all of the structural changes that have occurred in an interactive session. For example, objects that have been created then deleted and are no longer recoverable will not affect the future interaction. The user need not remember that they ever existed. Therefore, we can filter out such meaningless commands from the result of mapping *construct* onto the *history*. This completes our construction of the extraction function for this scenario. In summary, we would have

$$filter: C^* \rightarrow C^*$$

$$extract = filter \circ (map\ construct)$$

Another possible refinement to this model, which we will not expound upon in this chapter, would be to use attributes and signatures to arrive at a definition for the *extract*ing function.

We do not discuss the significance of the abstractions used to a psychologist except to say that we would hope that the concepts used in constructing extract may be useful in understanding the memory demands placed upon the user.

3.3. Conclusions

We conjecture that *extract* is a function whose appropriate definition (taking into account more global design implications) may be clarified by a psychological analysis. We can only conjecture at this stage and note that this bridge between psychology and computer science represents an important component of the AMODEUS project (Barnard & Harrison, 1989). This connection is further discussed in the next chapter. Design must capture assumptions about the system's environment (such as, the psychological make-up of the user population, work patterns, organisational architecture and so on). In chapter 2.3 we consider examples where user oriented view points cannot be folded into the system model. If we now turn our attention back to the example drawing package and the object selection method, we can analyse what memory demands the system design imposes on the user. In this case, the user needs to comprehend all three hierarchies which determine how the selection is made. The visual hierarchy is by its nature displayed while the structural and temporal hierarchies are not usually visible. Hence, for the selection procedure to be predictable, *extract* will be required to include the structural and temporal hierarchical information (such as an ordered tree of the graphical objects present). The extent to which users are able to reliably remember such information can be the subject of empirical analysis. In this way discrepancies between the mental competence assumed by a system and users' capabilities can be discovered. An analysis of this kind will determine whether it is necessary either to simplify the selection algorithm or to make explicit the structural and temporal information so that the user may derive from the display what has to be known in order to predict how the selection will work. In most graphics packages the selection algorithm is simplified. An appropriate solution to this design problem will depend on the way the system is used, so there may be cases when structural and temporal information should be made explicit as a more valuable alternative to simplification of the algorithm.

4. General conclusions

We have demonstrated in this chapter how a formal approach to system analysis can focus on the interactive aspects of the system and provide a means of elucidating principles to enhance interaction within the early design stages. We are concentrating on the use of such methods as a means of synthesis of interactive systems as well as a means of analysis that can be useful in demonstrating the use of interactive systems in practice. As a result we are concentrating our efforts now on incorporating the results of the abstract interaction models within more constructive software engineering and formal notations (Abowd, 1990; Abowd & Harrison, 1990; Duke & Harrison, 1993) with the aim of convincing practitioners in formal methods of the value to be gained from extending existing techniques into the realm of these requirements. Properties such as the ones described here are functional properties that are useful in showing that the system satisfies requirements that are traditionally understood to be non functional. Using the models analytically clarifies our understanding of how consistencies in the design of interactive systems may be employed to assist the use of such systems. We see the presentation of properties like layer conformance and history based predictability as contributions which can be used in the evaluation and assessment of the usability of interactive systems. It remains for behavioural scientists to assess the value of these properties in a general context.

Acknowledgements

We would like to thank members of the AMODEUS consortium for providing the initial descriptions of the scenarios used in our analysis. Discussions with other, more psychologically oriented modellers have continually led us in the direction of incorporating relevant psychological principles in our formal interaction models. We would also like to thank the members of the York HCI group for their constant support of our work and the stimulating discussions that they have provoked.

References

Abowd, G.D. (1990) Agents: communicating interactive processes. In D. Diaper et. al., (eds), *Human-Computer Interaction —INTERACT'90*, Elsevier, pp. 143-148.

Abowd, G.D. & Harrison, M.D. (1990) *On a constructive approach to applying formal methods in HCI*. University of York, Department of Computer Science, Technical Report YCS (151).

Barnard, P.J. & Harrison, M.D. (1989) Integrating Cognitive and System Models in Human Computer Interaction. In A. Sutcliffe & L. Macaulay, (eds), *People and Computers V*, Cambridge University Press, pp. 87-104.

Duke, D. & Harrison, M.D. (1993) Abstract Interaction Objects. In *Proceedings Eurographics '93*, Barcelona. Springer.

Hammond, N.V. & Myers, K. (1990) *Definition of scenarios for M1 Workshop*. ESPRIT BRA project 3066 (AMODEUS) Internal Report RP8/IR6.

Harrison, M.D. & Dix, A.J. (1990) A state model of direct manipulation in interactive systems. In M.D. Harrison & H. Thimbleby, (eds), *Formal Methods in Human-Computer Interaction*, Cambridge University Press, pp. 129-152.

Harrison, M.D., Roast, C.R. & Wright, P.C. (1989) Complementary methods for the iterative design of interactive systems. In G. Salvendy & M.J. Smith, (eds), *Designing and Using Human-Computer Interfaces and Knowledge-Based Systems*, Elsevier, pp. 651-658.

Nicholls, J. E. (ed.) (1992) *Z User workshop*, Springer Verlag.

Payne, S.J. & Green, T.R. (1986) Task-action grammars: a model of mental representation of task languages. *Human-Computer Interaction, 2:2*, pp. 93-103.

Young, R.M. & Whittington, J. (1990) Using a knowledge analysis to predict conceptual errors in text-editor usage. In *CHI'90 Human Factors in Computing Systems*, ACM, pp. 91-97.

Computers, Communication and Usability:
Design Issues, Research and Methods for Integrated Services
P.F. Byerley, P.J. Barnard and J. May (Editors)

Chapter 2.3

Integrating user requirements and system specification

Ann Blandford†, Michael D. Harrison* and Philip J. Barnard†

†MRC Applied Psychology Unit, 15 Chaucer Road, Cambridge, UK—CB2 2EF

*Dept of Computer Science, University of York, Heslington, York, UK—YO1 5DD

1. Introduction

If software engineers are to be alerted to human computer interaction concerns early in the design process then they must work with models of the system that incorporate user perspectives. In this chapter, we present a case for modelling the interactive system (involving computers and users) in a neutral way as a means of developing an understanding of the relationship between user and computer system concerns. In this section we outline a case for the use of formal models of interactive behaviour in the early stages of the system design process. In the following sections, we present the notation which will be used in the remainder of this chapter and discuss the mechanisms by which a neutral perspective based on a notion of agent may be translated into computer system terms through the formal expression of interactional requirements. This perspective will be referred to as *agent-neutral* in what follows.

Designing an interactive system involves a translation from informal requirements, which may be ambiguous, into precise constraints which can be shown to hold for an implementation of an interactive system. Requirements are typically imprecise, and are expressed in terms that can be understood by clients, while specification is designed to be as unambiguous as possible so that a computer program can be produced which reflects the requirements accurately. A challenge then, for software engineers, is to improve the precision of requirements so that the translation from requirements to specification may be made as precisely as possible and so that the 'formality gap' (Dix and Harrison, 1989) can be moved to an earlier stage of the design process.

In practice, we also wish to achieve greater precision to help to understand the implications of user requirements. We require precision so that there is a means by which the system can be analysed with respect to human centred models prior to the completion of the implementation. For example, we may wish to identify the situations where workload may be unacceptable or where additional problems are likely to arise.

In order to be more clear about the sorts of objects we are dealing with here we distinguish between an *interactive system* and a *computer system*. An interactive system may involve computers, people, paper and many other media and processes that assist the organisation of work tasks to achieve the goals of the interactive system. We shall call many components of the interactive system 'agents'. These are semi-autonomous processes that interact with the rest of the system. Agents are capable of some level of internal choice and therefore, from the point of view of at least one of the other agents, they will appear to be non-deterministic. Hence paper will not itself be an agent but may be considered to be part of the state of the human agent, whereas a computer system, or a process within a computer system, (which is designed to be deterministic, but may not appear so to a user) is treated as an agent. We are particularly interested in the properties of the agents which comprise the computer system, in terms of developing a better understanding of design issues relating to usability.

Another way of looking at the general design problem of concern here is that we are designing a system of agents. In order to design such a system more effectively we will be concerned with optimising the interactive behaviour of the agents, and in particular the usability of the computer system. When translating usability requirements into properties that can be used constructively in design it will become clear that they may cross boundaries between the different agents of which the interactive system is constructed.

Usability requirements typically refer to the ease with which the computer system may be used within the interactive system. Ease of use may include predictability, ease of learning, error visibility and error tolerance. In order to interpret usability requirements, some level of understanding of human cognitive processing is needed as well as a particular understanding of the task environment in which the system is to be used.

1.1 Interactive systems rather than computer systems

In practice, usability requirements place constraints on many, if not all, of the agents of the interaction. Despite this apparently obvious remark, the focus of the usability analysis of an interactive system is usually based on an analysis of the computer system. A number of recent techniques developed to apply or validate usability requirements tend to focus on the computer system.

- Informal techniques that extend traditional software *walkthroughs* by subjecting the system design to checklists drawing from understanding of where difficulties in cognition are likely to arise (Wharton et al, 1992) apply the checklists to some representation of the computer system.

- Usability metrics which in some informal sense can be used to assess whether a design is successful in relation to a particular implementation (Shackel, 1991) are metrics defined in terms of the performance of the completed computer system.
- The measuring of the complexity of interactive systems by analysing the typical performance of tasks by means of some simulation using a planning type cognitive architecture (Kieras and Polson, 1985) requires a full understanding of the characteristics of the computer system.
- Techniques that treat the finished implementation as a *micro theory* about how the system will be used in which claims about use can be tested empirically in a context similar to actual use of the system (Carroll et al, 1991) require the system as an encapsulation of the theory.
- Using abstract models of interaction as a first step in the development of a formal specification of the system and applying interaction principles and structures that the user is likely to understand as properties or artefacts of the interaction model (Dix, 1991; Harrison, 1992) requires an initial understanding of the computer system.

All these approaches assume that usability can be expressed in terms of a representation of the computer system under development. For example, the user's cognitive perspective is recruited informally as a means of criticising the system representation. Usability metrics give quantifiable measures that can be related to the use of the system. The system model captures assumptions about how the system will be seen by a user or users.

We argue that it is not sufficient to express usability requirements as properties of the computer system alone. Any notation for describing usability requirements should be capable of taking into account all the agents to the interaction. A framework is required that can be used as a starting point for addressing cognitive and systems aspects of usability in design as well as other wider issues such as domain and environment. Requirements may then be translated into the implementation either in terms of training or operating procedures, or in terms of the computer system design (Harrison and Barnard, 1993). Our concern in this chapter will be with the design of the computer system and we will focus on the way that requirements can be expressed in ways which will inform the computer system design; however, our focus will be guided by an understanding of the properties of the interaction. In order to demonstrate an interactional view of user requirements and their integration, a number of examples of interaction requirements will be considered. The examples are not intended to be exhaustive, but to demonstrate an approach to modelling interactions. The structure of the chapter is as follows:

In section 2 we provide an event based notation that can be used to describe interactions and develop a modelling framework for analysing interactions, in terms of goals and events as well as actions and states of the interactive system.

In section 3 this framework is applied to an exemplar drawing package. The example used here is one of those discussed in the previous chapter, and we explore what a system requirement such as predictability means in interactional terms.

In section 4 we consider the first of three examples that illustrate the demands of an interaction framework. This system is a walk-up-and-use system in which there are two relevant agents: a computer system and a user.

In section 5 we discuss a variant of the classic 'unselected windows' problem in which requirements are implemented in terms of an attentional perspective.

In section 6, we extend the discussion to a situation where more than one user agent may be involved by considering a computer based table tennis game, and interactional requirements that relate to whether or not the game is rich enough to be enjoyable.

Finally we make further observations about dynamic systems of many agents in general.

We wish to exemplify the demands of interactional requirements, demonstrate a notation which will express these requirements in a framework which accommodates various modelling perspectives, and describe computer system models that will be useful in specifying interactive properties of computer systems.

In the next section, we present a preliminary notation for representing interactions.

2. Adopting a means of representation

An interactive system is designed to do work that can be described in terms of goals—that a particular state of the interactive system should be achieved given some initial state. A goal may be described as a predicate on the state.

Goals are achieved by performing tasks. Tasks are procedures that employ the interactive system to achieve the work goals (this contrasts with the focus on user tasks in chapter 3.1). It is important to recognise that since goals may relate to the whole system, not just the state that is of concern to one of the agents, tasks will involve actions from more than one of the participating agents. Since tasks may be shared by agents we shall be concerned with issues such as which agent takes the initiative and which agent makes each of the decisions which influence the outcome of the interaction. If an interactive system involves a user and a computer agent then we may wish to describe issues related to initiative in the performance of any task.

Usability requirements will apply to those aspects of the system that involve interaction between a computer system and a user agent. We wish to investigate the means by which the requirement, at whatever level, is converted into constraints over the computer based agents of the broader interactive system. Having achieved this objective, the next step is to translate the agent description into a specification of the system using a standard software engineering notation such as CSP (Hoare, 1985) or Z (Spivey, 1989) so that we can achieve a precise consideration of design in the absence of implementation clutter.

Properties of systems will be described as properties both over interacting agents and within agents. Since we are concerned with properties over interactions we shall be concerned with the information that is exchanged. These information exchanges will be described in terms of *events*. In order to describe behaviour within agents we will begin with an initial model of agents which consists of states and actions. This model contains some of the components of a computer system design and is intended to be a first step in the direction of refinement towards implementation (Dix, Harrison, Runciman & Thimbleby, 1987).

2.1 Agents, states and actions

In the description of an agent it is often useful to describe an initial state (or 'point of departure'); then the interactive system is capable of producing a variety of states starting from this initial state. The state is transformed by the execution of actions. Viewing the interactive system as just that—a system 'in a box' without outside influences—all actions may be regarded as internal actions which update the state of the whole system. In this chapter a number of logical formulations will be presented—these will be bordered to distinguish them from the text:

The initial state is s_0 and the set of all possible states of the system is S. The system is capable of producing a variety of states $s \in S$ starting from this initial state. If the set of all possible agent actions is A then each action $a \in A$ and $A \subseteq S \rightarrow S$ transforms the state.

If s_i is the ith state and a_i is the ith action then the system goes through a sequence of states $<s_0,\ldots,s_n>$ where s_0 is the initial state and $s_i = a_i(s_{i-1})$. A sequence of actions, a^*, in general, starting at state s can be written $<a^*,s>$.

The nth state of the system can also be expressed as

$$s_n = \text{compose}\langle a_1, a_2, \ldots, a_n \rangle s = a_n(\ldots(a_1(s))\ldots).$$

Recognising that the system is made up of more than one agent, we have to account also for the interactions between the different agents. When agents communicate through action, we call the interaction that takes place an 'event'. An *event* occurs when a pair of agents both engage in actions that cause a communication of information. So, for example, if a computer system agent updates the screen display and a user agent observes and interprets the change then an event has occurred, whereas if the user fails to notice the change then one has not. Events may arise because two agents have synchronised on an interaction, but in general we shall not assume this; for example, with an auditory interface the agent actions of emitting sound and hearing it are often synchronised, whereas for a display-based interface, the system action of displaying visual information and the user action of reading it do not need to be synchronised—an event is still considered to have occurred when the information has been communicated. When we consider the interactive behaviour of such systems of agents we shall be interested in the sets of events and their causal relationships.

The model of agents can be used for two purposes. Firstly, it may be used constructively as a means of representation within the design process. Secondly, the agent-neutral representation may be used to capture requirements over 'interactional trajectories'—that is, the collection of events, and their ordering, involved in achieving some task. Analysis of the interactional trajectories will precede consideration of the implications of the requirements upon the computer based agents in the system. Requirements may have implications for different types of system models that emphasise different aspects of behaviour; for example: performance, security, or reliability. We are concerned that these models be used analytically to assess the interactive properties of a design. Here analysis is seen as complementary to construction; an iterative design process involves production of a version of the system which is subsequently subjected to analysis for verification, and experiment for validation, possibly followed by production of a modified version of the system.

2.2 Events and Tasks

We need to have a clear idea of what an event is because we aim to reason about sets and sequences of events, and use the orderings of events to apply analytic techniques from other modelling approaches.

Viewing an interaction from the perspective of one of the interacting agents, an agent is associated with a set of events (its *alphabet*). Events may be thought of in two ways; they are the effect of interaction that is the combination of actions that are invoked from different agents, and it is also useful to think of events as triggering actions. For example, an event which consists of a user issuing a command to a computer system and the system noting that command will trigger system actions which cause the system state to change.

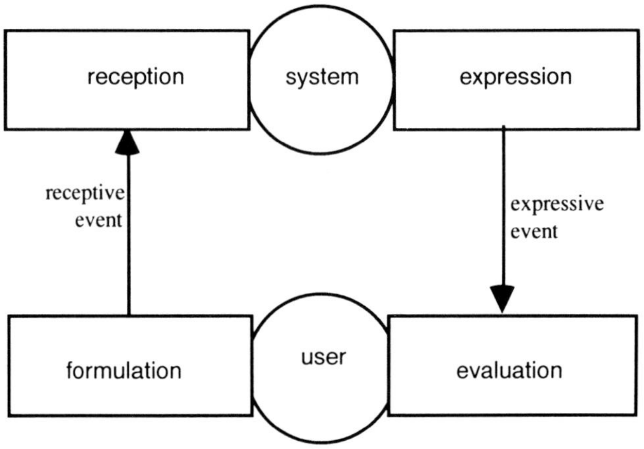

Figure 1: representation of the two-agent control loop

In order to understand a property of the interactive system in event terms we first consider the interactive system as a control loop (Norman, 1986). From a user's perspective, an interaction is an interleaving of *formulations* and *evaluations*. The user is considered to be the agent who carries out the operations of formulation and evaluation. On the other hand, the computer system has *receptive* and *expressive functions*. The *receptive function* accepts the formulation and acts upon it, while the expressive function reflects the state of the system resulting from the action—see figure 1.

It makes sense therefore to follow Norman and see an interaction from the point of view of one of the agents as a sequence of expressive and receptive events. Here the terms *expressive* and *receptive* merely indicate the direction of the interaction when the events synchronise with other events.

The interactive system is designed to support the execution of a set of tasks. A task is performed by sets of events, where we may view an event as a perceivable action produced by agents in the interaction as they communicate with each other.

In addition to an understanding of event relationships it will also be necessary to connect the events to computer system, cognitive or other models. The event formulation itself is merely a vehicle that can be used to structure and integrate the analysis that must be provided from the other modelling approaches.

A task has associated with it a set of performances. A *performance* is an ordered set of events (an event *trace* (Hoare, 1985)), where the events are generated by communication through action of the agents. Typically, there are a large number of possible event traces which achieve the goals of any given task.

> Let E be a trace of events that can be engaged in by the agents.
>
> The set of tasks is $t \in T$. E is a *performance* of task t. A task may have a number of possible performances associated with it. We shall refer to a particular performance as $t \prec E$.
>
> A partial performance of a task—that is, an event sequence which is engaged in while performing a task, but which does not achieve the goals of the task—is denoted by $t \prec\prec E$.
>
> An event sequence may be written $E = \langle e1, e2, e3 \ldots \rangle$, and a trace of events followed by a specified event as $E' = \langle e1, e2, \ldots e6 \rangle \triangleright e7'$. The length of trace E is $\#E$.

To illustrate the use of this notation, imagine five possible performances of a task, t, as shown in figure 2. Any one of these performances, here labelled E, E', E''' F and F', achieves the goals of the task. Choices between the alternative possible event traces—for example, the choice between events $e7$ and $e7'$ after $e6$—are influenced by various factors, as discussed later. Typically, if the even numbered events are receptive then the odd numbered ones are expressive, or vice versa. The length of a trace is the number of events of which it consists; for example, the length of trace $E'' = \langle e1, e2, e3, e4'', e5'' \rangle$ is 5, while that of $F = \langle f1, f2, \ldots, f9 \rangle$ is 9. Any incomplete traces which do not achieve the goals of the task yet are considered to be partial performances; for example, $t \prec\prec \langle e1, e2, e3, e4'' \rangle$, or $t \prec\prec \langle f1, f2, f3, f4 \rangle$.

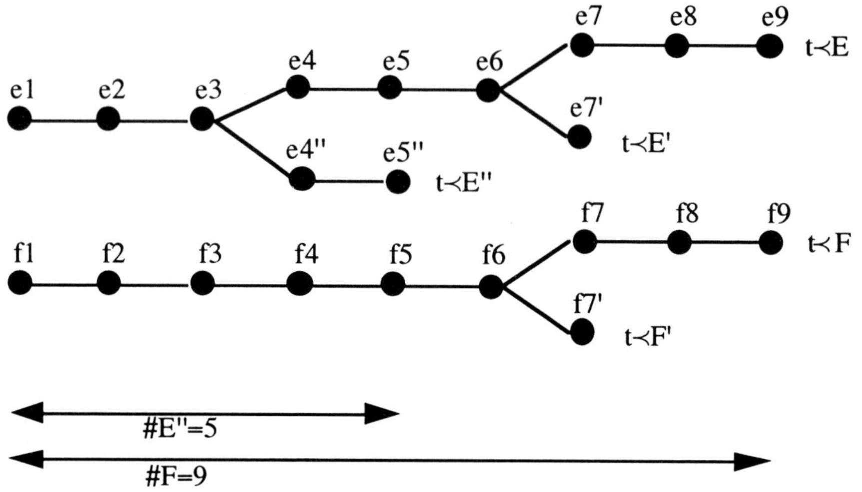

Figure 2: example event traces

The formulation in terms of events may be contrasted with that expressed in terms of actions and states, for which the basic notion is that the task goal is a predicate on the state, so a series of actions has to be performed by the interactive system to transform the state from the current state to one in which the task goal is achieved. In this case, rather than a *performance* of a task, we refer to an *implementation* of that task.

An implementation of a task t will take the system through a sequence of states $<s_0,\ldots,s_n>$ where s_0 is the initial state and $s_i = a_i(s_{i-1})$ and the sequence achieves the goals of the task. If $t \in T$ we say that $<a^*, s_0>$ *implements* t if $a^* \in imp(t)$ where $imp(t)$ is the set of all sequences of actions that satisfy the goals of the task. Implementation by a sequence a^*, in general, starting at state s shall be written as $t \lhd <a^*, s>$.

These action sequences will include all the possible ways that a task objective may be achieved including sequences that will occur perhaps as a result of an error or an afterthought—for example, action sequences which involve the use of 'undo' to restore the state to that which existed before the last action.

If a has an inverse a', that is $a'(a(s))=s\ \forall s \in S$, then if $t \lhd < a^*, s>$ for some $t \in T$ it will be the case that $t \lhd <<a_0,\ldots a_{i-1}, a, a', a_i, \ldots, a_{n-1}>, s>$.

The concepts of *performance* and *implementation* described above beg a number of assumptions about the nature of the system:

- It is assumed that system tasks can be conceived at design time. More exploratory systems may not be as closed as this and may subsequently be used in a variety of work domains not originally envisaged by the designer. A solution to this problem may be to consider the design of these systems in terms of *cognitive tasks;* that is, micro-tasks that capture the essence of the potential tasks and coincide with the basic patterns that are understood by the user.
- It is assumed that at the point of forming the intention to carry out the task, the appropriate sequence of actions is understood and can be performed without further modification as a result of external circumstances. This sort of system should be distinguished from systems where some tasks are also being performed autonomously by the system. For example, we may be talking about a process control system where many parameters are relevant to the performance of a task and many external factors may change the nature of the task during its execution.

In the description of the task we must provide more than an ordered set of actions. It will also be necessary to include the conditions that allow the refinement of the task to a particular action ordering, to understand what features of the interactive system and environment influence the choice between alternative possible action sequences. So tasks are implemented by sequences of actions, and agents are involved in deciding which specific action sequences are most appropriate to a particular performance. An interaction described in these terms can be seen as a *trajectory*.

Tasks will necessarily involve interaction between agents as they co-operate to implement a particular task. The effectiveness of the interactivity of a system of agents will be a property of these interactions, or events. Interaction requirements will relate to properties of such sets of events. In the remainder of this chapter we shall be concerned with task performance.

2.3 Multiple agents

Given this apparatus how can we express usability requirements? Usability requirements capture the essence of important properties of the interaction between the user agents and the other agents of the system. In a multi-agent system, we are concerned in particular with the subset of these interactions that involves the mechanisms of communication between user agents and system agents. These requirements should be expressed as properties of the interactive system which would enable the construction of computer agents that reflect these important properties.

Initially, we will be concerned with a system of two agents, user and computer system. In general, there is no need to limit ourselves in this way; the next step is to separate the interactive system into two groups, one consisting of just one agent and the other consisting of the agents that implement the rest of the interactive system (see figure 3). We will briefly discuss more complex systems in section 7. From the perspective of any one agent, we may view the interaction as consisting of events initiated by that agent and events initiated by the rest of the interactive system. In general, if we consider the perspective of any one agent (whether user or computer system based), the remaining agents in the system may be regarded as engaging in events with that one agent which are receptive (receiving communication from that agent) and expressive (expressing changes in state to that agent).

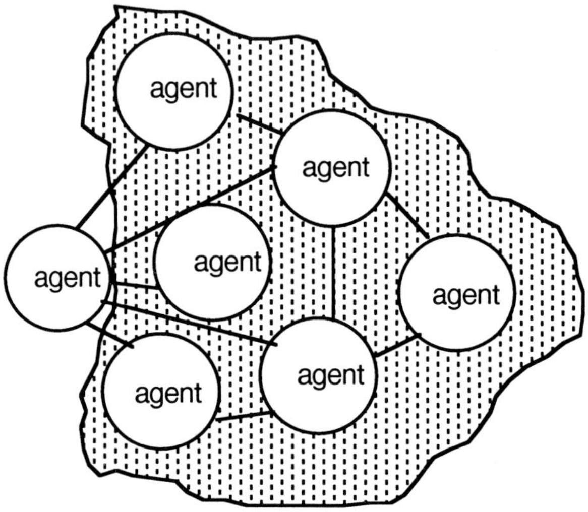

Figure 3: splitting the interactive system

When we take account of the agents within the interactive system, we recognise that an action will have, in addition to an attribute that prescribes its position in an ordered sequence of actions, a designation of the agent with which it is associated. Since actions are initiated from different agents we shall have to accommodate this in our notation.

These different autonomous perspectives shall be referred to generically by introducing an explicit set of agent names $\alpha \in Agent$. The model described above is extended to associate an agent with each action: $a^* = \langle\langle a_1, \alpha_1\rangle, \ldots, \langle a_n, \alpha_n\rangle, \ldots\rangle$.

We can redefine *compose* to include agents:

$$\text{compose}\langle\langle a_1, \alpha_1\rangle, \ldots, \langle a_n, \alpha_n\rangle\rangle s = a_n(\ldots(a_1(s))\ldots)$$

Similarly, for events, the model may be extended to incorporate the notion that an agent is associated with an event, i.e. (in general) is the initiator of the event. It will often be necessary to consider a trace that is restricted to a particular agent.

If the set of agents is $A \subseteq Agent$ then the trace of events, E, may be associated with the agents in the set, $\alpha_i \in A$: $E = \langle\langle e_1, \alpha_1\rangle, \ldots, \langle e_n, \alpha_n\rangle\rangle$. A trace that is restricted to a particular agent may be represented:
$E \uparrow \alpha$ restricts the events to those only that relate to the agent $\alpha \in A$.
$E \uparrow A$ restricts the events to those that relate only to agents in A.

At this stage, we will not explicitly consider properties of interactive systems which are particular to multi-agent systems, but will return to this topic in section 7. For now, it is sufficient to note that any interactive system can be viewed from the perspective of one agent, whether that agent be a user or implemented in the computer system.

2.4 Rendering

We can make a further connection between the event description and the description in terms of states and actions. So far events are the means of communication and are assumed to be transitory, in the sense that each communication episode happens, and is then past. Nothing has been said so far about how the state of one agent (or sub-system of agents) is made visible or, to be more general, *perceivable* to another agent in a persistent way which might allow another agent to acquire information from it over a period of time. Those parts of the state of the agent that are perceivable are modelled as a separate layer that we shall call a *rendering*. Perceivable components of a computer system might be features of the display (for example, a clock icon) or may involve other communication channels, for example an alarm bell or a smell of smoke. The important properties of the perceivable components, so far as the model is concerned, are that they bear some relationship to the underlying state. In some systems the rendering will be quite separate from the state of the system. For example, the instruments and displays in an aircraft cockpit will be entirely separable from the actual state of the engine or the flight surfaces. However, for many it will be a direct reflection of the state of the system. Duke and Harrison (1993a) further generalise the notion of rendering. The 'render' function here is analogous to the 'view' function in the previous chapter, and also shares many properties in common with the object representation dimension of the 'service relationship' introduced in chapter 1.1.

The *render* mapping r projects each state into a perceivable rendering $r{:}S \rightarrow P$. P is a collection of the perceivable components that represent the state.

It will often be the case that the rendering will provide an expressive function for the agent which can be identified with an expressive event. We can go further than this. It is possible that only task relevant components of the rendering, or agent relevant components of the rendering, are required or indeed are noticed. Expressive events can be associated with what we shall call an agent rendering. An agent, for example, may have an associated display that contains the information that contributes to particular expressive events, or a user agent may express task relevant aspects of their state by pressing keys or moving the mouse. A psychological analysis based on the event oriented notion may recognise what a user agent is likely to notice in interacting.

Only certain components of the state and the rendering, or renderings, will be relevant. Moreover, it is possible that another subset of the rendering, not including all the information that is required, is *noticed* in performing the task. In the context of a task we shall not only be concerned with implementations but also with those aspects of the state that are perceivable and relevant to that task. The rendering that is noticed in performing the task will be derived through analysis of what users are capable of perceiving and by experiment.

> Considering an agent oriented rendering, each agent has an associated perspective $r \in R$ such that $r_\alpha : S \rightarrow P_\alpha$ where P_α is the set of perceivable attributes for agent α.
> Those aspects of the state that are perceivable and relevant to task t will be written as $r(t) : S \rightarrow P$. We might also define $r_\alpha(t) : S \rightarrow P$ as the rendering that is noticed by agent α in performing the task.

As far as the task is concerned, it is important to notice that it achieves a goal that is relevant to part of the state, and may be expressed as a predicate on the state. During the performance of the task, various rendering templates will be relevant to checking that the user understands that the task is appropriately achieving goals and objectives.

Loosely speaking, an expressive event consists of one agent (typically a computer system agent) engaging in an action which causes a change in the rendering of its state and another agent (typically a user) viewing that change and interpreting (evaluating) it in the light of the current task goals.

3. Object selection: predictability revisited

In order to begin the discussion of usability requirements, we consider the principle of predictability that has been discussed in the preceding chapter and elsewhere (Harrison & Dix, 1990; Sufrin & He, 1990; Abowd, 1990). Expressed in agent-neutral terms, predictability is a general property of interactive systems that concerns the ease with which one agent can predict the effect of an action on another agent. The principle requires that the agent should not be 'surprised' by the effect of an action, given that the agent knows the general effect of the action and the agent engages with the current event. In this context, the action is a component of the current event, and the acting agent is the initiator of the event. From the perspective of that agent, the system is predictable if the effect of an action initiated by that agent can be predicted, and if there are no events which are unexpected—for example, initiated by the rest of the system. In event terms, the predictability is based on the current expressive event (i.e. the event in which the rest-of-the-system expresses its current state to the agent and the agent interprets that expression), and not on any memory of the history of the interaction.

Let us consider again the drawing package described in section 3.2 of the previous chapter. When several graphical objects overlap, there are various possibilities about which of the overlapping objects is selected when the user clicks on them. For example, objects may be stacked up in the order in which they were created or accessed, or in accordance with some sort of object hierarchy. Our concern here is to establish what features of the computer system design will make it predictable to a user agent.

In the example we are considering here, objects are accessed in reverse order of creation (i.e., most recent first), and the selected object is marked with 'handles'. Let us consider the alternative means by which the user can select the target object when presented with a diagram containing overlapping objects. The user might depend on memory of the order of creation to determine how many times to click on the overlapping objects to select the required one, or might rely on the feedback from the screen—i.e. the appearance of the handles. The choice will depend primarily on the user's memory of the order of creation, any feedback from the screen about the order of creation, and the quality of the feedback from the screen about the currently selected object. These are the factors on which the user's ability to predict the effect of an action on the system (a receptive event) is based.

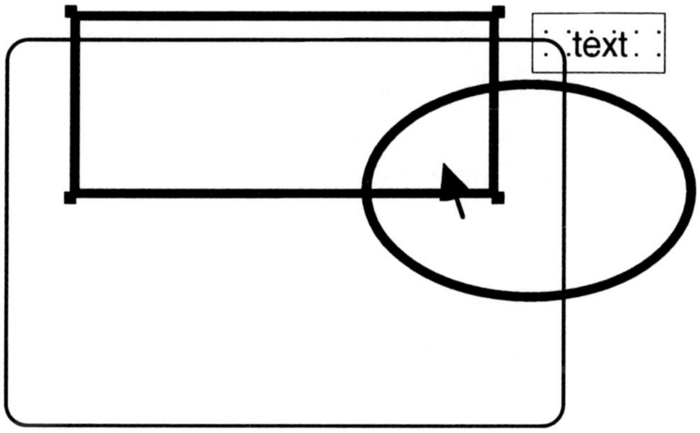

Figure 4: overlapping objects

This system is predictable in the sense that it is possible to predict the effect of selection if it is known how the drawing was constructed. Predictability is dependent on the preceding interaction history, certain features of which determine how a drawing was constructed.

For this small scenario the important aspects of the interaction state are:

the target object,
the currently selected object,
the information from the display about the currently selected object,
the interaction history (order of creation of objects),
the user's memory of the interaction history,
the information from the display about the interaction history,
the depth of the target.

If we consider what is happening from the point of view of goals, actions and events then the task goal is to achieve a state in which a particular object within the display picture is selected. An ordered set of events has led to the current *expressive event* (here the display is tied to an expressive event). The user has to gather enough information from the current display to make it possible to select appropriately. The design might be satisfactorily constrained by three possibilities. The first is that there is enough information imparted in the expressive event to enable the user to predict the effect of the selection command. The second is that the user can remember enough about what happened before to be able to predict the effect of the next action. The third is that further exploration will enable the necessary selection. The question then is which of the properties we wish to design to; we can use predictability as a property which assists the generation of the design.

If the system is to be predictable, in the sense discussed in the previous chapter, then there should be enough information in the display to demonstrate not only what object is currently selected (and this is done in many systems by displaying handles on the diagram as shown in figure 4) but also to present predictive information about how to select further objects. Therefore the challenge of event predictability is to define a display that removes the unpredictability of an ambiguous display. Here the principle is being used constructively to recognise the significance of certain design options.

What happens if the display cannot be designed to be adequate? In this case we will have to apply more relaxed criteria. For example, we may wish to apply a notion of predictability that presumes certain things about how the user agent can recall the history of the interaction as was done in the previous chapter. We therefore extend the notion of predictability to requiring that not only are the expressive events the same in the two cases but also elements of the trace of events that led up to the current event are the same.

3.1 Agent-neutral analysis and its relation to the system specification

In this section, we focus on an event based notion of predictability, and contrast it with the state based account given in the previous chapter. In describing predictability we are concerned with the nature of the interaction between one agent and the rest of the system.

For an agent to be predictable to another agent (in this case, for the computer system agent to be predictable to the user agent) there must be enough information in the current expressive event (e.g. the user agent must be able to infer enough information about the state of the system on the basis of the information about its state which the system agent is currently displaying) to allow the user to predict the effect of his or her next action. Put another way, predictability means that if the information exchanged in a current expressive event is the same for two different interactive systems, then further sequences of receptive and expressive events will not detect any difference between the systems. This, of course, reflects one of the qualities of the notion of an event: what is of significance in an event is not only what one agent imparts but also how the other agent interprets that information.

In agent-neutral terms, given any initially ordered set of events (the trace that generated the current situation) if we consider any pair of event traces E_1 and E_2 that both lead to a current expressive event e then if the same receptive event follows, the subsequent expressive events will be identical in the two cases; the systems are indistinguishable from the point of view of the single agent—see figure 5. (In this context, the terms 'expressive' and 'receptive' relate to the rest-of-the-system, outside the single agent which is assessing the predictability.)

For a predictable system, $\forall E_1, E_2$ which are possible traces of the system, if$E_1 \triangleright e$ and $E_2 \triangleright e$ are the only possible traces of the system, then given any receptive event r, both traces lead deterministically to e'. I.e. $E_1 \triangleright e \triangleright r \triangleright e'$ and $E_2 \triangleright e \triangleright r \triangleright e'$ are both the only possible traces from E_1, E_2 for any given receptive event, r.

It must be remembered that in this event oriented definition of predictability, the nature of the events depend on the properties of the interacting agents; an event depends not only on the action of the initiating agent but also on the interpretation of that action by the other agents in the system. This property of predictability captures the important point that it is the information which is communicated (and interpreted by the receptive agent) that is significant; it is this that is missing from the computer system oriented notion.

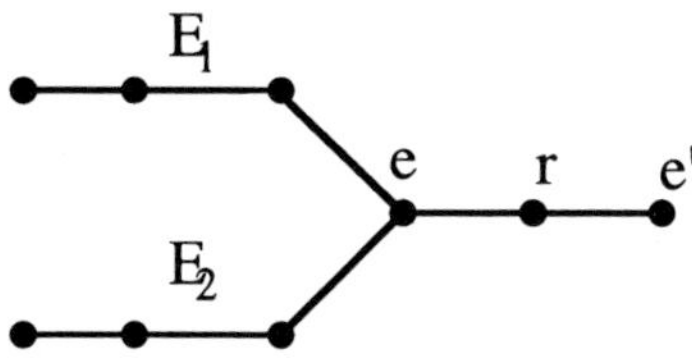

Figure 5: event trajectories

The general property of predictability is highly constraining. In practice we are concerned with the relationship between predictability and task. One way of relaxing this constraint is to limit the nature of the equivalence between the two interactive systems generated by the different event traces in various ways. Hence the requirement that the next user generated formulation (receptive event) will produce a predictable expressive event can be changed to be contextualised by task.

Taking any trace, E_1 that is a prefix for a current event, r, then any further event that continues to be a performance of the task will provide an expressive event that is also a performance of the task. In other words, the requirement that an agent should be predictable with respect to a task can be expressed as a statement that if the receptive event is consistent with achieving the task goal then the resulting expressive event will also be consistent with achieving that goal.

$$t \prec\prec E_1 \triangleright r \Rightarrow t \prec\prec E_1 \triangleright r \triangleright e$$

The question now is: how does this property of predictability translate into a property of individual agents? Before considering this in detail let us consider what an implementation of a task might be in terms of the agent constituent parts. An *implementation* is the system level description of what a task does that corresponds to the notion of *performance* at the event level. An implementation of a task will take the system through a sequence of states that will finally achieve the goals of the task (perhaps the final state produces some result which satisfies the task goals). There is a set of all sequences of actions that satisfy the goals of the task. It should be noted that it cannot be assumed that the task always starts at the same initial state because the task may be implemented from any state.

We have described three possible properties of the drawing package. The next step is to consider how these properties are mapped into properties of the system. Once these properties have been expressed appropriately in terms of computer systems then they may be used either to analyse an existing system or to construct a new one. The notion of predictability that we have mentioned has been expressed similarly though not identically in terms of a system model.

How do we express the weakened (task oriented) notion in terms of the system? The property can be expressed, as it was in a slightly different form in the previous chapter, as:

> A system is *observational predictable* if for any pair of states $s, s' \in S$ such that $r(s')=r(s)$ then $f(a(s))=f(a(s'))$ for any $a \in A$ where f is a function that extracts some features of the state hence: f may be the identity in which case the states may be required to be equivalent, or f may be the rendering r in which it is only required that the renderings are equivalent.

We can weaken this notion—for example, by acknowledging that what the user has done in the past is typically relevant to an understanding of the behaviour of the system. Hence we might define some kind of *history template* which extracts details of the important actions from the action sequence. In the example being considered here, the way in which a diagram is constructed influences the way a select action works. Hence when a system is predictable, on the basis of an ability to recall important components of what has recently happened, the user knows how to map task goals into actions reliably. It is not only the action sequence that is relevant; we might also include in this some components of the renderings that have led to the current state. In other words, what is relevant is the components of the event history which relate to the performance of the task, and here we start to see the relationship between the agent-neutral event notation and the computer system oriented notions of system state and rendering of the state.

In describing predictability we are concerned with the nature of the interaction between one agent and the rest of the system.

> A *history template* $ht{:}A^* \rightarrow A^*$ extracts details of the important commands from the command sequence. For all a_1' and a_2' such that $ht(a_1')=ht(a_2')$ then if $s_1=compose(a^*_1)s_0$ and $s_2=compose(a^*_2)s_0$ and $r(s_1)=r(s_2)$ then $f(a(s_1))=f(a(s_2)) \forall a \in A$.

We have now mapped the performance of a task directly into the system domain. What we have not yet done is to describe how this notion may be mapped to the decision making capabilities alluded to above. We can now interpret predictability in terms of a single agent.

Instead of considering the trace of events up to the current position we shall assume that some sequence of actions has been carried out and the system has arrived at a state s. We shall also assume that certain components of the state are relevant to the current argument and therefore assume a mapping, call it f, with domain S that extracts the important parts.

The notion of predictability then requires that if $f(s_1)=f(s_2)$ then any further action a that is available to the interactive system will not discriminate them; that is $f(a(s_1))=f(a(s_2))$.

What is the connection between this notion of predictability and the event based one? There is an important difference between event based predictability and state based predictability. It is clear that a computer system may be predictable in a formal sense but in fact something about the display is such that the user cannot predict the effect of the next command. For example, if the display rendering cannot be interpreted by a user then, although the information is there and is unambiguous, it cannot be used. On the other hand, in the case of the event oriented notion of predictability it is assumed that an interaction is carried out *between* the user and the rest of the system. In order for an event to take place the user and the rest of the system of agents must have engaged in a communication action. Although we say nothing about the cognitive demands that are put upon the user agent, the interactional perspective gives us hooks upon which to hang this analysis. The event however assumes that some communication takes place between the different parties.

3.2 Discussion

The relationship between the state and the rendering indicates the likely effectiveness of the interaction—that is, how well the user can be expected to assess the progress of the interaction. If the rendering is directly connected to the state, the user may readily realise the context of the predictability of certain commands. When there is no rendering, it is more difficult for the user to realise the context of predictability. Informative renderings allow the user to ascertain the values of important features of the state, so assessment of previous interaction can be performed. A system's output is adequate if whenever its state changes this is adequately reflected in the subsequent rendering.

For example, in practice, the drawing package provides feedback of a selection operation by showing the selected object's handles. The only effect of a selection operation is to change the selected object and this change can be interpreted from the display. However, this information is not sufficient to allow the user to predict the effect of the next action in all cases without a (task relevant) memory of the interaction. In event terms, we need the partial task performance notion of predictability. In practice, users of such packages often use a 'click until required object selected' strategy, which obviates the need for them to predict accurately the effect of an action. In terms of design, it is important to understand the degree to which a particular computer system design is likely to be event predictable, even if a strict implementation of system predictability is not a requirement of the design.

In this example, an analysis of possible user strategies in selecting a target object in a graphics system has led to a consideration of the predictability of the computer system and of how the notion of predictability may be formalised. In revisiting the concept of predictability, originally discussed in the previous chapter, our intention has been primarily to develop the links between an agent-neutral view of an interactive system and a computer system oriented view which is of more immediate relevance to design. Our argument is that taking an agent-neutral view of interactional requirements is a valuable first step to developing a deeper understanding of the relationships between the different agents in the interactive system, and of the properties of those agents which influence the course—and effectiveness—of their interaction.

4. Matching interface to expertise

We will now consider a class of computer systems, often referred to as 'walk up and use' systems, which are designed to be used by a wide spectrum of people with different educational backgrounds and varieties of experience of such equipment. A suitable system should provide enough information so that a complete novice can use the system but at the same time provide an interface that will not obstruct an expert user by requiring more interaction than is strictly necessary. Examples of systems that fall on one side or other of the expertise spectrum are some menu based systems that require hierarchies of interaction that are supportive to novices but obstruct experts, and command based systems that provide considerable command optionality for experts but are inscrutable to novices. In order to solve the expert novice conflict a requirement is needed that captures properties of the united behaviour of computer system and user that cannot be captured as a property of only one of the agents. The example we will use in our discussion is that of an Automatic Teller Machine (ATM), which will typically be designed to support domain tasks such as cash withdrawal, balance enquiry and the lodging of requests for cheque books or bank statements (figure 6).

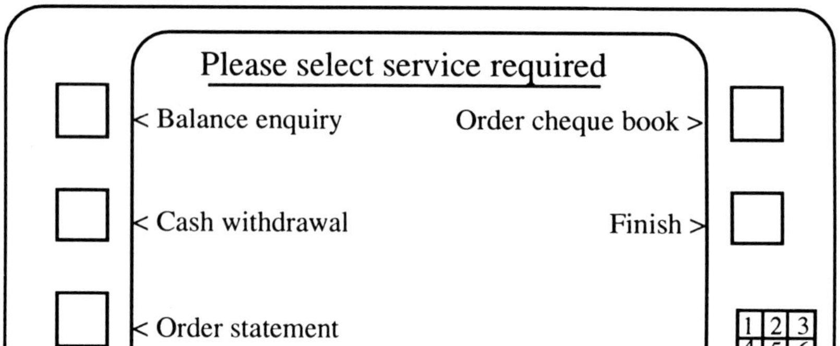

Figure 6: typical ATM screen

An important feature of all system-initiated (expressive) events in this interaction is that the communicative effect of the event is that it should be clear to any user agent what the interaction possibilities at this point in the interaction are.

Assuming that the prerequisites to an interaction are satisfied (e.g. the central computer is functioning and there is a user who wishes to perform transactions), the course of the interaction depends on factors including:

the cash supply in the machine,
the user's domain goals, in terms of banking requirements,
information about the user's status, such as their cash withdrawal limit, and
information of possibilities as displayed on the ATM screen.

An important feature of the overall interaction trajectory is that it is not entirely within the control of either of the interacting agents—some decisions about what event to initiate next are made by the user, and some by the system (based on factors defined as part of the system design). One of the consequences of this is that it might not be possible for a user to achieve her domain goals; for example, the ATM might not give cash when required (either because the user has exceeded her cash withdrawal limit, or because there is no cash available in the machine). This illustrates a difference between domain goals and task goals in terms of the interactional task; although the domain goal (e.g. that the user should obtain the required cash) may not be satisfied, in terms of the interactional task the task goal (e.g. completing the negotiation about one of the banking services supported) should always be completed; indeed, the requirement that all task goals be achieved is a particularly important requirement for banking and other secure systems.

Closely related to the requirement stated above that an expert user should not be required to engage in more interaction than is strictly necessary is a requirement that no user should be expected to engage in extended interaction when the prerequisites for achieving domain goals are not satisfied; for example, if the user requests a cash withdrawal but this is not possible then the user should not be asked how much cash is required. Both of these requirements can be simply restated as requirements that the length of the interaction trajectory should be kept as short as possible. It should, however, be noted that such requirements may be of secondary importance in some circumstances; for example, as a security measure, a UNIX system will request a password even if the user name given is not valid; usability requirements are only one of the sets of criteria which influence the design of the interactive system.

4.1 Agent-neutral analysis and its relation to the system specification

As outlined above, the two main requirements of a 'walk up and use' system are that the computer system should provide enough information to allow it to be used by a complete novice, while also providing an interface which an expert user can use efficiently. In terms of events, these can be stated simply as requirements that each event contains sufficient information to allow the agents to continue to perform the task, and that the length of the trace of events should be as short as possible, consistent with achieving the goals of the task. We note that the expressive event reflects changes to the system and that these system attributes are interpreted by the user. In addition to this model of receptive and expressive events, it is assumed that the system is designed to implement a set of task goals—i.e. to support the user in performing a set of tasks. When describing the properties of a walk up and use system we are concerned with whether the system provides enough information so that an incomplete performance (in the sense that further events must take place to achieve the goals of the task) can be completed.

> A system is *walk up and use* relative to task t if:
> • the current interaction is e such that $t \prec\!\prec E' \triangleright e$ where e is an expressive event;
> • then a receptive event r is generated that continues to be an appropriate performance for t. In other words a 'typical' interpretation of e produces an r such that $t \prec\!\prec E' \triangleright e \triangleright r$. Taking a user's perspective the system expression that is triggered by the trace ending in e can be interpreted by a suitable range of users.

This formulation in terms of events represents a requirement on the system that it is adequate to enable a typical user to generate a sufficient response; that is, the next user formulation should either continue to be a performance of the task or should recover from some error situation. We might wish to be more specific than this and discuss the nature of the *instructional* part of the expression. Support for this requirement might be produced by experiment. It should be noted that this is not a specific property of the user or of the system. The property is an *invariant* that relates the expression to the formulation in terms of events. The attributes of the expressive and receptive functions of the system should be designed so that for a large enough class of user agents this property should hold. Further analysis in terms of models of typical users or a model of the computer system would be required, driven by this invariant. The interaction requirement provides an integrating mechanism for linking the modelling of system and users.

The formulation so far omits the requirement that the trace should be as short as possible, consistent with it being a performance of the task:

> • in addition, the length of the trace which achieves the task goals should be as short as possible; minimise $\#E$.

Expressed in this way, the requirement is as much a requirement on the user agent as on the computer system agent. We return to the topic of the length of traces again in section 6.

4.2 Discussion

In this section, we have used the example of an ATM, illustrating how the user agent must be able to make decisions on the basis of the information available from the screen display and their own banking requirements. In this case, the analysis of the interaction has been used as the basis for a discussion of what it means, in formal terms, for a system to be 'walk up and use'—i.e. usable by a broad spectrum of people without extensive training.

Constraints do still typically exist on the class of users who can make effective use of such machines; for example, the positioning of ATMs often imposes limitations on the height of users (such that people who are unusually short and people who are confined to wheelchairs often cannot access them), and the use of text imposes the requirement that the user must understand the language in which the system is implemented. This issue, and others which are relevant to this example (such as that of how resources are assigned, and how the interaction can be described at different levels of detail), are covered in more detail elsewhere (Barnard et al, 1991).

One final point which it is worth noting in relation to this example is that system expressive events may involve not only information displayed on the screen of a transient nature, but also information which is printed on or near the machine, and which is read and interpreted by a user in the context of the performance of the task.

5. Reciprocal focus

Our next example, the unselected window problem, relates to the mutual focus of requirements for both system and user.

Consider the following variant of the unselected window problem. A user is editing a document with an electronic mail window also open but inactive. The user goes off to lunch and on return attempts to reply to some electronic mail that has come in over lunch. Their initial typed input reply goes into the document they were editing—a classic unselected window error.

As a first attempt at describing the 'unselected windows error', it can be seen that the problem arises over the inability of the computer system agent to detect certain user activity—i.e. going away, and looking at a particular window on the screen. As there are no fixed constraints to enforce synchronisation at this point, disengagement is possible. The user's next command is interpreted by the system agent as referring to the current focus (as perceived by it)—e.g. 'add the text being typed to the document being edited'. Actions by the agent—either autonomous ones (such as walking away) or communicative ones (such as interpreting the latest input / output)—cause that agent to update its focus. These have to be kept in synchronisation for there to be coherent interaction.

It can be seen that from a system perspective, there are user activities which cannot be detected by the system agent—they are not, in interactional terms, events. Whether input is interpreted as referring to the editing or the mailing task is based on the system agent's belief about the current focus. The user's perception of what the current task is depends on their focus, which is influenced by the affordances of the screen and the user's task goals.

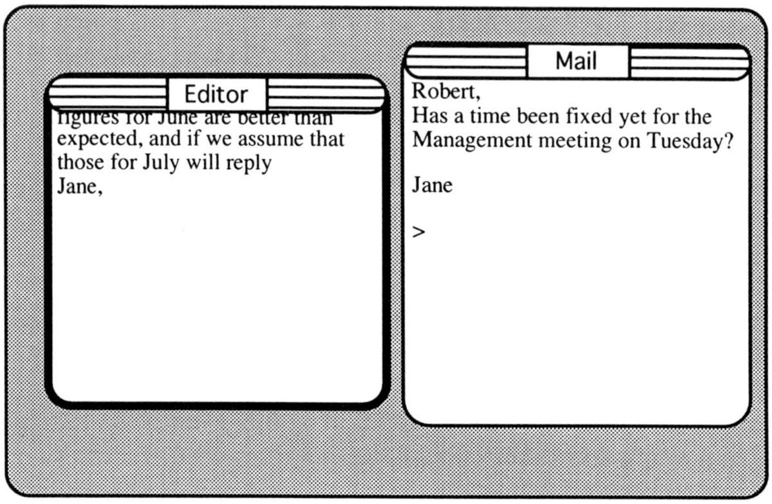

Figure 7: the 'classic' unselected window problem

The user's decision about whether to click to indicate focus shift is again dependent on the screen display—on how 'loudly' the system agent is declaring its current focus.

Disengagement might also occur when the user accidentally activates another window. Here we also have a synchronisation problem; in this case the system agent interprets the opening of a new focus as also denoting the closing of the previous one, which clashes with the user's intention. In this case, although an action by the user agent has been noted and interpreted by the system agent, causing the system's focus to shift, it is not an interactional event because the user agent does not engage with the event. The outcome is again a lack of synchronisation of focus as a result of the disengagement of one of the agents to the interaction. This illustrates that there is a sense in which the phenomenon can arise independently of the properties of the individual agents.

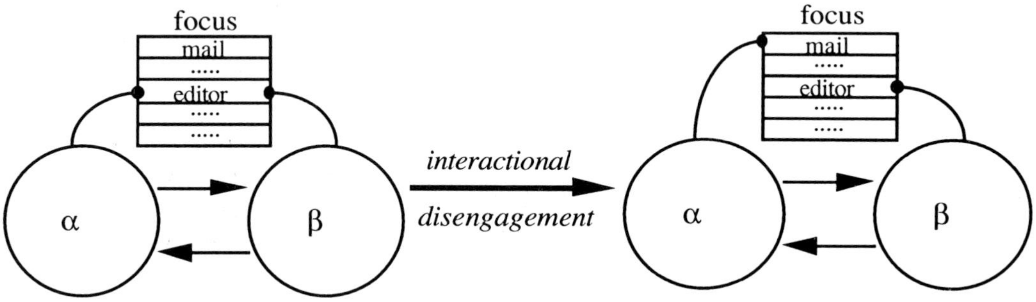

Figure 8: the effect of interactional disengagement on the focus of the agents

The interactional events immediately following the shift in focus of one of the agents constitute a class of *transitional events,* in which one agent communicates to the other its shift in focus. If interactional engagement is maintained, then this is matched by a second transitional event in which the other agent acknowledges and follows the shift in focus. In cases of interactional disengagement, this situation does not hold.

The most salient aspects of the interaction state for the purposes of analysing this problem are the user task goals and features of the screen which give the user cues about the current focus of the system agent.

5.1 Agent-neutral analysis and its relation to the system specification

We wish to specify an interaction requirement over traces that will support the designer's attempt to find a solution to this problem of reciprocal focus. It is important to recognise in this case that the problem of interactional disengagement occurs because we are attempting to interleave a number of performances. The essential property of traces that we shall require is that both the receptive party and the expressive party make appropriate transitions to new agent foci in a synchronised way.

Let U be defined as the set of transitional events $\bigcup_{\alpha \in Agent} \{te_\alpha, tr_\alpha\}$ where *Agent* is the set of agents involved in the interaction. Hence te_α causes a transition to the focus of agent α, and is an expressive event for α, while tr_α also causes a transition to the perspective of α, and is a receptive event.
A system *supports interactional engagement* relative to tasks $t_\alpha \in T$ if for any trace of such a system, E:
- the trace restricted only to the transitional events should consist of pairs of transitional events $\langle u_{\alpha 1}, \bar{u}_{\alpha 1} \ldots u_{\alpha n}, \bar{u}_{\alpha n} \ldots \rangle$ where u and $\bar{u}$ are related so that if u is expressive then $\bar{u}$ is receptive and vice versa.
- agent α's task is performed by the trace restricted to the events which are tied to α, $t_\alpha \prec E \Rightarrow t_\alpha \prec E \uparrow \alpha \;\; \forall \alpha \in Agent$

How does this specification over traces map to system and cognitive modelling? A system modeller would be faced with the problem of user indeterminacy. A user modeller would highlight the attentional problems of signalling window activity. This formulation of the interaction requirement in terms of events provides a means of bringing these together by defining the additional events associated with agent transitions. The cognitive modeller may note that there is a high probability of interactional disengagement when either:

- on resumption after a pause the user's focus changes so that they are no longer ready to engage in the interaction that the system is expecting, or

- discontinuity of input, as occurs for example when the mouse is dislodged, means that the user's focus is clear but now the computer system presents the wrong window as the foreground activity.

The requirement here is that the system should support interactional engagement, particularly in the face of user disengagement.

5.2 Discussion

In the light of these analyses, various solutions to the problem which arises when the user leaves the interaction for a while suggest themselves:

- the first possibility is to block the functionality in a situation where there is potential of interactional disengagement. In this design the user must explicitly select the task stream that is required before any further interaction is possible.
- a second possibility is to reinforce the expression of the task stream selected if there is potential for interactional disengagement. This might be achieved in various ways—for example, by flashing the appropriate window, giving the active window a fizzy border or greying out the non-focused windows. This is a design decision which would be based on the designer's beliefs about the efficacy versus the risk of distraction of the communicative options.

In both cases the design stress is on choosing ways of pushing the user into re-evaluating the expression before choosing the next user formulation, hence forcing a transitional event if necessary. A slightly different set of design solutions, including maybe brief flashing of a newly activated window or blacking out of the previously activated window, would be appropriate to addressing the other source of disengagement discussed here.

The event based notation provides a framework within which experiments can be conducted, and the results fed back into design. A recent experiment has provided some direct empirical support for the argument that an effective approach to the first type of interactional disengagement might be to reinforce the task stream when there is potential for interactional disengagement (Barnard and Harrison, 1992).

Lee (1992) carried out an experiment in which he asked two groups of users to perform a task of buying and selling shares with the aid of a multi-window system. The buying or selling price was displayed in one window and the actual transaction of buying or selling was carried out in another. During the interaction all users were regularly required to deal with a secondary task that they had to perform on a second computer. In the complete task environment, users therefore not only had to move between the windows for the main task, but also had to re-orient their activity from one computer display to another.

One group of subjects used an interface that provided only static feedback about the state of a particular window. In this case, the active window was marked by the presence of a thicker window border. The other group of users received context-sensitive dynamic feedback. In this condition, the dynamic condition, the border for the active window automatically changed from a passive thick border to an active 'fizzy' one following any period of inactivity in excess of 25 seconds. In practice, this meant that when subjects serviced the interruption on the second computer they would almost invariably return to the main task with the active window in this dynamic alert state. This dynamic feedback was transitory. As soon as they engaged in interactions with the active window, its border changed back from the fizzy state to the passive 'thick' border. The 'fizzy activity' marked the interactional state of disengagement. The state of transitional resumption occurs when the user re-orients to the main task and the outcome of this state is the return to a steady state passive border.

The basic result from this experiment was that on those blocks of trials when subjects were being interrupted, those in the dynamic feedback condition made significantly fewer unselected window errors than those in the static condition. There was no difference in the level of these particular errors on blocks of trials when they were not interrupted. This is not surprising because when users in the dynamic condition were not interrupted they received precisely the same feedback as those in the static condition (the dynamic state only occurred after 25 seconds of inactivity).

Closer examination of the data suggested that the advantageous effects of dynamic feedback were not simply confined to cases immediately following re-orientation. Users in the dynamic feedback condition seemed to be making relatively fewer unselected window errors at other points in the interaction when switching between windows. Since these are circumstances where both groups were receiving exactly the same feedback, the results suggest that dynamic feedback may lead users to develop different mental representations relevant to the interaction as a whole. The users of the dynamic feedback system may be more likely to incorporate into their mental representation a constituent explicitly marking the transitional step of activating a window than those in the passive border condition.

This experiment provides evidence to support the suggestion that this design solution is indeed an effective means of addressing one of the possible causes of interactional disengagement. However, the preceding analysis, which is symmetric in terms of the properties of the agents involved, yields insights about a more abstract property of the interaction—that of interactional disengagement—which may be used to inform computer system design.

6. Complexity of interaction

As illustrated by the examples already discussed, an interaction may be viewed as having a point of departure (or initial state), then proceeding through a sequence of events, and having an outcome. The final example we address is a simple dynamic system, a game of computerised tennis. Here, the overall shape of the interactional trajectory might be something like:

Point of Departure: player(s) has/ve money, device available;

1. Engaging the interactional setting, via a specific event (e.g., money in the slot);
2. Establishing constraints (level or difficulty, number of players);
3. Playing the game:
   - ball 1 in play;
   - ball 2 in play;
   - ...
   - last ball in play;
4. Terminating the interactional setting.

Outcome: player satisfaction/dissatisfaction, highest score changed/unchanged.

Such a sequence can obviously be described in increasing detail. So, for example, having a ball in play can be 'unpacked' in terms of its possible points of departure, more detailed properties of the interaction, and its possible outcomes. Although this is largely a setting involving continuous dynamic change, the more detailed analysis can effectively be reduced to significant events, like strikes or the ball crossing the base line and so on.

We will focus the analysis on step 3 above—that of playing the game. Such a game can be described as a series of rallies followed by finishing. The game may be viewed as involving three agents: the two players and the ball. We envisage a design such that either one or two human players can be involved, so the set of agents might be two user agents and one system agent, or one user agent and two system agents. Our analysis will be independent of the properties of the agents.

At a high level of abstraction, the requirements on a game are that it should be varied, challenging, and fun for the participants. In terms of a tennis game, this might be interpreted as a requirement that rallies should be neither too short nor too long, and that neither of the players should totally dominate the game. When considering the interaction, we might choose to specify requirements that the length of rallies should ideally be in the range 3 to 10 strokes and that the players should be well matched (i.e. neither much better than the other).

6.1 Agent-neutral analysis and its relation to the system specification

Representing the tennis game as three agents: two players and a ball, it may be assumed that the ball can engage in actions that generate two events: strikes that originate from each player ($strike_\alpha$,$strike_\beta$)—figure 9. In describing the processes thus we have conveniently ignored components of the interaction that are not needed in the formulation of the interaction requirement in event terms that the rally length should lie in the range 3 to 10; for example strokes that miss or the ball landing out of court are assumed here to be terminating conditions.

The rules of the game require that any rally is an alternation of strikes from the two players (it is no good for one player to hit the ball twice in a row). This rule can be expressed in terms of a property of event sequences quite straightforwardly. We can also express the fact that the length of a trace is between certain limits in the main.

> If an event sequence E partially satisfies the goals of the task ($t \prec\prec E$) and $t \prec\prec E \triangleright \text{strike}_\alpha$ then $t \prec\prec E \triangleright \text{strike}_\alpha \triangleright \text{strike}_\beta$ or $t \prec E \triangleright \text{strike}_\alpha \triangleright \text{strike}_\beta$, $\alpha \neq \beta$. This property expresses alternation.
> $t \prec E$. Then there is a high likelihood that $6 \leq \#E \leq 20$.

These constraints on traces, although derived from some understanding of the game and the capabilities of players and their frustration tolerance, tell us nothing about the capabilities of the individual agents. However it does allow us the possibility of mapping to the interdisciplinary modelling techniques from this relatively neutral perspective.

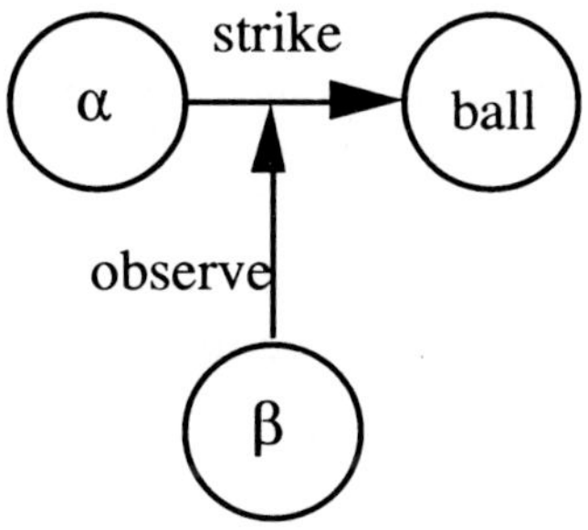

Figure 9: the significant events we are considering

For example, trace specifications can be used as means by which the various modelling approaches can be plugged together and therefore seen as integrated. A cognitive modelling analysis or assessment of perceptual motor skill requirements could be analysed in the context of this description by employing a device model to define the mapping between device inputs (in terms of manipulation and movement of the bat) and the events of the trace. This mapping and the trace specification can be used to analyse what skill or cognitive overheads are involved. In terms of a perceptual motor skill, the rally requires an alternation where user behaviour will be influenced by trajectory tracking and prediction, followed by a movement time to an estimated position (Fitt's Law). In cases more complex than tennis, such traces may serve to determine or bound the forms of cognitive uncertainty that must be resolved within the mental representation that the user must build to govern task execution.

The events can be used as a basis for understanding or implicitly defining the state transitions that the ball must undergo. These events in themselves tell us nothing about the state of the device; it is assumed, for example, that when the ball is out of play no more events can be engaged in as part of the current task, without specifying what 'out of play' might mean. The definition of traces, constrained by interaction requirements, therefore triggers an understanding of the behaviour of the device perhaps by mapping events to state transitions (Sufrin and He, 1990; Duke and Harrison, 1993b).

Rallies may be constrained by further interaction requirements. So, for example, two experts may have consistently high rally lengths. The basic event based notation provides a conduit between the modelling techniques for assessing the capabilities of the agents to the interaction and the modelling and specification techniques that are required to understand and implement the system.

6.2 Discussion

Since this game is particularly simple, we can provide straightforward illustrations of both an interactional principle and, as a requirement on implementation, its potential bridging role for integrating over domain, user and system concerns in an evolving design. Designing a successful game may involve not so much designing for error free performance, but designing a interactional space in which novice, intermediate and expert users may all achieve a positively motivating balance between outcomes on a dimension of satisfaction/ dissatisfaction. We would not want to design an interactional space in which a high proportion of intended users could be expected to achieve a mean stroke value of one or less per ball in play, and neither would we wish to design one in which most rallies were seemingly endless. Rather, we might want to specify an interactional principle, as discussed above, that the average ball in play activity should involve more than three but less than ten strikes for each contributing player.

This is primarily a requirement on the interaction that can be understood *without reference to* either a system model, user model or domain model. It can be formalised in terms of the events that contribute to the interactional state. Its refinement can also be modelled using psychological techniques, such as Fitt's Law, and system techniques, dealing with the computation of the ball's velocity and flight path, to assess whether the requirement is likely to be met by a specific design proposal.

7. Multi-agent systems

Most of these examples have either dealt with two-agent systems, or have treated multi-agent systems in a very simple way. In particular, the important aspects of each interaction have been captured by considering only two autonomous agents at a time. The term *agent* may be used to describe different potentially autonomous perspectives, for example:

- different workstations each involving a user;
- different windows within a single workstation each supporting a thread of activity;
- a collection of processes, some of which are users, while others are systems. An example of such a system would be a power plant.

These different autonomous perspectives shall be referred to generically as a set of agents. The model referred to earlier may be extended to incorporate the notion that an agent is associated with an event, i.e. (in general) is the initiator of the event. It will often be necessary to consider a trace that is restricted to a particular agent as outlined in section 2.3.

An event based notation will only give a limited ability to describe the sorts of interaction requirement that are likely to be encountered; for example, if we take an aircraft flight deck, it is clear that many parallel tasks are being performed. We can associate tasks with restricted sub sequences of the traces of the system; however, it will not be possible to describe the multiple causalities in terms of a simple ordering. When considering multi-agent systems more generally, it will be necessary to define interactions as partially ordered sets based on a causal relationship.

Events that are incomparable in terms of the partial ordering may be considered to have been perceived as independent of one another and are possibly occurring in parallel. Events that are ordered with respect to each other may be considered to have a causal relationship.

> The set of tasks is $t \in T$. A performance of a task t will be $\langle E, \leq \rangle$ where $\leq$ is a partial ordering that defines a causal relationship between events. Parallel events will be incomparable according to the ordering. A particular performance may be denoted as $t \prec \langle E, \leq \rangle$.

All the definitions we have presented so far will carry over to the case of multiple agents with parallel streams of activity.

We have indicated that we may consider a system as a potentially large collection of agents. We leave the problem of analysing this situation to future work, but notice here that if, from the perspective of one agent (figure 3), the rest of the system is deterministic, then the required properties of that system can be reduced to those of a two-agent system. Whether or not the effect of interaction by an individual user is deterministic depends on what else is going on at the same time within the system. An agent's activity, in relation to an intention, will be deterministic if the activity can be carried out irrespective of what else is going on in the system as a whole. In the context of some intention we may require that the user's implementation of the intention is unaffected by what else is going on. This might include the fact for instance that other users are trying to perform another implementation of the same intention.

It is possible to define properties that are agent centred in the sense that they refer to the ways that one agent can carry out activities without concern about what is going on in the rest of the system.

> An interaction is *agent deterministic* with respect to agent α, given some initial interaction u and an agent $\alpha \in Agent$, if for all traces E of the system after it has engaged in u $t \prec u \triangleright E \Leftrightarrow t \prec u \triangleright (E \uparrow \alpha)$.

This definition of agent determinism captures the idea that an agent's use of the system is not affected by the activities of other agents of the system. We might describe an *opaque* system as one which is agent deterministic for all agents in the system. This property is almost achieved by the interactions of multi-access systems such as UNIX (if no files are shared and electronic mail is not used).

> An interactive system is *opaque*, given some initial interaction u if for all traces E of the system after it has engaged in u $t\text{-}\prec u \rhd E \Leftrightarrow t\text{-}\prec u \rhd (E \uparrow \alpha)\ \forall\ \alpha \in Agent$.

Our concern here is to find properties of systems involving multiple tasks. In practice interactive systems in which there is some co-operation are of more interest. The issue here is what can be said when some mutual interference is possible and indeed essential.

In practice, when we consider multi-agent systems, we are concerned with interactions that are much more co-operative than the ones we have considered in this chapter. For example, a pilot in an aircraft is interacting with a large number of agents to maintain a safe flight envelope and to reach the required destination at the required time of arrival with appropriate fuel consumption. The pilot, as one agent in this interaction, is concerned with perturbing the system of agents in a variety of ways in order to achieve the goals. Rather than being involved in an interaction which is entirely deterministic, in which the initiative resides entirely with one of the agents, we have a situation which is more like the resolution of a variety of interfering interactions.

8. Conclusions

All these examples provide only the simplest form of illustration of an approach to understanding computer system requirements in terms of properties of the interactive system. In this chapter, we have illustrated how taking an agent-neutral view can lead to a better understanding of the interactional significance of system properties such as being predictable, 'walk up and use' or 'supporting interactional engagement'. We have also illustrated some requirements, such as those on trajectory lengths, which cannot be readily expressed solely in terms of computer system requirements.

This chapter generates some ground work. It demonstrates that many properties which are widely considered to be features of the computer system have interactional significance and that developing an understanding of this can give a better understanding of these properties, which can be 'fed down' into system design. It also demonstrates that there are properties of interactive systems which cannot be easily understood in terms of only one of the agents to the interaction; particularly for this class of requirements, a good understanding of the relationships between computer system and user modelling concerns is needed. The simple trace model demonstrates how we might describe these interaction requirements as 'invariants' over system and user models. The examples demonstrate that we may need to examine user and system more closely in order to understand requirements more clearly. The trace model provides a framework upon which we can hang these more detailed analyses of the components to the interaction. Finally, we have shown how requirements of interaction involving only two parties is unduly limiting and extended the perspective to multi-agent systems.

References

Abowd, G.D. (1990) 'Agents: communicating interactive processes' in Diaper, D., Gilmore, D., Cockton, G. & Shackel, B. (eds.) *Human-Computer Interaction—INTERACT'90*, pp.143-148, Elsevier Science Publishers.

Barnard, P.J. & Harrison, M.D. (1992) 'Towards a framework for modelling human-computer interactions' in Gornostaev, J. (ed.) *Proceedings of the East-West International Conference on Human-Computer Interaction, EWHCI'92*, pp.189-196, ICSTI, Moscow.

Barnard, P.J., Harrison, M.D., Krening, S.M. & Abowd, G.D. (1991) 'Notes on a methodology (with worked examples) for an integrated design of interactive systems based on behavioural scenarios' *Amodeus Project Document D12*, Applied Psychology Unit, Cambridge.

Carroll, J.M., Kellog, W.A. & Rosson, M.B. (1991) 'The Task-Artifact Cycle' in Carroll, J.M. (ed.) *Designing Interaction*, pp.174-102, CUP, Cambridge.

Dix, A.J., Harrison, M.D., Runciman, C & Thimbleby, H.W. (1987) 'Interaction models and the principled design of interactive systems' in Nicholls & Simpson (eds.) *European Software Engineering Conference*, Springer Lecture Notes.

Dix, A. J. (1991) *Formal Methods for Interactive Systems*, Academic Press.

Dix, A.J. & Harrison, M.D. (1989) 'Interactive system design and formal development are incompatible' in McDermid, J.A. (ed.) *The theory and practice of refinement*, pp.12-26, Butterworths.

Duke, D. & Harrison, M. D. (1993a) 'Rigorous Development of Interactive Systems' *Amodeus2 Project Document SM/WP11*, Applied Psychology Unit, Cambridge.

Duke, D. & Harrison, M. D. (1993b) 'Abstract Interaction Objects' *Proceedings Eurographics '93, Barcelona*, Springer.

Harrison, M.D. (1992) 'Engineering human-error tolerant software' in Nicholls, J. E. (ed.) *Z User workshop*, pp.191-204, Springer.

Harrison, M.D. & Barnard, P.J. (1993) 'On defining requirements for interactions' in Finkelstein, A. (ed.) *Proceedings of IEEE International Workshop on Requirements Engineering*, IEEE, New York.

Harrison, M.D. & Dix, A.J. (1990) 'A state model of direct manipulation in interactive systems' in Harrison, M.D. & Thimbleby, H. (eds.) *Formal Methods in Human-Computer Interaction*, pp.129-152, Cambridge University Press, Cambridge.

Hoare, C. A. R. (1985) *Communicating Sequential Processes*, Prentice Hall International.

Kieras, D.E. & Polson, P.G. (1985) 'An approach to the formal analysis of user complexity' *International Journal of Man Machine Studies*, 22, pp.365-394,.

Lee, W-O (1992) 'The effects of skills development and feedback on action slips' in Monk, Diaper & Harrison (eds.) *People and Computers VII*, Cambridge University Press, Cambridge.

Norman, D. (1986) 'Cognitive Engineering' in Norman, D.A. & Draper, J.W. (eds.) *User Centered System Design*, pp.31-62, Lawrence Erlbaum, Hillsdale NJ.

Shackel, B. (1991) 'Usability—Context, Framework, Definition, Design and Evaluation' in Shackel, B. & Richardson, S. (eds.) *Human Factors for Informatics Usability*, pp.21-37, Cambridge University Press.

Spivey, J.M. (1989) *The Z Notation: A Reference Manual.*, Prentice-Hall International.

Sufrin, B. & He, J. (1990) 'Specification, analysis and refinement of interactive processes' in Harrison, M.D. & Thimbleby, H. (eds.) *Formal Methods in Human-Computer Interaction*, pp.153-199, Cambridge University Press, Cambridge.

Wharton, C., Bradford, J., Jeffries, R. & Franzke, M. (1992) 'Applying cognitive walkthroughs to more complex user interfaces: experiences, issues and recommendations' in Bauersfield, Bennett & Lynch (eds.) *CHI'92 Conference Proceedings*, pp.381-388, Addison Wesley.

Computers, Communication and Usability:
Design Issues, Research and Methods for Integrated Services
P.F. Byerley, P.J. Barnard and J. May (Editors)

Chapter 2.4

Developing the Design Space with Design Space Analysis

Allan MacLean†, Victoria Bellotti† and Simon Shum*

†Rank Xerox EuroPARC, 61 Regent Street, Cambridge, UK — CB2 1AB

* Department of Psychology, University of York, Heslington, York, YO1 5DD, UK

1. Introduction

This chapter reports work carried out as part of the AMODEUS project (Esprit BRA 3066). The goal of the project is to develop interdisciplinary approaches to studying human-computer interaction and to move towards applying the results to the practicalities of design. We describe one of the approaches taken to represent design — Design Space Analysis. One of its goals is to help us bridge from relatively theoretical concerns to the practicalities of design. Design Space Analysis is a central component of a framework for representing the design rationale for designed artifacts. Our current work focuses more specifically on the design of user interfaces. A Design Space Analysis is represented using the QOC notation, which consists of Questions identifying key design issues, Options providing possible answers to the Questions, and Criteria for assessing and comparing the Options. In this chapter, we give an overview of the Design Space Analysis approach and describe a model of the design process which helps guide its use.

Amodeus (Assimilating Models of Design, Users and Systems) is a project which brings together teams from different disciplines with three general objectives (1) to extend the scope of modelling techniques to provide analytic leverage on the problems of user-system interaction: (2) to bridge the conceptual gaps between behavioural and computing disciplines: and (3) to bridge from theory to the practicalities of designing software artifacts. The first two objectives have been addressed by the previous chapters in this section; this chapter reports work aimed at supporting the third objective. It has two goals. The first is to develop a technique for representing design decisions which will, even on its own, support and augment design practice. The second goal is to use the framework as a vehicle for communicating and contextualising more analytic approaches to user-system interaction into the practicalities of design. We have made significant conceptual progress towards both of these goals in the course of the Amodeus project. However, an important pre-requisite for their application is to understand how the basic approach can actually be used in practical settings. That is the main focus of this chapter. We describe a model of the design process which acts as the basis for a simple method to support the systematic development of a design space representation. We continue by first describing the Design Space Analysis framework and the QOC notation. We go on to describe the process model. We then give a more concrete illustration of the process model in use by showing how a design space is developed starting from a description of a design problem. We conclude by discussing how the broader range of approaches represented within Amodeus are likely to fit into this model and the further benefits which that promises.

1.1 Design Space Analysis

Design Space Analysis is an approach to representing design rationale (MacLean et al., 1989, 1991a). It is essentially an argumentation-based approach to design (see Shum and Hammond, 1993). It is a central part of a long term project in which we are interested in helping software designers to reason about design (individually and in groups) and to produce an output which can help others understand why the resulting design is the way it is. Although our intent is to develop a framework for design applicable to a variety of domains, much of our work to date has been aimed at HCI audiences for three main reasons. First, the design domain in which we are most actively working is user interface design. Secondly, we believe that HCI expertise is a crucial ingredient in developing the approach itself as the usability of the tools and techniques we produce will ultimately be critical in determining the acceptability of our work to its users—i.e. the designers of computer systems. Third, and perhaps most importantly, HCI design is much more informal, more open-ended and more poorly understood than some other domains. Reasoning is not always clear and developing the 'knowledge base' by cumulating design rationale promises to be a powerful way to compile contextualised design knowledge.

A key characteristic of our approach is that the output of design is conceived of as a design space rather than a single artifact. The approach therefore contrasts with the traditional conception of design which assumes that the end product is simply a specification or artifact. The final artifact, although embodying the designer's decisions, does not preserve any of the thinking and reasoning which went into its creation. We use a semi-formal notation (called QOC, for Questions, Options & Criteria) to represent the design space around an artifact being produced. This design space is an explicit representation of alternative design options, and the reasons for choosing among those options.

Figure 1 illustrates the relationships between the main concepts we use to represent the design space. Questions highlight key issues in the design space. Options can be thought of as 'answers' to Questions. Criteria are the reasons that argue for or against possible Options. We find that considerable power can be gained simply from regarding relationships between Options and Criteria as relatively positive (solid line between Option and Criterion) or negative (dotted line) — i.e. arguing for or against the Option. Options may spawn off consequent Questions which allow more detailed aspects of the design to be addressed.

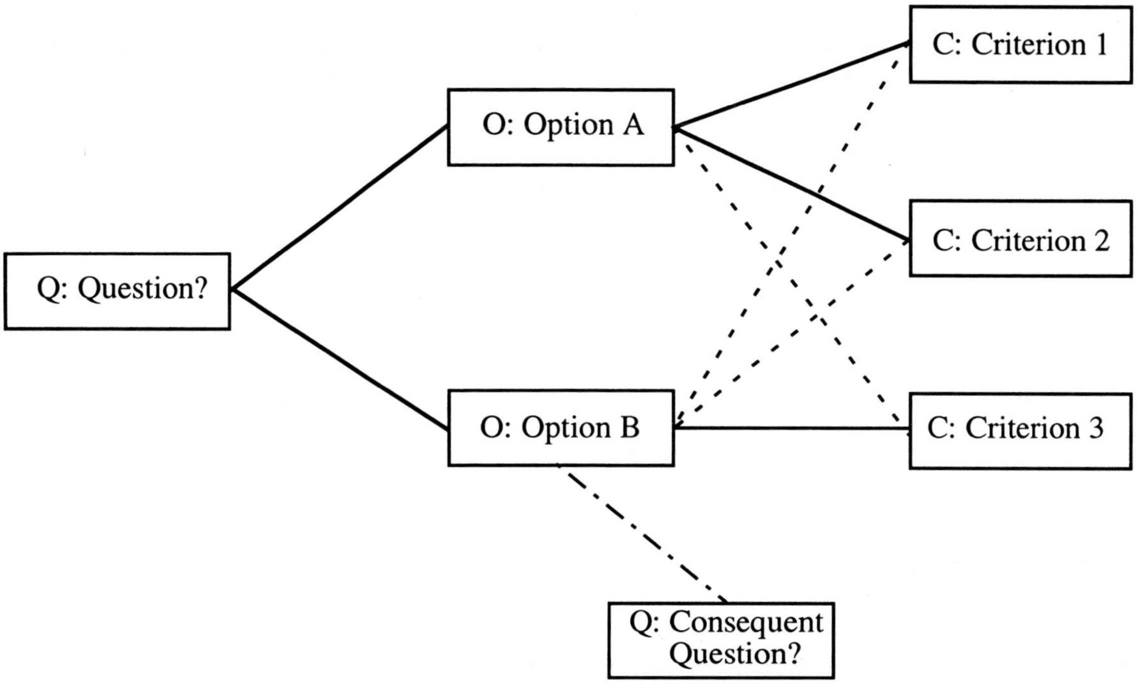

Figure 1. The components of a design space using the QOC notation

This representation of a design space provides a succinct rationale for the final design by placing it in a broader context which highlights how it might be different and why it is the way it is. Such a representation should be able to support communication between people with different backgrounds and goals, for example between members of a design team working on an initial design (see McKerlie and MacLean, 1993a, 1993b), between the original designers and designers of a later generation system who want to re-use parts of the original design, and even between the designers and users of a computer system.

Exploring some of these claims is still a future part of our research strategy. However, Design Space Analysis provides a framework for design which we have already found useful to help better understand design issues (MacLean et al., 1989, 1990, 1991a, b). This theoretical approach drives such activities as understanding the design process and how it can be improved (MacLean et al., 1990); requirements for tools to support the creation of a QOC representation (and thus the design process) (MacLean et al., 1991a); and the integration of other approaches, such as cognitive and system HCI modelling techniques, into software design (MacLean et al., 1991c).

Such motivations, arguments and examples are important for developing the kind of approach we are taking. However, to get practical advantage we need to understand how to actually *produce* a design space representation.

1.2 A model of the design process

As a way of articulating the process of carrying out Design Space Analysis, we have developed a model of the design process which focuses on gathering, organising and reasoning with design information. The model has two components. First, there is a sequential model which consists of five phases ranging from gathering and organising given information, through structuring it into a QOC design space, augmenting the design space and finally making decisions (see figure 2). The second component is a declarative model based on the use of design heuristics which focus on sub-parts of the design representation to assist the designer in developing the space (see figure 3).

Phase 1: Identify relevant information
  Activities:
    Get a feel for the main issues
    Work out what information provided is relevant (& classify as Q, O, and C if possible)

Phase 2: Structure material into rough QOC
  Activities:
    Structure and make sense of the information available
    Find good Questions

Phase 3: Flesh out design space
  Activities:
    Use current understanding of design to help generate new ideas
    Generate new Options
    Generate new Criteria

Phase 4: Reformulate design space to tidy it up.
  Activities:
    Tidy up description and make it more coherent
    Reword Q, O, C if necessary
    Reformulate Questions (and reorganise O, C) to improve decomposition

Phase 5: Make design decisions
  Activities:
    Evaluate and select Options
    Use Criteria to evaluate Options
    Represent decisions by drawing a box around selected Options

Figure 2. The five phase sequential process model and the activities underlying each phase.

The goal of the sequential model in figure 2 is to structure and highlight the main activities. Clearly design is not carried out in the strictly sequential way implied by this model. This is taken account of in two ways. First, the phases in figure 2 should not be interpreted as a strictly serial model which requires one phase to be finished before the next one is started. For example, it may be that some decisions are clear well before phase 5. Nevertheless, there is some logic to the ordering of the phases. For example, beyond the obvious point that you have to have information before you can work with it, we would like to encourage (though not enforce) designers to delay making decisions until they have an adequate basis for making a good decision.

Secondly, we complement the sequential phases with a declarative model based on heuristics derived from the QOC framework. Figure 3 shows some of these heuristics. They focus on sub-patterns within a QOC representation and suggest design "moves" which can be used to explore the design space. They encapsulate ways in which the QOC notation itself can provide leverage or 'cognitive scaffolding' for reasoning within a design space. Being declarative, these heuristics can be applied during any of the sequential phases and avoid the need for postulating extensive iteration between the phases. They therefore provide a mechanism for coping with the opportunistic aspects which we know are a characteristic feature of design behaviour (e.g. see Guindon, 1990). Their main role is to provide a mechanism for developing the design space in a systematic way. For example, *Use Options to Generate Questions* highlights focusing in on an explicit identification of the issue being addressed, which in turn helps to encourage the generation of alternative Options for its solution (*Use Questions to Generate Options*). *Represent Both Positive And Negative Criteria* encourages a balanced view of potential solutions and helps counter confirmation bias (Wason, 1968). If a pattern is observed of all Criteria which have been considered supporting a given Option, this heuristic encourages the designer to think about possible problems with the solution.

Use Questions to Generate Options

Use Options to Generate Questions

Use Options to Generate Criteria

Consider Distinctive Options

Look for Novel Combinations of Options

Represent Both Positive And Negative Criteria

Overcome Negative, But Maintain Positive, Criteria

Design to a Set of Criteria

Search for Generic Questions

Figure 3. Examples of heuristics which form the basis of the declarative component of the design model. (Based on MacLean et al., 1991a)

2. An example: Distributed multimedia services

Let us now look in more detail at the way in which a QOC design space is developed using the process model just presented. The example is taken from a 'design scenario' used in the Amodeus project. A design scenario is a description of a design situation and some of its surrounding context. The kinds of scenario we used could take a number of forms, corresponding to typical design situations. For example, they may highlight a particular problem to be solved, may ask for advice on a particular problem, may ask for assistance in choosing the best of a number of possible solutions, or may simply present a design solution to be critiqued. Such scenarios served as a common focus to allow the variety of approaches represented in Amodeus to be exercised on the same material to enable comparisons to be made among them. The set of scenarios ensured reasonable sampling over a variety of different situations.

2.1 The initial design problem

The particular example used here comes from the 'production interface' developed as part of the RACE project GUIDANCE (further details about the nature and development of this design can be found in chapter 3.3). The starting point is from a situation in which two different designs for providing information about users and work objects in a distributed multi-media CSCW system are being considered. The objects are displayed in a desktop style in both alternative designs, providing information about their type and usage by other people in the system. The main contrast between the designs is that one has people represented in the main display along with the various work objects, and the other has people displayed in a separate window from the work objects. We will refer to these as design A and design B respectively. An important issue which has been identified is the need for users to be able to find out what objects other people are currently working on. The design scenario produced from this situation is reproduced in Appendix A. It provides details about the nature of the design project, its current state, and some of the surrounding context.

The designers are looking for advice on choosing between the alternative designs or suggestions for improving on them. They are also concerned with the implications of dealing with a larger number of work objects or people than can comfortably fit on the screen and would like to know how scrollable displays would affect the situation. They are open to the possibility that the issue identified may not be the best way of looking at the problem.

2.2 Developing the design space

The approach taken by Design Space Analysis in this kind of situation is first of all to identify the relevant information which is available and try to understand its implications by organising it into a design space using the QOC notation. This then acts as the basis for further reasoning over the available information to expand the design space. The models presented in figure 2 and figure 3 above help guide this process. The first step is to identify the relevant information which is known, not necessarily worrying about classifying it in terms of QOC at this stage.

2.2.1 Phase 1: Organise initial information

The analysis was begun by looking at the people, objects and tasks involved in the scenario. The work objects, users and their tasks are the same for both design alternatives in the scenario, but some of the more detailed system operations are different. They can be summarised as follows (see Appendix A, Figure 8 and Figure 9 for more details):

People and Objects from the Scenario (the same for designs A and B)

Users are represented by name-labelled icons showing self (in black) current users (in white) and non-current users (in grey).

System/work objects are represented by icons which show their type including: documents, records, text, video snippets, audio, stills, and music.

The purpose of the system is to facilitate collaboration among users working on the preparation of multi-media documents. A variety of specific tasks can be identified from the description given.

Users' Tasks from the scenario (the same for designs A and B

Co-operative document planning (several users contribute to the same document(s) over time).
Editing and assembling of existing materials using integrated editor functions.
Communication directly between system users.

We summarise these three classes of tasks as:

An individual or group working on multiple objects to try to pull them together
Several people working on one object
Users communicating with each other directly

This may be viewed as communication between people, mediated through more than one, one, or no work objects.

When we look in more detail at how these tasks are carried out with the alternative interfaces, we now see differences between a and b. We can identify three specific activities which are necessary for carrying out the classes of task highlighted in the previous section and which are instantiated differently. These are:

Finding another user

Design A displays all users in the same area with the work objects.
Design B displays all users in a separate list (the ordering scheme is not clear).

Finding out what work objects a person is using

Design A displays the user inside rounded "busy-boxes."
Design B does not do this directly. The viewer has to select the user from the list to see what objects that person is using.

Identifying who is using a "busy" work object

Design A displays an icon representing the person inside the busy-box of each object they are using.
Design B requires you to select a user in the list and then puts busy-boxes around each work object that person is using.

2.2.2 Phase 2: Produce rough QOC

The next step is to represent the information, which phase 1 identified as being important, in terms of QOC. It should be emphasised that the choice of what to represent in QOC is strongly driven by the nature of the design problem. In the present case, the combination of the collaborative tasks to be supported and the contrasting design solutions between the two proposed designs define the focus we take. The pairs of solutions are contrasting Options in QOC terms. The different activities which we identified highlight dimensions along which the Options differ, and therefore form the basis of Questions. The resulting QOC representation is shown in figure 4. In this case, suitable Questions are fairly easy to derive, at least in part because having directly comparable alternatives clarifies the nature of the dimension along which they relate. The reason for not asking Questions about how to support the users' tasks in a more general sense is that this would expand the design space away from the focus which best contrasts the alternative designs.

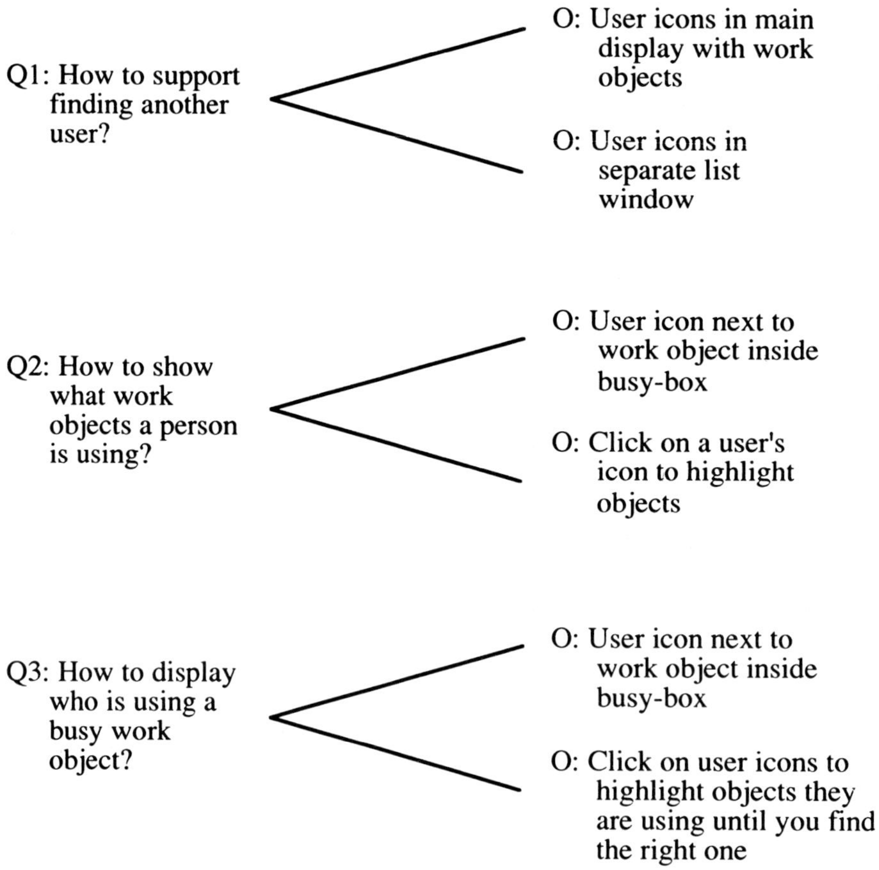

Figure 4. Initial QOC Questions and Options derived directly from the information given in the scenario. The upper Option for each Question is the solution found in design A and the lower Option is from design B.

It should be noted that in general identifying Questions can be one of the most difficult activities in producing QOC (see Bellotti et al. 1991; McKerlie and MacLean, 1993a; Shum (in press). The heuristic *use Options to generate Questions* is aimed towards helping in this situation. The idea is to encourage the designer to identify what is important about a particular Option—for example an aspect which is thought to be problematic for some reason.

2.2.3 Phase 3: Expand the design space

Now that we have a basic design space, the next step is to use the focus it gives us to reason beyond the information available in the description of the design problem. It was interesting to note that no explicit Criteria were given in the scenario description to help us understand why particular options might be good or bad. Finding some of the relevant Criteria is therefore the first thing we need to do for evaluating the alternative solutions.

2.2.3.1 Finding Criteria

Our analysis in phases 1 and 2, which began structuring some of the important issues at stake, leads to the generation of Criteria which, in this case, basically involves a process of going through the scenario designs and critiquing them. Three heuristics are relevant for this activity: that of *using Options to generate Criteria*, i.e. looking for advantages and disadvantages of Options which are being examined; *representing positive and negative Criteria* which aims to ensure that both advantages and disadvantages of Options are considered; and *designing to a set of Criteria* which encourages Criteria relevant for one Option to be considered for others (More generally it is used to help identify a set of Criteria which are globally important for a given design, but that will not emerge with a small example like this one). We will now consider each pair of Options and their implications to see what Criteria we can identify.

Q1: How to support finding another user?

The Option of putting *user icons in main display with work objects*, which represents design A, has *high information content permanently in the display*. This could be useful, for example, if the person looking for the other user also knows which object that person is working on. However, the alternative Option, *user icons in separate window*, representing design B, gets more support because this display structure better *supports an organised serial search* (because it is clear where to look for user icons—although it is not clear exactly how they are organised), there is *less display clutter* with only one icon per user (it is assumed that users of multiple files have multiple icons in A). In addition, a dedicated display of user icons means that the user does not have to search through other icons as well which makes this more *supportive of task relevant information for finding another user*.

Q2: How to display what work objects a person is using?
(Assuming person has been found)

The Option *user icon next to work object inside busy-box* (design A) is supported by a Criterion of *few steps to get information from the display*. Even this may be called into doubt if the person is using many objects (hence having multiple icons) and the display is very cluttered since visual search would have to include reading the name label of every active user icon, and would be prone to omission and repetition of inspection. In contrast, the Option *click on a user's icon to highlight objects* (design B) *supports task relevant information about a particular user*, i.e., what objects they are using. It also generates *less display clutter* than design A, which means that more can be fitted on the computer screen so less scrolling should be necessary. If we were to refine our analysis further in this direction, we might suggest ways of getting all the objects a person is using on to the screen when the user clicks on their name. This could require redisplaying the objects in another window, or making them move onto the existing desktop display.

Q3: How to display who is using a busy work object?
(Assuming object has been found)

In this case the design A Option, *user icon next to work object inside busy-box*, is supported by three Criteria of *low user memory demands* (because the user need not remember the name of the work object), *few steps to get information from the display* (because they only have to look, unless they need to scroll the desktop display) and *support task relevant information on object's status* (because all the information they need is immediately accessible). The design B Option, *click on user icons to highlight objects they are using until you find the right one*, comes off very badly in this assessment because the user has to remember the object while clicking on each of the user icons, which requires many mouse keypresses. Task relevant information is not supported at all. The only Criterion favouring this Option is that of *low display clutter*. Note that the same Option from A appears under Q2 and Q3 and the Option from B is also similar in both cases. This shows that the same solution may be a candidate for different design Questions, which, everything else being equal, might be a reason for favouring it. However, the analysis presented here suggests that everything is not equal, and that design B which was preferable for the other tasks is a failure here. Trying to overload the solution to cover a variety of tasks does not work in this case.

These Criteria and their assessments against the Options are summarised in the QOC diagram in Figure 5. The overall pattern is that design B is claimed to be better on consideration of screen clutter. It also wins on the main user-oriented task of displaying what work objects a person is using because the person is easier to find, and because a person may be using more than one work object. Although the structure of the display does not actively support the search for the objects, it is likely to be easy to pick out the rounded boxes which provide at a glance the task relevant information required. Design A, on the other hand, is much better on the work object oriented task of identifying an object's current user because, once the object has been found, the relevant user information is immediately visible, whereas Design B loses heavily because it requires a highly time consuming search. This failing seems more serious than the two failures with design A, as long as the desktop display area is not very large. If the display area does become large then design A begins to fare much worse on both of its weak points.

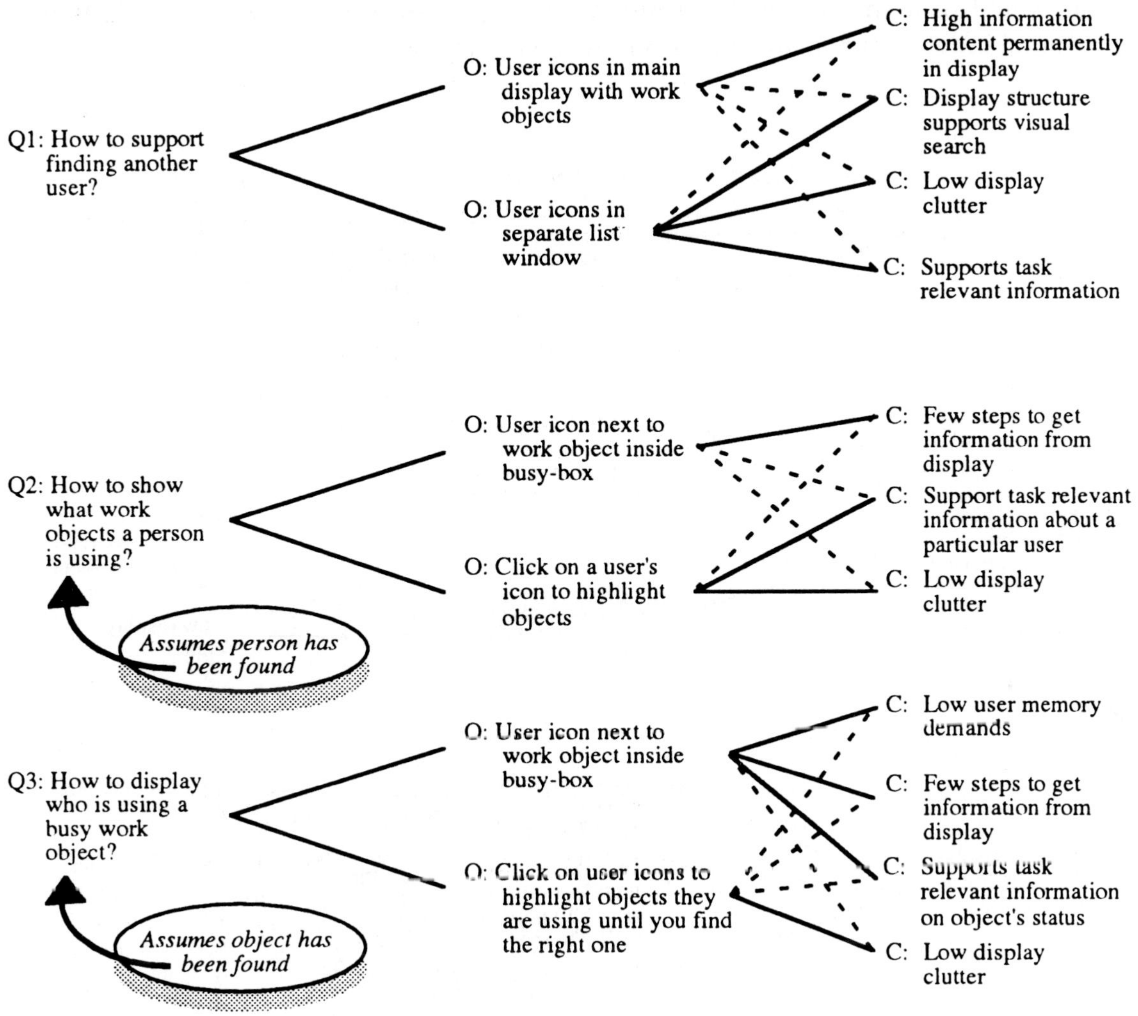

Figure 5. QOC design space with Criteria for and against the Options.

2.2.3.2 *Finding New Options*

It is clear that neither design provides a satisfactory solution for all three Questions, particularly if the desktop display area becomes larger and more cluttered. The next stage is to look for further Options which might lead to a better design. The Criteria just identified are important in this process. The heuristics which help here are *overcome negative, but maintain positive, Criteria* and *use Questions to generate Options.*

For the Question *how to support finding another user,* the *icons in a separate window* was the favoured Option, but although it was better than the alternative it was fairly weak on the Criterion about the *display structure supporting visual search*. One way of improving the Option would be to impose more structure on the user icons—for example placing them in alphabetical order. This new solution is represented in figure 6.

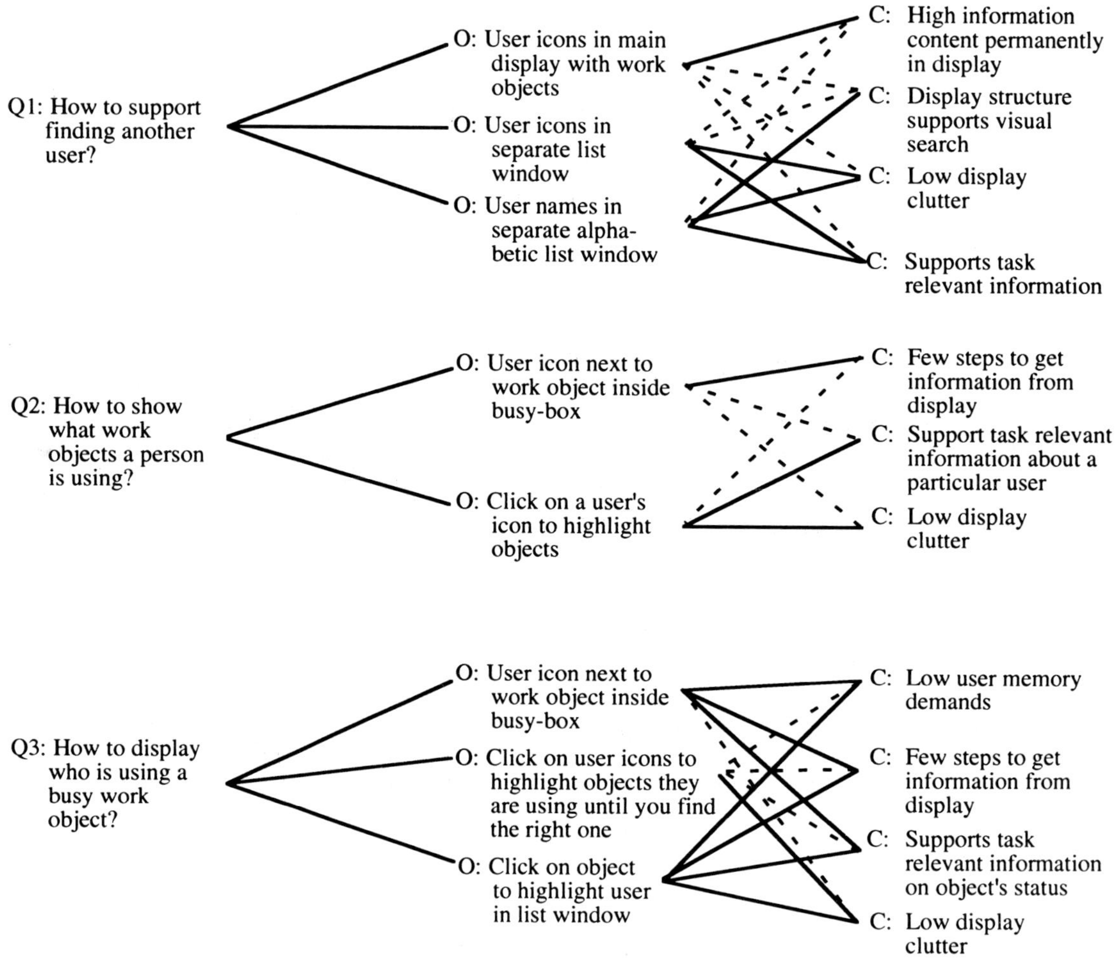

Figure 6. Expanded QOC design space. The new Options are the bottom ones for Q1 and Q3

Note that the assessment against the *display structure supporting visual search* Criterion has now been altered to reflect the identification of this better solution. Given the Criterion of *low display clutter* it is not clear that the icons make the best use of screen real estate. This is increasingly important when the number of users increases and may not all fit within the window. Removing the icons and having a list of user names only would improve the Option. Doing away with the icons here does not hurt the design as they are only really useful in the main display to distinguish people from objects (we have not represented this in QOC, but it could easily be done if the claim was felt to be contentious). In the list, status information could be displayed by the colour (or shade) of labels as easily as with icons. The main advantage is that the names alone are far more compact, thus making it more likely that the user you are looking for is on screen. Having an alphabetical ordering scheme means that, if you know the name you will find them, no matter how many there are. This can be summarised by a proposed Option to put *user names in separate alphabetical list window*. The above argument claims that such an Option would be assessed better than the original design B Option. The result is shown in figure 6.

A major difficulty highlighted by the QOC analysis was that neither original design was better overall in the part of design space examined. They support different activities reasonably well, but all of the activities are important ones for the final system. For Q3: *how to display who is using a busy work object?* an alternative Option might be to *click on the object to highlight name in a list window*. This could also enable you to go on and find out what else that person is using by clicking on the name, which might be useful task relevant information. Such a solution is supported by all of the Criteria for this Question, and is therefore a good example of overcoming the negative assessments against the original Options while maintaining the positive ones.

Further examination might reveal flaws in our new design proposals, but this would draw us away from the scenario issues. Therefore we will not seek to refine the analysis further, although we would hope that the design suggested is an improvement on the original proposals. In this particular case, further refinement of the QOC in phase 4—tidying up the QOC representation is not necessary. This is in part because of the nature of the design problem we have been addressing. Other kinds of problem where it is more relevant will be mentioned shortly. We will now move to the final phase in which we make decisions about the most appropriate design solutions based on our new understanding of the problem and of the design space.

2.2.4 Phase 5: Make decisions

Figure 7 highlights the design recommendations resulting from the analysis. It should be clear from the previous section how the argument has built up to support these choices. The Design Space Analysis helped highlight weaknesses in the original proposals, thereby allowing these to be overcome. In each case we have selected the Option which is supported by the largest number of Criteria. Things are not always this simple, since we may have chosen a solution which had little support from several Criteria but strong support from a very important one. We would then have designed ways to overcome its negative assessments. However, in this case we were able to find solutions by recombining the basic ideas already in the two designs A and B. Note that we do not typically make detailed distinctions about the weighting to be attached to Criteria or assessments in the notation. We typically rely on the supporting argument for that (i.e. the textual description in the present case) — see MacLean et al., (1991a) for further discussion of this point.

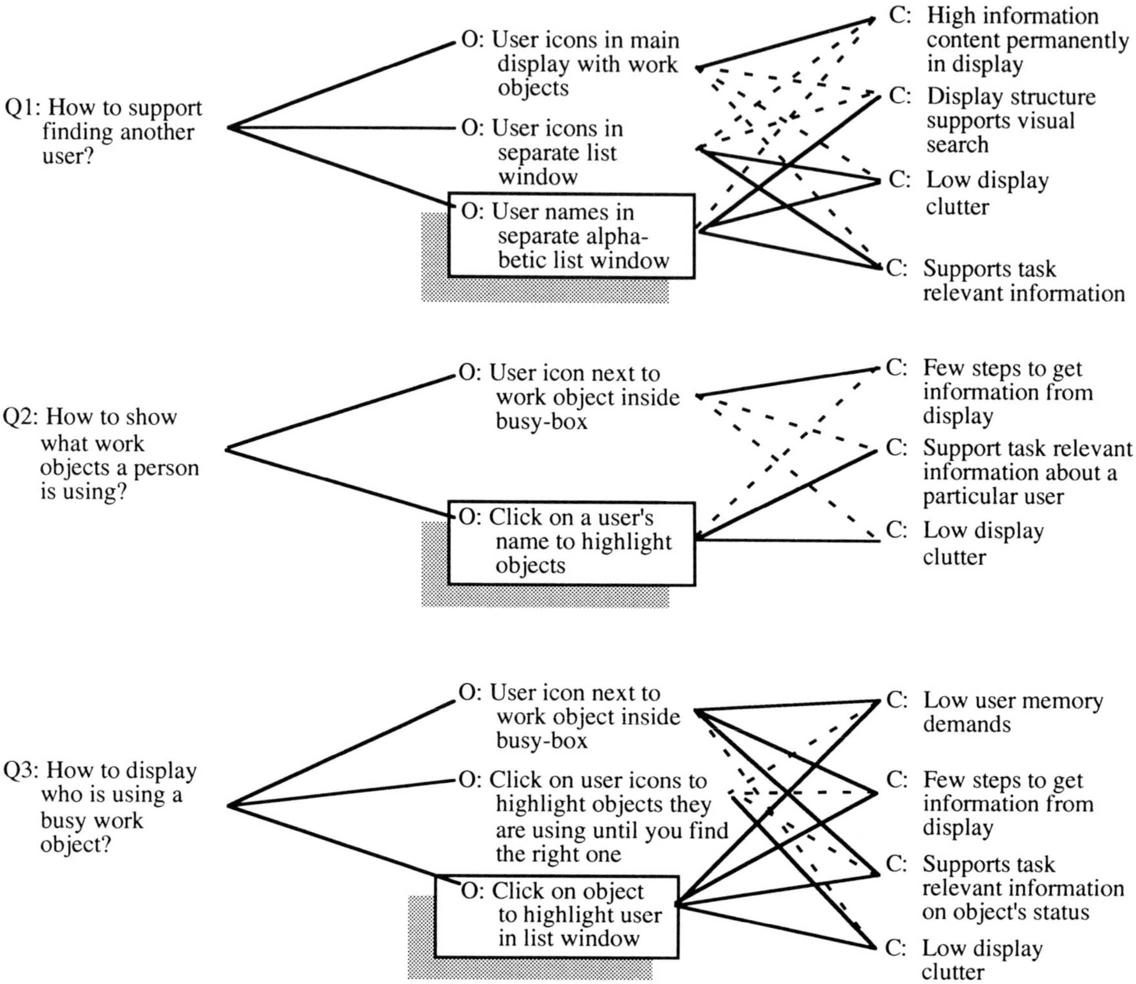

Figure 7. Phase 5; QOC Design Space Analysis and Design Recommendations (shown as boxed Options).

To summarise the proposed solution shown in figure 7, we suggest the provision of multiple ways of accessing both people and objects to allow the user to choose the means to suit their task. The user should then be able to access further information via the most salient object in the current task. Our conclusion favours a separate, compact and alphabetical list of users' names (which could be colour coded as are the icons in both A and B for status information). You should be able to click on a user's name to display the objects they are using, and you should be able to cause the list to highlight the user when you click on a busy work object. This design supports both user-oriented and work object-oriented tasks equally and reduces display clutter considerably, making any visual search that much easier. This resolves the trade-off between these two types of tasks which would have had to be made if one or other of the original designs was accepted.

3. Discussion

The reader should not over-generalise from this example. There are many different kinds of design problem which emphasise different kinds of design activities and are therefore supported by Design Space Analysis in different ways from that illustrated here. The starting point for the current problem was two alternative designs. In such a case, starting a design space is relatively easy as competing Options can be identified fairly readily to give a good start to the QOC representation. This allowed us to present the analysis in quite a lot of detail. Although this is not an uncommon kind of problem, other design problems can be much more open-ended. For example, if we start with a single solution, it can be harder to find an appropriate Question to identify how best to look at the problem. Sometimes in early design there may be no existing solution to improve upon, so still more open-ended exploration would be necessary.

Such situations will have implications for the way in which the process model is used. For example, phase 2 may have much rougher QOC than shown here. When exploring the design space it may be much less clear what are candidates for Questions, Options or Criteria, and fairly extensive re-working may be necessary to produce a coherent QOC. Indeed, for some kinds of problem a rough QOC may be sufficient to get a feel for the issues, and the later phases may be ignored. Similarly, phase 4, 'tidying up' the QOC, may be much more problematic than the example described here. Sometimes the Questions may not be giving adequate coverage of the relevant issues, or may not be organising the design space in an appropriate way (see Bellotti et al., 1991). Fairly significant re-shaping of the design space may be necessary in such circumstances.

Some more discussion of rough QOC can be found in Shum (1991a, 1991b), and some examples of rough QOC and re-organising design spaces can be found in McKerlie and MacLean (1993a). Neither of these references put so much emphasis on the systematic development of the design space as we have done here. However, McKerlie and MacLean (1993a) go into considerably more detail about uses of Design Space Analysis over a period of time in a collaborative project. This highlights benefits we have not discussed at any length here, such as the role the QOC notation can play in assisting problem solving, in providing a basis for sharing and collaboration within a design team and in preserving a record of design deliberations over time. The present chapter has presented the development of a design space in a fairly clinical way in an attempt to communicate how the process models work. The reality within a design project is rather more subtle, both because of the range of different problems which may have to be tackled, as discussed above, and because of the dynamics of the design process itself.

We should emphasise that the intention is not to produce a 'complete' description of a design space using QOC. Rather its value comes from supporting exploration of fairly focused parts of the design space. Some of the situations where producing QOC might be encouraged include poorly understood issues, critical issues, areas where there is disagreement among the design team or where there is an expectation that the solution chosen will not be obvious in the future (e.g. because of a particularly clever solution, or because a short-cut had to be taken). There are many pragmatic reasons for choosing to produce QOC, but as McKerlie and MacLean

(1993a) point out, the goal of design is doing design — not producing QOC. It is therefore important to place effort in areas where there is likely to be benefit, either for the immediate design requirements or for future comprehensibility of the design produced.

Another point worth emphasising here is that the QOC representation is not supposed to be a stand-alone representation. Rather, it summarises the structure of a design space in ways which linear text, diagrams of possible solutions or even prototypes cannot. The terse wording within the various elements is seldom comprehensible on its own. It generally requires further detail to be fully understood. For example, the text in this chapter has served to describe the various QOC figures presented. The value of the notation is to show alternative design possibilities, how they relate to each other and what determines their suitability, in a form which allows these relationships to be examined simultaneously. Compare this to a more traditional representation (e.g. the screen shots in figures 8 and 9). These show more clearly how the components of possible solutions fit together, but do not highlight the differences between these solutions, far less any reasons for these differences. A variety of complementary representations are essential for describing designs adequately (see also McKerlie and MacLean, 1993b).

The key to the technique we describe here for carrying out Design Space Analysis is to focus on identifying and organising relevant information. Design Space Analysis in and of itself has nothing to say about what domain information is relevant or should be represented. It simply provides a framework and a notation within which it can be represented. In the example described here all the initial information came from the description in the design scenario. In other situations it might come from a requirements document or the experience of the designers. Once some initial design information has been assembled, the approach uses that to 'bootstrap' the design space. The distinctions between Questions, Options and Criteria and the design process models presented in Section 1 help focus the designer on parts of the space which might fruitfully be augmented and give some guidance on the form that the augmentation might take — for example, in looking for better Criteria to evaluate the current state of the design, or in looking for new Options which meet the Criteria better.

The way in which Design Space Analysis assists with organising design information is valuable. However, it would be even more valuable to provide support for identifying what information *should* be present and for identifying the implications of information provided. That is the role of other contributions to Amodeus. The system modelling component is centred around work on formal specifications from the computer science tradition (e.g. chapters 2.2 and 2.3). The user modelling component comes from the psychological tradition and two approaches are represented in the project (e.g., chapter 2.1). Finally, CO-SITUE is a framework which provides a systematic basis for the categorisation of the factors that affect user performance (see chapter 2.5). In each case, these approaches emphasise a particular kind of information (to do with the system, the user, or the overall coverage which is required).

In a number of scenarios which have been analysed during the course of the Amodeus project, we have attempted to produce a QOC design space representation which combines this variety of perspectives within a single design space (see Bellotti, 1993 for an example). To oversimplify the conclusions a little, the major initial impact of the other Amodeus approaches on Design Space Analysis is seen in the Criteria they generate. For example, user modelling tends to emphasise Criteria to do with the cognitive processing of information or the knowledge required to use the system. In general, our attempts so far to combine these analyses have been post-hoc, taking the separate results of each individual analysis and relating them to each other using the QOC notation. It would clearly be more powerful to have the contribution of this variety of approaches embedded within the design process. If we consider how this might be done within the process model presented here, there are implications for both the sequential and declarative components. For example, in phase 1 of the sequential model, rather than rely only on the initial design information provided, we could suggest specific kinds of information which should be considered, and perhaps techniques for generating it. In the declarative model, specific kinds of information could perhaps be included. For example, the heuristic *design to a set of Criteria* could be expanded with guidance on how to develop an appropriate set of Criteria using the other Amodeus modelling approaches.

The approaches included in Amodeus promise to help select what available design information is important, and to help identify what other information is necessary but is not given. We believe that they can contribute to a solid information base on which to build a design space. The challenge now facing us is to take the next step beyond the process model presented in this chapter and make that a practical reality.

Acknowledgements

The work reported here was partly funded by the European Commission under the Esprit Basic Research Actions 3066 (AMODEUS) and 7040 (AMODEUS II). We would like to thank Phil Barnard, Jon May, Diane McKerlie and Tim O'Shea for helpful feedback on an earlier draft of this chapter.

References

Bellotti, V., MacLean, A. & Moran, T. (1991) *Structuring the Design Space: The role of Questions in QOC Design Rationale.* AMODEUS working paper, RP6/WP4.1.

Bellotti, V. (1993) Integrating theoreticians' and practitioners' perspectives with Design Rationale. In Proceedings *of INTERCHI'93: Human Factors in Computing Systems, CHI'93 and INTERACT'93, April 24 —29, Amsterdam, Netherlands.* New York: ACM.

Guindon, R. (1990) Knowledge exploited by experts during software system design. *International Journal of Man-Machine Studies,* Vol. 33, 279-304.

McKerlie, D. & MacLean, A. (1993a) *Reasoning with rationale: Experience with Design Space Analysis.* AMODEUS II working paper ID/WP7.

McKerlie, D. & MacLean, A. (1993b) QOC in Action: Using Design Rationale to Support Design. In*Video Program, INTERCHI'93: Human Factors in Computing Systems, CHI'93 and INTERACT'93, April 24 —29, Amsterdam, Netherlands.* (Available in ACM SIGGRAPH Video Review Series).

MacLean, A., Young, R., & Moran, T. (1989). Design Rationale: The argument behind the artifact. In *Proceedings of CHI'89: Human Factors in Computing Systems, April 30—May 4, Austin, Texas,* 247-252. New York: ACM.

MacLean, A., Bellotti, V. & Young, R. (1990). What rationale is there in design? In Diaper, D., Gilmore, D., Cockton, G. and Shackel, B. (Eds). In *Proceedings of INTERACT'90: Third IFIP Conference on Human-Computer Interaction.* (Aug. 27-31, 1990, Cambridge, England). Amsterdam: Elsevier North-Holland, 207-212.

MacLean, A., Young, R., Bellotti, V. & Moran, T. (1991a). Questions, Options, and Criteria: Elements of Design Space Analysis. *Human-Computer Interaction,* Vol. 6 (3&4). [Special Issue on Design Rationale: John M. Carroll and Thomas P. Moran (Eds.)], 201-250.

MacLean, A., Bellotti, V., Young, R. & Moran, T. (1991b). Reaching through analogy: A design rationale perspective on roles of analogy. *Proceedings of CHI '91 Conference on Human Factors in Computer Systems,* New York: ACM, 167—172.

MacLean, A., Young, R., Bellotti, V. & Moran, T. (1991c). Design Space Analysis: Bridging from Theory to Practice via Design Rationale. In *Proceedings of Esprit'91, Brussels, 25-29 November, 1991,* Office for Official Publications of the European Communities, Luxembourg, 720-730.

Shum, S. (1991a). Cognitive dimensions of design rationale. In D. Diaper & N. V. Hammond (Eds.), *People and Computers VI.* Cambridge, England: Cambridge University Press, 331-344.

Shum, S. (1991b). A Cognitive Analysis of Design Rationale Representation. *Doctoral Dissertation,* Department of Psychology, University of York, UK.

Shum, S., and Hammond, N. V. (1993). Argumentation-Based Design Rationale: Conceptual Roots and Current Use. *Technical Report EPC-93-106,*: Rank Xerox EuroPARC.

Shum, S. (in press). Analysing the Usability of a Design Rationale Notation, and its Interaction with Modes of Designing. In T. P. Moran and J. M. Carroll (Eds.), *Design Rationale.* Hillsdale, NJ: Lawrence Erlbaum Associates.

Wason, P.C. (1968) Reasoning about a Rule. *Quarterly Journal of Experimental Psychology,* Vol. 20, 273-281.

Young R. and Barnard, P. (1987) The Use of Scenarios in Human-Computer Interaction Research: Turbocharging the Tortoise of Cumulative Science. In *Proceedings of CHI+GI'87: Human Factors in Computing Systems,* Toronto, Canada. ACM, New York, 21-26.

Appendix A

The design scenario: Distributed multimedia services

A prototype CSCW (Computer supported co-operative work) environment is being developed by a manufacturer which wants to create a market for its integrated multimedia workstations. Among the applications being targeted is the production of multimedia travel brochures by the publicity and marketing departments of international holiday companies.

Task analyses have been conducted on the targeted users. The system should allow users access to a shared database of 'work objects' (elements of brochures, plans about brochures, and comments on both of these), and provide users with both real-time communication and a mail system. The task analyses have differentiated between the 'entities' in the worksystem (i.e. the people who can use it) and the 'objects' (the contents of the database that they work upon), and identified the need for users to have an overview of the worksystem so that they can get information about both entities and objects.

To provide this overview, the designers have decided to use a desktop-style presentation of the objects, with icons representing types of object in the worksystem. Only one person can access an object at a time, and so if an object is being used it is surrounded by an additional rounded border. People in the worksystem are to be similarly represented by stylised head and shoulder icons (that do not attempt to resemble the individual they represent), and which are coloured black if they are the user of the workstation themselves, white if they are other people currently using the system, and grey if they are people known to the worksystem but who are not currently using it. Users can individually arrange the objects on their desktop as they please, but the entities are positioned by the system (see system section below).

The Problem

The task analysis identified the need for users to be able to find out what objects other people were currently working on.

A) *People in object overview (Figure 8).* The person icons could be represented in the object overview, placed next to the icon of the object that they are working on (inside the rounded ' busy' border). When a user stops working on one object and starts working on another, their entity icon would be moved (by the system) next to the object and enclosed with it in a box. (Specifying which object to work on can be achieved with a single mouse click). Entities without proprietary boxes remain adjacent to the object they were last using, even when the user is represented by a white icon, indicating that they are now using some other facility such as mail or the videophone not represented as objects.

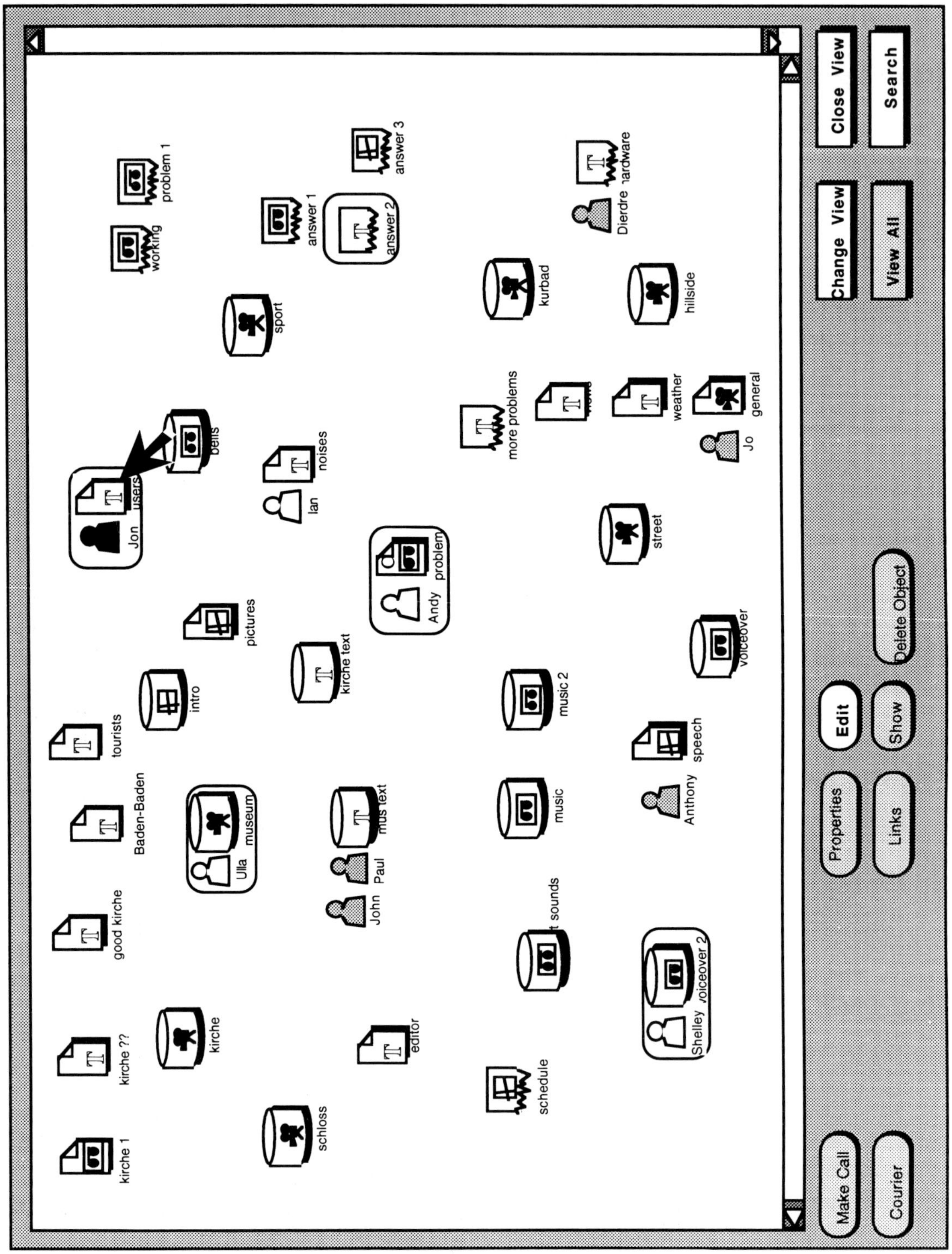

Figure 8 The suggested design with person icons included in the object window.

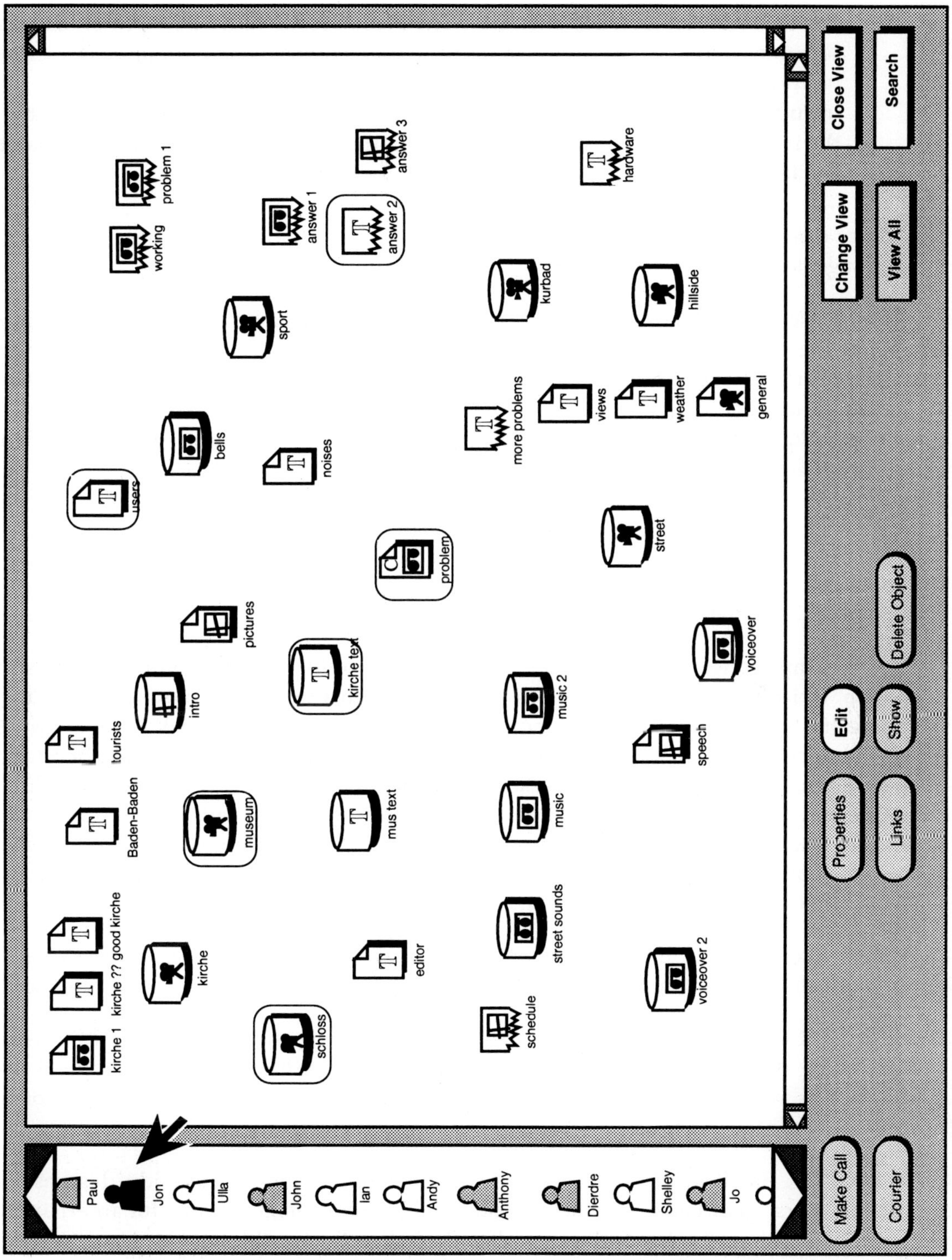

Figure 9 The suggested design with the person icons in a separate window.

B) *People in separate overview (Figure 9).* The person icons could be represented in a separate overview, and users would have to select them and click on a 'Properties' button to find out what they were working on (along with other information including their role and their 'do not disturb' status which would be editable by the individual user). This would mean that the information was not immediately available, and only one person's information could be seen at a time.

What advice can you provide? Are the designers focusing on the correct issues? Could your approach suggest any alternatives?

Subsequent problems: The designers have been working on the assumption that the desktop overview of the objects would be large enough to display icons for all of the objects in the worksystem. For large worksystems, this might not be the case, and the overview would have to be scrollable, as some objects may not be visible within the current view. There could also be too many people in the worksystem to represent in a separate workgroup overview without making that also scrollable. How would this affect your advice?

The Context

System: A prime consideration is that the system should use the company's existing workstation, which incorporates a large colour monitor with a top/centre mounted wide-angle lens camera (for video communication), mouse, keyboard, microphones, speakers, laser printer, scanner and document camera. In both cases, the user arranges the 'objects' on their desktop to suit themselves; 'entities' are located adjacent to 'objects' automatically by the system. The buttons located below the 'objects' windows appear white when available and grey when unavailable: these buttons are associated with a selected 'entity' or 'object'. For both versions this includes communications functions, but additionally status information for the second version. Simultaneous or multiple selection of 'entities' and 'objects' is not allowed.

Target users: Professional publicity and promotional staff, with high levels of task-specific skills in production and layout of paper-based brochures and videos using existing technology.

General tasks: Co-operative planning of documents; editing and assembling of existing materials using integrated editor functions. A typical task situation would involve a user requiring access to an 'object', for example to edit a section of videotape, but having to contact the current user to request access.

Design situation: Small group of designers with stated intention to provide designs that support the users' goal tasks. Product intended as flagship demonstrator for international shows and exhibitions in first instance rather than directly for the marketplace. No client other than the company that makes the workstations.

Open Issues

Information presentation. The applications screen layout is as defined in the figures.. Unconfirmed aspects include the presentation within the workstation environment of the editing windows for the different 'objects': including what control over them, for example in resizing and moving, that the user can exhibit. Also unknown is the form, positioning and precise effects of the "Properties" button mentioned in design B. Design questions which have not yet been addressed include how the system will position entity icons once a user has finished using the application associated with an object, and are succeeded by another user. Also unaddressed (for design A) is the question of what would happen if a user arranges the objects on their desktop in such a manner that when the system updates the entity positionings, another object is occluded by an entity.

Task. Little explicit detail about *why* the users would want to access the 'object' applications, and therefore no parameters about frequency of use or on the ramifications of user-user interaction or protocols for access.

Dynamics of the interaction. Underspecified in terms of the users' action strategies and system responses. We know in general terms what the user has to do, but not the overall structuring of this within the context of an ongoing interaction.

Computers, Communication and Usability:
Design Issues, Research and Methods for Integrated Services
P.F. Byerley, P.J. Barnard and J. May (Editors)
1993 Elsevier Science Publishers B.V.

Chapter 2.5

The structure of the design space

Niels Ole Bernsen

CCI, Risø National Laboratory, PO Box 49, DK—4000 Roskilde, Denmark

1. Introduction

Much has been said and written on the complexity of the problem facing designers of advanced information technologies and applied scientists of human-computer interaction (HCI) alike. The overall conclusion is that there is a huge gap between the complex, real life diversity of users, systems, work contexts, tasks, etc., on the one hand, and anything which has so far been proposed to provide principled, predictive support for the design of usable artifacts on the other. The dilemma for HCI is the following: either such support actually does address in detail part of the complex, real life diversity mentioned, but then its predictive scope turns out to be quite narrow and cannot easily, if at all, be generalised for broader coverage; or the support is maintained at a general level of description in order to maintain broader coverage, in which case its applicability for predictive purposes to concrete design processes is shallow.

This gap may well be unbridgeable. Prescriptive design guidelines will never bridge it and have lately fallen into disrepute exactly because bridging attempts based on them tend to result in an unmanageable complexity of rules and heuristics. The theories and systems available for predictively supporting the detailed design of usable artifacts each have a very limited domain of application and do not lend themselves easily to extensive generalisation (e.g., Card, Moran and Newell 1983, Kieras and Polson 1985, Payne and Green 1986). The basic science of cognitive psychology has proved of less use to HCI than many used to believe a few years ago. There are two main reasons for this. Firstly, for most usability engineering problems, the relevant support from scientific theory does not currently exist. Secondly, even if it did exist, the lessons from the relationship between physics and engineering teach us that applied science is much more complex than simple deduction from basic principles to some real life (design, engineering, or otherwise) problem at hand. Realistic problems exhibit complexity and variability far surpassing the conceptualisations of basic theory.

The physics/engineering example suggests, however, that more principled support for usable artifact design in information technologies might be achievable than is currently the case. To investigate ways and means of providing such support, many HCI scientists are turning towards deeper analyses of full-scale design processes (cf. the papers in J.M. Carroll et al. 1991). The understanding gained is intended to be fed back into requirements for new developments of the science base of HCI, new ways of representing the knowledge in the science base for the purpose of predictively supporting design processes and, ultimately, operational support tools based on science (Barnard 1991). This situation forms the overall context of the Esprit Basic Research project AMODEUS II. The applied science of HCI is no longer viewed as being a matter of directly importing the science base into the design process. In view of the complexity of realistic problems one has, at least for the time being, to be more modest. Predictive support for usability during early design may be a matter of combining designer craftsmanship with (a) advanced rapid prototyping methods involving empirical user testing, (b) focused application of science-based methods where available, be it through consultancy, walk-through support, development of computer-based tools or in other ways, (c) use of relatively low-level and robust conceptual frameworks which can be turned into computer-based tools, and (d) communication with designers through example design studies in which usability issues are analysed in depth, the idea being that 'intuitive generalisation' performed by designers themselves will do some of the work that scientific theory is currently unable to accomplish.

This chapter addresses points (c) and (d) above in presenting first steps towards the development of a robust, top-down framework for characterising the design space within which software information systems designers work, and starting at the coarsest level of description. It presents an idea that is still in development, rather than a mature theory. In section 2, an iteration is made over the issue of what is being designed and the CO-SITUE framework is presented as a coarse-grained analysis, for the purpose of supporting usability engineering, of a central part of the design space within which artifact design takes place. Section 3 presents the full set of components of the CO-SITUE framework and an empty CO-SITUE frame. Sections 4 and 5 describe the application of the CO-SITUE frame notation to the first phase of the design of a particular information service system, i.e., a spoken language dialogue system. Section 6 concludes the analysis and relates the work to another approach currently under development in AMODEUS, i.e. the Design Rationale framework for design space analysis using Questions, Options and Criteria.

2. CO-SITUE and the design space

The question 'What is being designed?' may appear trivial but in fact is not. One may want to say that, obviously, what is being designed are systems or, if one prefers, systems and user interfaces intended to facilitate human performance on specified tasks. As chapter 1.1 describes, this is a simplification. We have learned that normally if not always, the introduction of information technologies as intermediaries between users and their tasks and work domains change the tasks themselves. Systems and interfaces are not simply being tailored to pre-specified tasks. So what *is* being designed? Let us call what is being designed an artifact. What is an artifact? It turns out that artifacts are much more complex entities than suggested above and this fact seems to have implications for the understanding of design practice and for the nature of the design support approaches which are feasible in HCI.

In addressing the question of what is an artifact, let us look at CO-SITUE first. CO-SITUE stands for the following aspects of the artifact which is being designed:

C = Collaborative aspects.
O = Organisational aspects.
S = System aspects.
I = Interface (or more generally: system Image) aspects.
T = Task aspects including task domain aspects.
U = User aspects.
E = User experience aspects.

User performance on the designed artifact will be a function of all these aspects, i.e.:

UPERF = f(C,O,S,I,T,U,E)

Put an artifact designed for one organisational setting into another, different organisation and the artifact may not be used any more. Or change the user population from occasional users into full-time professionals and they may end up being highly frustrated when using the artifact. User performance is the decisive measure of the usability and efficiency of the artifact as a tool for human work. The basic claim behind CO-SITUE, therefore, is that in designing a 'system' what is actually being designed is not merely a system but something much more complex. What is being designed is a particular CO-SITUE complex. Or, what is being designed is an artifact and artifacts are CO-SITUE complexes.

It is important to note from the outset that there are other generic constraints on design tasks than those designated by CO-SITUE. The overall design space framework to be presented in this paper has three main components, i.e., CO-SITUE, the general design goal and a number of general constraints and criteria on the design process. The latter two components are presented in section 3 below. For the moment, let us consider the question whether artifacts are 'really' CO-SITUE complexes. The justification for claiming that this is at least approximately true is the following. Information systems artifacts have an *interface* at which *users* accomplish their *tasks*. In the case of interactive information systems, task accomplishment is interactively shared between user and system and this will be marked by the introduction of a distinction between *system's task* and *user's task* below. Intended users' *experience* is an important parameter in artifact design. *System* performance is a major parameter in early artifact design. Many information systems artifacts are being designed to fit into *collaborative* schemes of work and many artifacts, even though they are not intended for collaborative use, are designed to fit into specific *organisational* frameworks. Each of these CO-SITUE aspects obviously have their own intrinsic complexity. The claim here is that any analysis of design space structure, however coarse-grained, will at least have to include the CO-SITUE aspects. This is also true of design space structure analysis from the particular point of view with which we are concerned, namely, usability engineering. The aim of usability engineering is the optimisation of user performance and the artifact being designed is analysed with this aim in mind. This implies that, e.g., interface technicalities or system technicalities are only relevant to the analysis to the extent that they make a difference to user performance. Otherwise, they are left out of consideration.

During the design process, designers have to consider and actually do consider to some extent and at various levels of detail all the types of aspect relevant to artifact design summarised in CO-SITUE. The artifact is designed the way it is in order to satisfy multiple criteria and constraints an important part of which are derived from a consideration of the CO-SITUE aspects of the artifact. The claim here is not that designers do consider all possible CO-SITUE aspects of the artifact, or all relevant aspects, at appropriate levels of detail, or that they consider the aspects which they do consider in any systematic fashion to make sure that no relevant aspect has been overlooked and left unanalysed. In fact, designers currently have no way of making sure that this happens. Instead, they work with personalised stopping rules and evaluation functions (Goel and Pirolli 1992). The claim is rather that designers actually work within the conception of an artifact designated by CO-SITUE; that is, *CO-SITUE describes an important part of the overall shape of the design space around an artifact during design.* If that is at least approximately true, then CO-SITUE does link up with actual design reasoning and hence might provide common ground among designers and HCI scientists on which to pursue principled and predictive approaches to usable artifact design.

There are two main points involved in the claim that what is being designed is a particular (CO-SITUE) *complex*. Firstly, during the design process the various aspects of CO-SITUE constantly interact. This means that neither normatively nor in actual design practice is there such a thing as first specifying a system and then looking at user requirements, or vice versa, or first specifying a system and then looking into user tasks, or vice versa, or first specifying a system and then specifying its interface. In design reasoning, multiple constraints derived from very different aspects of the artifact to be designed are simultaneously and continuously brought to bear within the design space and user performance on the resulting artifact is a function of the sum of the design decisions made. The resulting physical artifact is an embodiment of a specific CO-SITUE complex. It represents a particular set of commitments within the design space. Which one, cannot be determined from the physical artifact itself through 'reverse engineering' because of the complexity involved. The CO-SITUE frame notation has been designed to capture the specific set of commitments which constitute the emerging artifact during design and are relevant to usability engineering (see below).

Secondly, there is an important sense in which designers design, not only systems and their interfaces but also collaborative and organisational schemes, tasks, users and the degree of experience of users. Usable artifact design is not a one-way traffic of optimising the way constraints from those domains influence the usability of the resulting artifact. It has already been noted by several that computer artifacts change users' tasks (e.g., Norman 1991). However, the point about change is more general than that. As viewed *from within* the dynamic and creative design process, all or most aspects of CO-SITUE are potentially subject to change as a result of design decisions. The reasons are that (1) computer artifacts change work conditions: they change not only tasks but very often also collaborative and organisational schemes of work; (2) artifacts 'change' users in the sense that the types of target end-users and the requirements on their knowledge and experience for dealing with the artifact are themselves to some extent variable design options; and (3) that, rather more obviously, during the design process the actual system to be designed and its interface are themselves variable design options.

In other words, when designing an artifact one constantly has to identify and select between options concerning not only the way the system is to be built, or the way the interface is to be built, but also concerning the possible ways to change the organisational and collaborative work schemes of users, the possible ways to change their tasks, and the possible ways to select the types of end-users and the knowledge and experience they will need for operating the artifact. The artifact which evolves during the design process is a result of design decisions which are normally concerned with all aspects of CO-SITUE. The design space is a field of possibilities or potential changes, and the scope of those changes is CO-SITUE as a whole. Thus, for instance, in an experimental design process where the goal is to generate ideas for concrete applications of a new technology, one may start by designing an artifact with which naive users can easily accomplish tasks a, b and c, and end up with an artifact with which only trained users can accomplish tasks b, c, d and e using an interface very different from the one which was originally intended and requiring a rather different collaborative and organisational context. It is true that most design processes are more constrained than that but this does not affect the general point.

One common, current idea for producing relatively low-level HCI science support of design processes is that of developing taxonomies for tasks, systems, interfaces or users. Such taxonomies might be useful, it is argued, since locating an artifact-to-be-designed within them might facilitate property inheritance from the set of properties characterising a specific type of system, interface, user or task within a particular taxonomy. However, if what is being designed during artifact design are CO-SITUE complexes, it follows that the research goal of constructing *independent* taxonomies for tasks, systems, users, interfaces or possibly also for collaborative or organisational schemes of work, represents a misguided approach to design-applicable HCI. In the design context, tasks, interfaces, users and systems can only be described and analysed with comprehensive reference to one another. It is, therefore, not surprising that no such taxonomies have yet been devised by HCI researchers nor proved successful to design practice. Indeed, the prediction is that such taxonomies are not only misguided, but impossible to construct unless they start combining features of users, tasks, interfaces, etc. (compare Brooks 1991).

The above claims should not be misconstrued as stating that every aspect included in CO-SITUE is a legitimate variable during the design process. CO-SITUE itself is not, nor are, for instance, general aspects of users' cognitive architecture. These are *constants* or *invariants* rather than variables which designers can only ignore at their peril. Bad design often results from overlooking such invariants, from users' limited working memory to the fact that doctors do not want to transfer responsibility to medical expert systems. The existence of such invariants opens up the possibility that further articulation of invariant CO-SITUE aspects of the design space might provide increased support for artifact design (see section 6 below).

If CO-SITUE provides an approximate, coarse-grained characterisation of a central part of the design space surrounding an artifact during design, the design process itself becomes one of making explicit the constraints relevant to the particular artifact being designed and trading them off against one another in handling the design options that become apparent. This way, CO-SITUE gives rise to the definition of interacting design commitments at increasing levels of detail which can be represented in a frame notation supported, whenever necessary, by relevant design documents. The following three sections describe how this process worked during the initial specification phase of a spoken language dialogue system prototype. The prototype is the first of two experimental prototypes of increasing sophistication designed for eventual product development by a Danish research consortium including the Speech Technology Centre, Aalborg University, the Centre for Language Technology, Copenhagen University and CCI, Roskilde. The system will allow users to book flights, change flight reservations and obtain information on Danish domestic flights using spontaneous spoken natural language over the telephone. The design of the first prototype has aimed at ensuring that a usable system can indeed be built given rather hard technological constraints. The design of the second prototype will aim at improving the naturalness of the system. A full account of the initial specification phase is given in Bernsen (1993). Project results are described in Larsen et al. (1992), Larsen et al. (1993), Dybkjaer and Dybkjaer (1993) and Dybkjaer, Bernsen and Dybkjaer (1993). The description below focuses on usability aspects and is analytic (or *post hoc*) rather than temporal and historical. It shows that the artifact is indeed a CO-SITUE complex and how the initial design specification phase can be represented as a limited set of interacting commitments at increasing levels of detail. CO-SITUE has yet to be applied *ab initio* to the detailed temporal structuring of design processes. The analysis reported in this paper was actually used as the basis for the subsequent knowledge acquisition phase of the design project and in defining the knowledge acquisition model (Bernsen 1993, Dybkjaer and Dybkjaer 1993).

3. Overall design goal and design constraints

As said earlier, the overall design space framework to be presented has, in addition to CO-SITUE, two more components. The design task has an overall design goal and a number of generic design constraints. The *overall goal* of the design task is to develop a state-of-the-art spoken language dialogue system prototype operating via the telephone and capable of replacing a human operator. The *generic design constraints* are the following: The design task is constrained, firstly, by limitations in cost, time, manpower, available machine power and so on. There may also be relevant legal, social and political constraints. These are *general feasibility constraints*. Secondly, the design task is constrained by *scientific and technological feasibility*. Thirdly, the artifact to be designed is subject to a number of important conditions concerning its *realism, usability and naturalness*. These conditions help shaping (or constraining) the initial design space within which subsequent design decisions are to be made. The distinction between realism, usability and naturalness can be viewed as a somewhat more structured alternative to the ill-defined notion of 'system usability' or 'system habitability'. Finally, it is valuable for the sake of making the design process transparent to include a fourth type of design constraint, namely, *designer preferences*. What happens during the design process, then, is that the generic design constraints are applied, i.e. interpreted and traded off against one another in a process of interpretation, discovery, justification, trade-off and decision-making, with respect to the overall design goal of designing a particular CO-SITUE complex. The implemented physical artifact is the eventual outcome of this process. Since CO-SITUE aims to support usability engineering rather than design reasoning in general, the constraints from general feasibility, scientific and technological feasibility, and designer preferences will in general only enter into the CO-SITUE representation when they are judged relevant to usability issues during design.

Here follows an empty version of the CO-SITUE notational frame. The distinction between general constraints and criteria (A) and their application within the design space (B) amounts to a distinction between general constraints and criteria, on the one hand, and interpretations of the general constraints and criteria and trade-offs between them with respect to the particular artifact under development (or with respect to the overall design goal), on the other. The frame includes the distinction already mentioned between system's task (T(S)) and user's task (T(U)).

CO-SITUE No. (0)

A. General constraints and criteria
  Overall design goal:
  General feasibility constraints:
  Scientific and technological feasibility constraints:
  Designer preferences:
  Realism criteria:
  Usability criteria:
  Naturalness criteria:

B. Application of constraints and criteria to the artifact within the design space
  C =
  O =
  S =
  I =
  T(S) =
  T(U) =
  U =
  E =

C. Hypothetical issues:

D. Conventions:
  C = Collaborative aspects.
  O = Organisational aspects.
  S = System aspects.
  I = Interface (or more generally: system image) aspects.
  T(S) = System Task aspects including task domain aspects.
  T(U) = User Task aspects including task domain aspects.
  U = User aspects.
  E = User experience aspects.

CO-SITUE No. ( ) indicates the number of the current CO-SITUE specification.

---

4. CO-SITUE exemplified

Let us apply the CO-SITUE framework analytically to the initial specification phase for a spoken language dialogue system. The notational frame itself is indifferent to the order in which information is entered into it. The reason is that, in general, no particular order can be normatively imposed on the way in which frames have to be filled during initial artifact specification. In the present case we are dealing with general, problem-structuring constraints on the design task. Design reasoning at this stage consists in discovering the constraints, developing their implications and making design decisions on this ground. Each decision represents the binding of a variable leading to a more specific conception of the artifact being developed. Let us start with a number of constraints which seem rather straightforward, such as the overall design goal and the following aspects.

4.1 Collaborative and Organisational aspects

Many artifacts are to be introduced into collaborative and organisational work contexts in which it is important to consider the possible changes needed in collaborative and organisational schemes of work in order that appropriate and efficient use be made of the artifact. If such changes are not properly analysed and evaluated, the artifact may eventually not be used at all or may not be used optimally by its intended users. Some artifacts, such as the one described here, may not initially appear to have any clear collaborative or organisational aspects to them. If analysis of the work domain shows this to be the case, then there is no need to take collaborative and organisational aspects into consideration during the design process. As for the current design task, a first analysis of this kind was conducted as a field study in a professional travel agency. A successor to the artifact being designed may eventually end up being used by the agency workers and it has already been established that its introduction would cause changes to their organisational and collaborative schemes of work. These changes must be taken into account eventually. However, it will be assumed in what follows that we are dealing with an 'individualistic' artifact (a SITUE complex).

4.2 Experience and Interface aspects

The *E* aspect of CO-SITUE can be comparatively simple. In the case of our spoken language dialogue system prototype, the design decision was to initially aim at novice users in order to create a fully self-explanatory system. Unless ways were found later to accommodate them, experienced users would therefore have to accept an artifact designed for novice users even though this might force them to listen to (for them) redundant information from the system during each occasion of use.

At the most general level of description, the *I* aspect of CO-SITUE is simple and fixed throughout the design process since we are dealing with spoken dialogue over the telephone. Thus we have:

CO-SITUE No. (1)

A. General constraints and criteria

Overall design goal:

Spoken language dialogue system prototype operating via the telephone and capable of replacing a human operator;

General feasibility constraints:

Scientific and technological feasibility constraints:

Designer preferences:

Realism criteria:

Usability criteria:

Naturalness criteria:

B. Application of constraints and criteria to the artifact within the design space:

C = *Null*

O = *Null*

S =

I = *Spoken telephone dialogue*

T(S) =

T(U) =

U =

E = *Novices*

C. Hypothetical issues:

- *How to accommodate experienced users?*

D. Conventions:

Italics indicate new elements in the design specification (n) as compared with the immediately preceding specification (n-1).

'Null' means that the artifact does not embody a certain aspect of CO-SITUE.

The *C, O, I* and *E* components of *CO-SITUE* have now been fixed to the extent possible at this stage. During the remainder of the design process, *E* fixed as *novice* continues to exert its influence on design decisions unless it gets changed to something else because of novel considerations which might help resolve the hypothetical issue of how to accommodate experienced users. If changes are made to *E*, the changed *E* will then exert its constraining influence on the design process. As a matter of fact such a change happened later during the early knowledge acquisition phase when a promising option was discovered concerning how to make the artifact adapt on-line to experienced users. This illustrates how, during the design process, E is itself being designed and re-designed. CO-SITUE aspect *I* having been fixed as spoken telephone dialogue, on the other hand, will remain fixed throughout the design process. This also illustrates how design commitments, or the assignment of values to variables in the design space, act as constraints on subsequent design reasoning and design decision. The issue about which design decisions are *really* fixed and which ones are more or less fixed or temporary is not important. The significant point is that the notational frame makes explicit all important design constraints and decisions relevant to usability engineering.

5. Criteria and preferences

During initial design specification we want to identify the type of artifact to be designed, specify important aspects of its performance, identify design tools and basic technologies for incorporation into the artifact, and develop the broad overall system architecture. We also want to prepare subsequent phases of the design process. As regards the focal issue of usability, the current design task requires consideration of the linked criteria of *realism, usability and naturalness*. These are general criteria to be used throughout the design process in evaluating different design options prior to the making of design decisions. The following description demonstrates a number of cases in which the criteria of realism, usability and naturalness support the making of early design decisions, sometimes in conjunction with other kinds of constraint on the design process. The net result is a gradual shaping of the design space from the top down.

5.1 Realism criteria

The artifact to be designed should be *realistic*, that is, it should demonstrate the use of the generic technology (i.e., spoken language dialogue systems) in a task domain in which such systems might actually be superior, equivalent or at least acceptably inferior to other known ways of performing similar tasks. Among the preferred task domains for experimental spoken language dialogue systems today are time table enquiries for flights and trains. It was decided to select one such domain, i.e., that of information on flight travels and reservation of tickets to be obtained over the telephone. This task domain is realistic in the sense that, in all or most countries today, the telephone is already being used extensively for the tasks indicated. In principle, the relevant tasks might be mechanised instead by having users perform long series of telephone keystrokes or through the use of screen, keyboard and mouse. However, the first option implies rapidly decreasing usability as the task-relevant dialogues grow in complexity, and the second option is not (yet) available to most potential users.

In summary, then, the realism of the artifact to be designed derives from the facts that it (a) addresses real and known user needs, (b) is preferable to other technological solutions currently available and (c) is assumed to become, from the users' point of view, at worst tolerably inferior in performance than humans performing the same tasks. It is possible that these realism criteria are valid for all artifacts designed to replace humans on some task or set of tasks. Criterion (c) is expanded in the usability criteria below. Note that criterion (a) makes an assumption about user needs. Note also that if the realism criteria cannot be met then the system will be practically useless. This means that realism criteria are normally basic to the decision whether to design a system at all. Finally, we have seen how the realism criteria were interpreted with respect to the task domain of the artifact. The general realism criteria themselves do not state anything specific about the task domain.

5.2 Usability criteria

Roughly speaking, usability criteria serve to ensure that the artifact *can do* the tasks done by the human it replaces or, speaking more generally, that the artifact can serve the tasks it is being designed to serve. If the usability criteria cannot be met then the system will be practically useless. Strictly speaking, usability in this sense sometimes is a matter of degree. In chapter 1.1 two generic usability criteria were presented, and these will be enlarged upon in chapter 3.1 and 3.2. However, it is also useful to think of usability criteria as specific conditions which *unconditionally* have to be met by the artifact, thereby contrasting them with naturalness criteria (see below). This way of conceiving of specific usability criteria emphasises the importance of being able to identify all usability aspects of the artifact during early design. It turns out that many specific usability criteria can be self-evidently derived from or interpreted with respect to the design goal. Usability criteria normally are basic to the decision of whether to design a system at all, unless we are dealing with pure research systems.

In view of the importance of usability to HCI, and to illustrate the actual contents of the design task, the interpretation of usability with respect to the design goal is presented rather comprehensively in what follows. A usable spoken language dialogue system should meet at least the following requirements, most of which are rather self-evident implications of the overall design goal in conjunction with assumptions about users and tasks:

- the system should *understand* all or most of the users' utterances in their appropriate task context as evidenced from its responses to users. In cases where the system fails to understand an utterance, it should be able to repair the dialogue through appropriate responses to the user. This feature of the system is often called 'robustness'. It is self-evident that if a system is not sufficiently robust in this sense it will not be usable;
- the recognised *vocabulary* should be large enough to encompass all or most terms relevant to completing the dialogues necessary for the chosen tasks. If the vocabulary is too limited, then users will have an unacceptably difficult time trying to get the system to do what they want;
- the system's *grammar* should be natural to users, i.e., the system should recognise and understand the varieties of syntactic forms that users find it natural to use during their dialogue with the system. It is an empirical fact that it is practically impossible for users to remain within the bounds of an unnaturally restricted syntax during natural dialogue. This point may be less self-evident than others on this list and hence might be overlooked by designers who are not familiar with the literature on earlier spoken language dialogue systems or with relevant principles of human language processing;
- the system's handling of *discourse phenomena* should be natural to users during their dialogue with the system. The principle behind this condition is the same as in the case of grammar above. The point about lacking self-evidence made in that context holds true here as well;
- the system should *respond to user input* in something not too remote from real time. This is evident from practical usability considerations even without invoking cognitive theories of attention span;

- the system should perform speaker-*independent* recognition of speech as usability in the chosen task domain would be damaged by requiring the training process which is otherwise needed for spoken language understanding systems to adapt to new users. In other related task domains (e.g., voice process control in the flight cockpit), speaker-independence is less important;
- the system should preferably perform *continuous speech recognition* as constraints on users' sentence pronunciation seriously affect the usability of the system in the task domain. Again, in other related task domains this requirement may be less important;
- the system should clearly communicate to users *which tasks* they can accomplish with the system in the chosen task domain and the system should possess the task *domain information* necessary for users to accomplish those tasks. Since the system's performance will be limited as compared to that of human operators, users obviously have to be told what the system can and cannot do. And when the system announces it capability of doing something, it should be able to do it.

Although this list is (mostly) self-evident and states conditions necessary to usability, there does not currently seem to exist any procedure for making sure that it is sufficient. For this reason, moreover, it is possible for different design teams, depending on their collective know-how, to derive somewhat different sub-sets of the necessary conditions for system usability. This, indeed, seems to happen all the time. This is clearly an area in which principled support for usability engineering would be beneficial. Note that the current design process is in some sense a limiting case since the *C* and *O* aspects of CO-SITUE have been set to null and the *I* aspect is extremely simple. In the general case, usability may be affected by all aspects of CO-SITUE.

Let us take a closer look at how the above commitments on usability are derived. The usability criterion is *minimal* in the sense that it states conditions necessary for the system to be at all able to replace a human operator in the task domain. The design commitments derived from it are derived from the overall design goal specification in conjunction with task domain information and information on user needs and users' cognitive capabilities and limitations. An important cue as to how this derivation takes place is the following. In general, when a system is usable, it *can do* the tasks done by the human operator it replaces. This may not be sufficient to ensure ultimate usability (cf. the point made about collaborative and organisational aspects above) but it certainly seems to be necessary. In other words, *to determine specific usability criteria for an artifact during early design, one has to identify conditions such that, if they are not met by the designed artifact then the artifact cannot do the tasks assigned to it given the users assigned to it*, or, in the general case, given CO-SITUE as a whole. CO-SITUE may thus provide some support for the derivation of usability commitments in the design space.

The 'meta-criterion' just stated for the identification of specific usability criteria for a given artifact may provide some support during early design. But this support is not sufficient even with respect to user and task-related aspects of usability. Since, as indicated, important assumptions about the cognitive capabilities and limitations of users may be relevant, the expertise needed for deriving a reasonably complete set of usability criteria during early design is not necessarily present in the designer team and strong penalties may have to be paid later if some crucial usability criterion has been overlooked. Some way of testing, as early as possible in the design process, the cognitive assumptions about users underlying the design of a particular artifact is therefore mandatory. The same applies to other aspects of the artifact denoted by CO-SITUE. CO-SITUE is intended to provide support for the derivation of usability commitments and identification of usability issues which may be in need of empirical testing or other forms of investigation, and a further articulation of the basic structural properties of the design space within which system design takes place may yield additional support (see section 6 below).

CO-SITUE No. (2)

A. General constraints and criteria

Overall design goal:

- spoken language dialogue system prototype operating via the telephone and capable of replacing a human operator;

General feasibility constraints:

Scientific and technological feasibility constraints:

Designer preferences:

Realism criteria:

- *The artifact should meet real and/or known user needs;*
- *The artifact should be preferable to current technological alternatives;*
- *The artifact should be tolerably inferior to the human it replaces, i.e., it should be at least usable (to be expanded in the usability criteria);*

Usability criteria:

- *Make sure that the artifact* can do *the tasks done by the human it replaces;*

Naturalness criteria:

B. Application of constraints and criteria to the artifact within the design space:

C= Null
O= Null
S= *Understand user utterances in task domain;*
Ability to repair if understanding fails;
Large enough task-related vocabulary;
Natural grammar;
Appropriate semantics;
Natural discourse handling;
Close-to-real-time response;
Speaker-independent recognition of continuous speech;
Complete task domain information;
Clear and comprehensible communication of what the system can and cannot do;
I = Spoken telephone dialogue
T(S) =
T(U) =
U= *Need for specific types of flight information;*
Need for natural grammar and discourse handling;
Communication failure in case of delayed response;
Need to use continuous speech in task domain;
Risk of communication failure in case of lacking task domain information;
Risk of communication failure in case of lacking knowledge about what the system can and cannot do;
E= Novices

C. Hypothetical issues:

- How to accommodate experienced users?

In CO-SITUE (2), the realism and usability criteria and their interpretation have led to a number of design decisions on system performance partly based on assumptions about users. It is clear that the commitments made so far do not completely specify a number of artifact properties relating to them. For instance, the task domain has not yet been fully specified; the task-related vocabulary recognised by a usable system might still consist of as little as one or two words; the notion of close-to-real-time response is somewhat vague and so is the notion of speaker-independent recognition. Note also that some of the most self-evident user needs are absent from the list. There is no need to complicate the CO-SITUE description of the evolving design space with trivial claims such as that users need to be understood by the system. This of course introduces an extra element of judgement into the CO-SITUE analysis. However, judgements are traceable from the analysis as long as the *S*-specification contains elements which, because of assumed self-evidence, are lacking their corresponding elements in the *U*-specification. In general, the *U*-specification provides justification for elements elsewhere in the CO-SITUE analysis.

5.3 Naturalness criteria

Usability is a basic standard indeed for the design of interactive computer systems. Arguably, a usable system corresponding to the above commitments on usability could be developed by having a dialogue with users which is totally controlled by the system so that the user will only have to respond by 'yes' or 'no' throughout. However, such a dialogue would not be *natural*, i.e., it would not even remotely correspond to the way in which spoken language dialogues are normally conducted in the task domain. In other words, a natural dialogue requires that users can use language (i.e., vocabulary, grammar, semantics and discourse) to some considerable extent freely during their dialogue with the system.

It is interesting to note that in the case of 'pure' spoken or written natural language interfaces to computers, we actually do have an objective standard of naturalness. We know what natural spoken language communication is like and can easily recognise unnatural limitations imposed on it. This is not the case with respect to most other types of human-computer interfaces. With respect to the current design effort, a fully natural dialogue cannot be allowed given the task domain and the state of the relevant science and technology; so there have to be some constraints on naturalness. In other words, *naturalness can be* - and in fact has to be, in this case just as in most or all other cases of contemporary systems design - *traded for system feasibility* (general as well as scientific and technological). However, just as importantly, *the constraints on naturalness should themselves be principled and natural* in the sense that they can be easily assimilated and practised by users, possibly based on some initial advice communicated to them by the system itself. The trade-offs have to be principled and natural because user experience has already been set to 'novice' and because of the assumption that novice users are able to handle only principled and natural limitations on system performance. Constraints on system naturalness may affect at least the tasks the users can perform with the system, the task domain covered by the system, the mode of user-system interaction and the types of users who can operate the system. The resulting design decisions led to significant reductions in the interactive tasks of user and system:

CO-SITUE No. (3)

A. General constraints and criteria

Overall design goal:

- Spoken language dialogue system prototype operating via the telephone and capable of replacing a human operator;

General feasibility constraints:

Scientific and technological feasibility constraints:

Designer preferences:

Realism criteria:

- The artifact should meet real and/or known user needs;
- The artifact should be preferable to current technological alternatives;
- The artifact should be tolerably inferior to the human it replaces, i.e., it should be at least usable (to be expanded in the usability criteria);

Usability criteria:

- Make sure that the artifact *can do* the tasks done by the human it replaces;

Naturalness criteria:

- *Maximise the naturalness of user-interaction with the system within the scope of feasibility;*
- *Unless a naturalness criterion cannot be met for feasibility reasons, it should be incorporated into the artifact being designed;*
- *Constraints on system naturalness resulting from trade-offs with system feasibility have to be made in a principled fashion based on knowledge of users in order to be practicable by users;*
- *Constraints on system naturalness have to be clearly communicated to users;*

B. Application of constraints and criteria to the artifact within the design space:

C= Null
O= Null
S= Understand user utterances in task domain;
Ability to repair if understanding fails;
Large enough task-related vocabulary;
Natural grammar;
Appropriate semantics;
Natural discourse handling;
Close-to-real-time response;
Speaker-independent recognition of continuous speech;
Complete task domain information;
Clear and comprehensible communication of what the system can and cannot do;
Intelligible, practicable and principled limitations on natural system performance;
I = Spoken telephone dialogue
T(S) = *Make system limitations clear to users from the outset;*
Provide information on and allow booking of flights between two specific cities;
T(U) = *Obtain information on and perform booking of flights between two specific cities;*
U= Need for specific types of flight information;
Need for natural grammar and discourse handling;
Communication failure in case of delayed response;
Need to use continuous speech in task domain;
Risk of communication failure in case of lacking task domain information;
Risk of communication failure in case of lacking knowledge about what the system can and cannot do;
Need for principled limitations on natural system performance;
E= Novices

C. Hypothetical issues:

- How to accommodate experienced users?

What has happened here is that:

(1) More fixed usability commitments have been introduced.

(2) Distinction has been made between the tasks to be performed by the artifact and the tasks to be performed by the user during user-artifact interaction. This distinction comes very natural in the case of natural language dialogue systems but may prove useful to the analysis of other types of user-artifact interaction as well.

(3) A further assumption about users has been introduced.

(4) Trade-offs between naturalness and feasibility have led to initial specifications of system and user tasks.

An important point illustrating the claim that the design process is a process of designing a CO-SITUE *complex* is that questions of feasibility have led to design decisions on user and system tasks. In other words, it will not do in general to assume that we have well-defined tasks to start with and then simply design a system and an appropriate interface in order to support or perform those tasks.

5.4 Feasibility constraints

Let us finally add to the map of the evolving design space some important general feasibility constraints, designer preferences and scientific and technological requirements and constraints, and their trade-offs with the design criteria mentioned already. From a usability engineering point of view, such requirements and constraints should be incorporated into the CO-SITUE notation only if and when they interact with issues of realism, usability and naturalness. There is no need, therefore, to include details of, e.g., the speech recogniser or the semantic frames used to represent the meaning of natural language expressions. The CO-SITUE frame came to look like that shown as CO-SITUE (4). In the current context, the interesting new points in CO-SITUE (4) are the following:

(1) Trade-offs between the naturalness criteria and feasibility have led to a change in system performance specification (*limited* speaker-independent recognition).

(2) A change in the users' task (use single sentences) has been decided. And

(3) a hypothesis has been made about the size of the sublanguage vocabulary needed in the task domain.

(1) Limited speaker-independent recognition was adopted for general feasibility reasons. Given the amount of time available for completing the prototype, it was agreed that *limited* speaker-independent word recognition might be acceptable in the first prototype. This would save an amount of effort in training Markov word models, which might preferably be invested in other aspects of artifact development. Although limited speaker-independence adds to the fragility of system performance, it was decided that limitations in this respect were not of a principled nature, the degree of speaker-independence being largely a function of the amount of work spent on training the system's word models.

CO-SITUE No. (4)

A. General constraints and criteria

Overall design goal:

- Spoken language dialogue system prototype operating via the telephone and capable of replacing a human operator;

General feasibility constraints:

- *9 person/years available for the first version of the prototype;*
- *Limited machine power available;*

Scientific and technological feasibility constraints:

- *Limited capability of current speech and natural language processing;*
- *Open research questions;*

Designer preferences:

- *Use of the Dialogue Design Tool (DDL);*

Realism criteria:

- The artifact should meet real and/or known user needs;
- The artifact should be preferable to current technological alternatives;
- The artifact should be tolerably inferior to the human it replaces, i.e., it should be at least usable (to be expanded in the usability criteria);

Usability criteria:

- Make sure that the artifact *can do* the tasks done by the human it replaces;

Naturalness criteria:

- Maximise the naturalness of user-interaction with the system;
- Unless a naturalness criterion cannot be met for feasibility reasons, it should be incorporated into the artifact being designed;
- Constraints on system naturalness resulting from trade-offs with system feasibility have to be made in a principled fashion based on knowledge of users in order to be practicable by users;
- Constraints on system naturalness have to be clearly communicated to users;

B. Application of constraints and criteria to the artifact within the design space:

C= Null
O= Null
S= Understand user utterances in task domain;
Ability to repair if understanding fails;
Large enough task-related vocabulary;
Natural grammar;
Appropriate semantics;
Natural discourse handling;
Close-to-real-time response;
Limited speaker-independent recognition of continuous speech;
Complete task domain information;
Clear and comprehensible communication of what the system can and cannot do;
Intelligible, practicable and principled limitations on natural system performance;
500-600 words vocabulary;
I = Spoken telephone dialogue
T(U) = Obtain information on and perform booking of flights between two specific cities;
Use single sentences (or max. 10 words);
T(S) = Make system limitations clear to users from the outset;
Provide information on and allow booking of flights between two specific cities;
U= Need for specific types of flight information;
Need for natural grammar and discourse handling;
Communication failure in case of delayed response;
Need to use continuous speech in task domain;
Risk of communication failure in case of lacking task domain information;
Risk of communication failure in case of lacking knowledge about what the system can and cannot do;
Need for principled limitations on natural system performance;
E= Novices

C. Hypothetical issues:

- How to accommodate experienced users?
- *Is a vocabulary of 500-600 words sufficient to capture the sublanguage vocabulary needed in the task domain ?*

(2) It was decided that, during each turn in the user-system dialogue in which users address the system, they should use brief utterances, and preferably single sentences in doing so. This will increase the likelihood of utterance recognition and understanding by the system. At the same time, this constraint on the user-system dialogue seems to be *principled* and thus meets the naturalness criterion. The assumption is that the brevity constraint can be easily understood, assimilated and respected by the intended users during their dialogue with the system.

(3) The close-to-real-time and naturalness design constraints jointly led to the design decision that a vocabulary of 500-600 recognised words should be sufficient for the selected user tasks. Fewer words recognised would constitute an unnaturally restricted sublanguage for the task domain whereas a larger vocabulary would endanger the close-to-real-time design constraint. This design decision clearly rests on a *hypothesis* about the nature of the task domain sublanguage and therefore has been made explicit in CO-SITUE as a crucial hypothetical issue. The hypothesis was tested and confirmed during the knowledge acquisition phase. This testing conformed to the idea that artifacts are CO-SITUE complexes, since the two parameters of user task and system task were manipulated until the resulting user sublanguage vocabulary was smaller than required.

6. Conclusions

CO-SITUE (4) represents both the artifact and the design space surrounding it after completion of initial artifact specification. The representation is the result of making explicit a number of generic constraints and criteria on the artifact to be designed and applying, through a process of interpretation, discovery, justification, trade-off and decision-making, those generic constraints and criteria to all aspects of the evolving conception of the artifact. Each general constraint or criterion, and each design commitment resulting from applying these to the artifact within the design space represent additional constraints on the subsequent artifact specification process.

There is no reason to expect that a series of CO-SITUE frames will be filled in any systematic fashion. Design reasoning, *qua* dealing with open-ended and ill-structured problems, may use a limited commitment-mode control strategy (Goel and Pirolli 1992) under which different aspects of the design problem are being addressed concurrently and interchangeably and no one problem needs to be completely solved before other problems are being addressed. Instead, designers are taking advantage of, often linked, multiple problem-solving contexts in their generation, evaluation, re-consideration and modification of design solutions. As a consequence, successive CO-SITUE frames are generally expected to be incrementally filled in a seemingly random, across-the-board fashion.

In the design process described, subsequent artifact specification was done through two parallel approaches. One was a more detailed specification of system architecture, the other was empirical knowledge acquisition through *T(S)* simulation using the Wizard of Oz methodology. Interestingly, the *S, I, T(S)* and *T(U)* aspects of the artifact resulting from initial artifact specification allowed a direct derivation of an optimal knowledge acquisition model. Knowledge acquisition for the artifact should be done through a simulation of an artifact which reflects the design commitments represented in CO-SITUE (4). If empirical knowledge acquisition is carried out according to this model, there is an optimal likelihood that the results obtained will be valid for the implemented artifact (Bernsen 1993).

A noteworthy aspect of the CO-SITUE notation is its compactness. Despite describing a relatively substantial artifact development effort with a focus on usability issues, most of the important design decisions (excluding only the overall system architecture) and the generic constraints and criteria on which they are based can be represented rather compactly. This no doubt is a virtue of the notation although it should be emphasised that it has been realised through excluding a large amount of design process information. What matters, however, is that what has been included is viewed by consensus in the designer team as constituting the most important criteria, constraints and design decisions adopted so far which relate to usability engineering issues. It can safely be claimed that use of the CO-SITUE notation has been helpful in keeping track of designer consensus during our development of the first prototype dialogue system, and that the existence of the compact design space representation in CO-SITUE (1)-(4) will make a difference to our work on the following, second prototype.

The virtues of CO-SITUE in its current form can be summarised as follows:

- CO-SITUE offers a framework for making explicit the most general aspects, criteria and constraints characterising the design space surrounding computer artifacts and hence enforces a consideration, during artifact design, of each of those from the point of view of usability engineering;
- maintaining a numbered series of CO-SITUE frames during the design process provides a means of explicitly and succinctly capturing the most important design decisions made as artifact design evolves as viewed by designer consensus. When the designer team is large and/or organisationally or geographically distributed, CO-SITUE can be useful in keeping track of designer consensus and the reasoning behind it, including the ultimate justification of the specific design commitments made;
- CO-SITUE in effect provides a series of 'snapshots' of the evolving design process including design space and artifact specification.

Some drawbacks of CO-SITUE in its present form are:

- the CO-SITUE notation does not incorporate the design reasoning linking up CO-SITUE (n-1) with CO-SITUE (n). Instead, design reasoning has been represented above in the form of brief discussions linking each consecutive pair of CO-SITUE frames. Ways should be sought of providing more structure to the representation of design reasoning (see below);
- CO-SITUE in itself offers little in terms of scientific theory. This is unfortunately true of all current and realistic approaches to artifact design support. What CO-

SITUE offers so far is an analysis of artifacts and the design space surrounding them at the highest level of generality, and a notation for analytic and dynamic capture of design decisions combined with informal representations of the reasoning behind them.

Some open issues are:

- how does CO-SITUE work when its comes to representing the details of tasks and interfaces of the artifacts being designed?
- how much extra overhead on the design process does use of the CO-SITUE notation imply?

More open issues will be pointed out below. It is the aim in the context of the AMODEUS project to address these drawbacks and open issues of CO-SITUE as it stands. The following two sub-sections describe the relationship between CO-SITUE and the Design Rationale approach to design space analysis using Questions, Options and Criteria (DR/QOC for short) and the issue of a 'level 2' articulation of CO-SITUE, respectively.

6.1 CO-SITUE and Design Rationale

DR/QOC represents an approach under development in AMODEUS to providing additional structure to design reasoning. The idea is that Design Rationale can be captured through design space analysis in terms of Questions, Options and Criteria (QOCs), which helps provide explicit analytic structure to the design space surrounding an artifact (MacLean et al. 1990, MacLean et al. 1991; and chapter 2.4 of this book). CO-SITUE and DR/QOC are complementary approaches and the current aim is to combine them into a framework for usability engineering which can be turned into a computer-based tool. The main similarities, differences and complementarities between CO-SITUE and DR/QOC in their present forms are presented below.

6.1.1 Similarities

Both DR/QOC and CO-SITUE represent forms of design space analysis which, generally speaking, is a natural style of reasoning to designers. Both approaches are therefore potentially useful to the structuring of design discussions. Both are analytical accounts of design processes rather than historical documentations of them. Both representations provide useful documentation for supporting effective communication within design projects, with future design teams responsible for extending or maintaining the artifact, with experts in science-based approaches to artifact evaluation, and possibly also with end-users. The representational formats involved are simple enough to be understood by a variety of people. On the other hand, both types of design space analysis face the challenge of having to offer more in cost-benefit terms than familiar design representations such as generating scenarios, sketching possible solutions, listing required attributes, using rapid prototyping and producing prototypes. Finally, the fact that design space analysis is a natural style of reasoning to designers of course does not imply that their reasoning maps directly into the CO-SITUE and DR/QOC notations and their underlying conceptual structures. More research is needed on this issue which is crucial to the eventual adoption by designers of CO-SITUE and/or DR/QOC.

6.1.2 Differences and complementarities

The main strengths of DR/QOC as compared with CO-SITUE are that DR/QOC-style design space analysis explicitly aims at supporting, during the design process, the development of an organised space of alternative and innovative designs or possible solutions, as well as facilitating the analysis of weaknesses of proposed solutions. Such DR/QOC representations might be directly incorporated into the CO-SITUE frame notation. However, it seems preferable to develop DR/QOC representations in the form of separate links between the individual CO-SITUE frames in a series representing a particular design process. The design process would then be analytically represented for usability engineering purposes as the series:

CO-SITUE(1)→DR/QOC(1)→CO-SITUE(2)→DR/QOC(2)→...→CO-SITUE(n)→DR/QOC(n)

The representation DR/QOC (n) would provide the design reasoning which justifies the design decisions made between CO-SITUE (n-1) and CO-SITUE (n). Furthermore, DR/QOC (n) would not necessarily have to represent a comprehensive analysis of every design decision made between CO-SITUE (n-1) and CO-SITUE (n) but might focus on the most important decisions for which it would offer a detailed analysis, whereas other decisions might be discussed in the informal style used in the present paper.

CO-SITUE, on the other hand, attempts to make explicit the overall semantic structure of design spaces, i.e. to systematically conceptualise the aspects of the artifact being designed which need to be taken into account from a usability engineering perspective. DR/QOC does not represent overall design space structure. Linked to that, and in contrast to most DR/QOC work done so far, CO-SITUE represents a top-down approach to design processes in that it aims at capturing those constraints and criteria, and their gradual application within the design space, which are most important from a usability engineering point of view. Outside of this unfolding context, it can be claimed, DR/QOC representations make little sense as an aid to design. The unfolding design context involving general criteria and constraints and the design commitments based on these are potentially relevant to any subsequent design decision to be made during the design process. This context provides the design space structure which often carries the specific style of the artifact and is used in justifying the more specific design arguments which are applied to the solution of specific design problems. It follows that the general criteria, constraints and design commitments dealt with by CO-SITUE necessarily have to be represented also in a DR/QOC-style approach.

Linked to these two points, again, is the fact that CO-SITUE explicitly captures design constraints which do not directly express usability issues but which, nevertheless, affect usability issues by enforcing trade-offs with the criteria of usability and naturalness. It may be claimed that capturing such constraints is necessary to the analytic capturing of design reasoning even though they sometimes trade off *against* the naturalness of a particular design. Indeed, the apparent opposition between purely historical design process accounts and purely justificational accounts is fictitious. Any historical account will contain justifications, and any justificational account, however analytically expressed, will contain history. Since design problems are ill-structured and do not have unique and correct solutions, what matters is to provide as much analytic structure to them as possible in order to support the justificational strength of their solutions. Finally, it is an important question for further work whether a CO-SITUE representation of the design process can facilitate identification of those design issues for which a detailed DR/QOC analysis could provide the most benefit. This is an as yet unsolved problem within the DR/QOC approach, the problem being that a 'complete' DR/QOC design space analysis of medium- and large-scale design processes would probably, for most types of design processes, require too much effort to be realistic in cost-benefit terms.

The similarities, differences and complementarities between the DR/QOC and CO-SITUE approaches noted above would seem to justify the attempt to combine them into an explicit and analytic representational framework for design processes.

6.2 Level 2 Articulation of CO-SITUE

If CO-SITUE as presented above is considered a general level articulation, or a *Level 1* articulation, of the structure of design spaces, the question arises whether a *Level 2* articulation of design space structure is possible. If approximately valid, CO-SITUE captures the most general aspects of design spaces and artifacts. These aspects are thus invariants involved in any computer artifact design process. As remarked above, such design processes have to reckon with many more invariants characterising artifacts and design spaces. The problem is to articulate these invariants in a systematic fashion which both maintains a sufficiently high level of generality and is directly relevant to practical design reasoning. Such invariants are potentially of at least two kinds:

(1) *A taxonomy of CO-SITUE complexes* describing design-relevant combinations of aspects of organisations, collaborations, systems, interfaces, interactive tasks and their domains, and users. As argued above (section 2), independent individual taxonomies of these aspects are probably neither relevant to artifact design nor possible in the first place. It is an empirical question to what extent a taxonomy of CO-SITUE complexes is feasible. Their articulation would probably have to proceed via:

(2) *The identification of general invariant structures* of organisations, collaborations, systems, interfaces, interactive tasks and their domains, users and features of the novice-to-expert development process. This would constitute a Level 2 articulation of CO-SITUE and forms the

subject of ongoing work. In the particular context of AMODEUS, the Cognitive Subsystems framework developed by Barnard is an obvious candidate for articulation of the *U* aspect of CO-SITUE (see chapter 2.1).

It is an open question whether much further useful articulation of the general criteria and constraints of realism, usability, general feasibility, and scientific and technological feasibility is possible or worthwhile. However, it seems likely that the naturalness criterion might profitably become further articulated. There clearly is work to be done on how to define naturalness with respect to multi-modal interfaces, for instance.

We cannot expect systematic HCI theory to ever overcome the gap between general constraints and criteria, and general CO-SITUE-like aspects of design spaces and artifacts, on the one hand, and the immense complexity of detail met with when interpreting constraints, criteria and aspects with respect to particular artifacts or design goals, on the other. However, some further systematic, top-down articulation of constraints, criteria and CO-SITUE aspects might prove useful in practical design work as one possible way of making the design process more rational, or more systematic, or more transparent, or more amenable to receiving input from an appropriately developed basic science. The CO-SITUE framework is a candidate approach along these lines but it is clearly too early to make strong claims about its capabilities to deliver.

Acknowledgements:

Many thanks are due to Phil Barnard, Hans Dybkjaer, Laila Dybkjaer, Allan MacLean and Tove Klausen for helpful comments on earlier drafts. The work was done as part of the Esprit Basic Research project 7040 AMODEUS whose support is gratefully acknowledged.

References

Barnard, P. (1991): Bridging between basic theories and the artifacts of human-computer interaction. In J.M. Carroll (Ed.): *Designing Interaction. Psychology at the Human-Computer Interface.* Cambridge: Cambridge University Press.

Bernsen, N.O. (1993): Design of a Spoken Language Dialogue System. A Study of the Initial Specification Phase. *Esprit Basic Research Action AMODEUS Working Paper* RP3-WP3.

Brooks, R. (1991): Comparative Task Analysis: An Alternative Direction for Human-Computer Interaction Science. In J.M. Carroll (Ed.): *Designing Interaction. Psychology at the Human-Computer Interface.* Cambridge: Cambridge University Press.

Card, S.K., Moran, T.P., and Newell, A. (1983): *The Psychology of Human-Computer Interaction.* Hillsdale, NJ: Lawrence Erlbaum Associates.

Carroll, J.M. (1991): *Designing Interaction. Psychology at the Human-Computer Interface.* Cambridge: Cambridge University Press.

Dybkjaer, L., Bernsen, N.O. & Dybkjaer, H. (1993): Knowledge acquisition for a constrained speech system using WoZ. In *Proceedings of the EACL.*

Dybkjaer, H. & Dybkjaer, L. (1993): *Wizard of Oz Experiments in the Development of the Dialogue Model for P1.* Report 3, Spoken Language Dialogue Systems. STC Aalborg University, CLT Copenhagen University and CCI Roskilde University.

Goel, V. and Pirolli, P. (1992): The structure of design problem spaces. *Cognitive Science* 16, 395-429.

Kieras, D.E., & Polson, P.G. (1985). An approach to formal analysis of user complexity. *International Journal of Man-Machine Studies,* 22, 365-394.

Larsen, L.B., Bernsen, N.O., Broendsted, T., Dybkjaer, H., Dybkjaer, L., Music, B., Povlsen, C., and Ravnholt, O. (1992): *Spoken Language Dialogue Systems. A Survey of the State-of-the-Art.* Report 1, Spoken Language Dialogue Systems. STC Aalborg University, CLT Copenhagen University and CCI Roskilde University.

Larsen, L. B., Broendsted, T., Dybkjaer, H., Dybkjaer, L., and Music, B. (1993): *Overall Specification and Architecture of P1.* Report 2, Spoken Language Dialogue Systems, STC Aalborg University, CCI Roskilde University, CST University of Copenhagen.

MacLean, A., Bellotti, V. and Young, R. (1990): What Rationale is There in Design? In Diaper, D., Gilmore, D., Cockton, G. and Shackel, B. (Eds.): *Proceedings of INTERACT '90: Third IFIP Conference on Human-Computer Interaction.* Amsterdam: Elsevier North-Holland 207-212.

MacLean, A., Young, R., Bellotti, V. and Moran, T.P. (1991): Questions, Options, and Criteria: Elements of Design Space Analysis. *Human-Computer Interaction* Vol. 6, pp. 201-50.

Norman, D.A. (1991): Cognitive artifacts. In J.M. Carroll (Ed.): *Designing Interaction. Psychology at the Human-Computer Interface.* Cambridge: Cambridge University Press.

Payne, S., & Green, T. (1986). Task Action Grammars: A model of the mental representation of task languages. *Human-Computer Interaction,* 2, 93-133.

Section 3

Synthesis for current design practice

Jon May

MRC—APU, 15 Chaucer Road, Cambridge, UK—CB2 2EF

The first two sections of this book have set out the need for requirements in design, and described a number of ways in which research is providing systematic ways of assessing requirements. In this final section we wish to show how requirements can be successfully used to guide the design process.

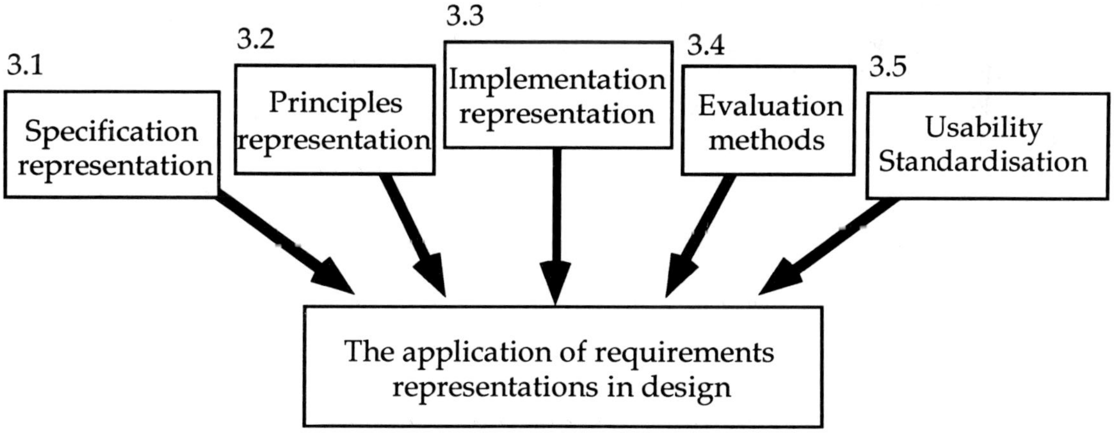

The first three chapters describe the Enabling States approach to design, which was developed by the RACE project 'Guidance'. This approach embodies the 'internal perspective' of the eight-step method presented in chapter 1.1. The latter four steps are represented as a sequence of 'specification tables' which allow requirements to be recorded in a structured fashion as they develop from a consideration of the end-users' task context—chapter 3.1 covers the conceptual background to the approach and describes the structure of the specification tables. Support in completing these tables is provided in the form of design principles, which allow human-factors knowledge to be targeted to the appropriate decision points in the design process. The derivation, content and application of these principles is covered by chapter 3.2. The third chapter reports the use of the method and the principles in the implementation of two prototype multimedia systems, one of which also allowed multiple users to collaborate. The experience of the specification and implementation teams was central to the development of the method, and this chapter copntains examples of completed specification tables to show how the requirements feed into implementation.

Once a prototype has been implemented, it must be evaluated in order to assess how well the requirements have actually been met, and chapter 3.4 reviews evaluation approaches in order to highlight general issues for the practitioner, and makes recommendations on the optimisation of evaluation practice. Finally, chapter 3.5 describes the relationship between usability and standardisation, showing how the results from RACE and Esprit are being taken up by the international standards bodies. This chapter also debates the usefulness of standardisation as an approach, examining the way that designers actually use standards. A comprehensive glossary of standards bodies and their inter-relationships is given to help practitioners unfamiliar with the area unerstand the often complex path that usability infomation must take before it is disseminated as a 'standard'.

Together these chapters show how research has enabled requirements to be generated in a structured fashion, together with principled human factors support, and used to drive the implementation effort. A coherent set of requirements benefits evaluation, and can be used to enhance the value and impact of usability standardisation.

Computers, Communication and Usability:
Design Issues, Research and Methods for Integrated Services
P.F. Byerley, P.J. Barnard and J. May (Editors)

Chapter 3.1

The enabling states method

Jon May[1], Paul F. Byerley[2], Ian Denley[3], Becky Hill[3], Shelley Adamson[4], Pat Patterson[4] and Leif Hedman[5].

[1]MRC—APU, 15 Chaucer Road, Cambridge, UK-CB2 2EF

[2]Alcatel-SEL AG, Lorenzstraße 10, D-7000 Stuttgart 40, Germany

[3]Ergonomics Unit, University College London, 26 Bedford Way, London UK—WC1H 0AP

[4]Roke Manor Research, Roke Manor, Romsey, Hampshire UK

[5]Telia Research, Box 85, S—20120 Malmö, Sweden

1. Introduction

This chapter describes, step by step, an approach to design that has been developed to provide specific design support to the designers of Integrated Broadband Communication (IBC) systems, and which can be applied more generally to the design of interactive systems. The method focuses on the internal design perspective that is described in chapter 1.1, beginning with an analysis of the usage context and the human end-user's goal tasks, and produces a specification of the operational requirements for a user-centred design. At each stage in the method specific design support is available in the form of parameterised principles (these principles and their application are discussed in detail in chapter 3.2).

The chapter begins by considering the problems of integrating two or more services into a single design, and then describes a distinction between the tasks that an end-user wants to do with a design and the tasks that the design forces them to do. This distinction, and the concept of the 'enabling state', form the basis of two 'criteria of usability' that must be met if a design is to be usable. The design method that is proposed has been derived to provide support to designers in meeting these criteria.

2. The goal task—enabling task distinction

Chapter 1.2 has shown how IBC will allow users to achieve much more than just talking to other users—they will be able to communicate audio, video or data information to other people, or to machines; and they will be able to do so in two-way conversations, or by sending messages, or by retrieving stored information. It also described how the very potential and power of IBC will cause many problems for the design of usable systems. End-users will face problems caused by lack of integration between services, and these problems may make the integrated systems' usability even worse than existing independent services.

To solve the problems of integration described in chapter 1.2, it is essential to remember that people become 'users' of systems because they have something to do: that is, they have some *goal task*. The raison d'être of the system that they turn to is to assist them in performing this goal task. Whatever support the system provides to the user, however, it is almost inevitable that the user will have to communicate with the system to give it information about their goal task—perhaps they will have to tell it what their goal task is, or exactly how they want to carry it out, or they might decide to change their goals once they have started to use the system. In fact, the greater the functionality of a design, and the wider the range of possible goal tasks that it supports, the harder it will become for a system to make assumptions about the support it is to provide, and the more information will be required from the user.

The problem that this raises for designers is that users will not be interested in providing this information, since it apparently does not directly help them achieve their goal. From the user's point of view any tasks that they must perform that involve telling the system what to do, configuring it, making choices about courses of action that it suggests, or even telling it that the goals have been achieved and that it can stop, are not goal tasks. They are tasks that must be carried out to enable the goal tasks to be performed, and it is because of this that we must make the distinction between *goal tasks* and *enabling tasks*.

Consider a system as simple as a bank's Automated Teller Machine (ATM). A person approaching this might simply have the goal of walking away with a sufficient quantity of money withdrawn from their account. The design of even the simplest ATM in common use in the UK, though, requires them to:

- push a button to specify the amount of cash that they want
- to insert a plastic card into a slot to identify their account
- to type in a security code to identify themselves
- to take their card out of the slot again
- to wait while their cash is counted and a receipt slip printed
- and only then can they take their cash from a second slot and walk away.

While each of these steps can be considered as tasks at one level of granularity or another (and we will discuss the problem of task decomposition in section 5.2) only the last in the list actually corresponds to their goal task—the other five are all enabling tasks that, while quite necessary from the point of view of the function of the system, the user would be quite happy not to have to perform. If any of these tasks could be omitted from the overall task sequence, the ATM would become 'easier to use' as far as they are concerned. If the interaction is represented using the conventions presented in chapter 2.3, to show the sequence of actions and interactions that are involved in this transaction, the distinction between the enabling tasks and the goal task is clear (figure 1). Since the emphasis in chapter 2.3 was on describing the machine or system side of an interaction, the enabling tasks are represented in much more detail than the goal tasks (and to avoid confusion, it should be made clear that in this chapter we are using the term 'goal task' to describe *only* the user's goals, whereas in chapter 2.3 the 'goals' were those of both the user and the machine, in their interaction).

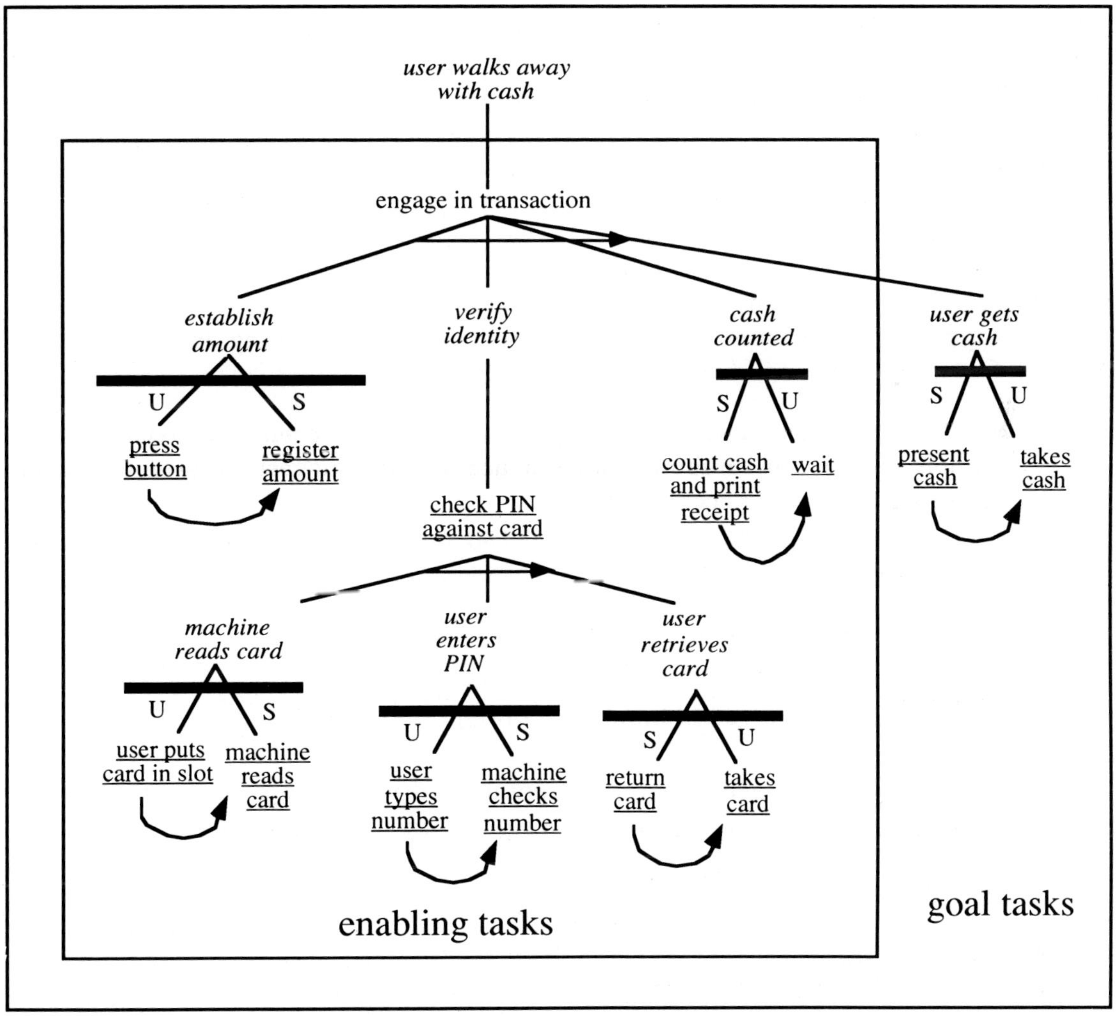

Figure 1: the ATM interaction represented using the conventions of chapter 2.3, with the goals of the user and system (i.e., machine) in italics, and actions underlined (S=system, U=User).

The ATM task sequence described here is taken from a 'rapid cash dispenser' whose operational requirements must have emphasised speed and security of transaction above all else. The designers have faithfully (and quite successfully) followed the specifications that they were given, and have provided a design whose enabling tasks have been cut down to a minimum. Furthermore, even the performance of the enabling tasks has been simplified as far as possible, with the most common amounts of cash that people request preset and printed in big characters on large red buttons—making it easier to specify the amount than it would be with a fiddly numeric keypad and an enter button.

However, this streamlining of enabling tasks is not without cost in terms of functionality, for it is not only impossible to carry out any goal tasks other than cash withdrawal, but it is not even possible to request an amount that is not printed on one of the six buttons. In other words, the enabling tasks that have been designed only support a very limited range of goal tasks: the withdrawal of one of six predetermined amounts of cash. The person who wants to take out a different amount of money, or who wants to request a statement, obtain a balance of their account, or to talk to the bank manager, must use a more complex machine or contact the counter staff. The consequence in terms of usability is that the rapid cash dispenser is highly usable for most people most of the time, but extremely unusable for some people the rest of the time.

This simple example has been used to illustrate the distinction between the tasks that people want to do and those that they have to do to enable the performance of their goal tasks. It might seem a long way from the complexities of IBC design, multimedia conferencing, computer supported co-operative work or multimedia document editing, but it shares some important features that are common to all of these design areas:

- the system can support all of the goal tasks that have been specified as appropriate to the context within which it will be used (and none that it has been decided that it need not support)
- any enabling tasks that must be performed are wherever possible carried out by the machine (i.e., reading the account number off the card instead of having the user type it in; and assuming that they would like a receipt instead of giving them the option of not having one)
- when the user must perform an enabling task, it is made as simple and as obvious as possible (i.e., large red buttons and bright prompts that light up)

The next two sections will expand upon these points to describe the concept of the *enabling state* and two criteria of usability that a design must meet in order to be usable.

3. The enabling state

Imagine an ATM design in which the user did not have to press any buttons, insert any cards, or do any waiting, but could simply walk up and take their money. Somehow the machine knows their account number, the amount of cash they want and who they are, and has the correct amount all ready for them to pick up. They would find this machine extremely usable, since the only task that they would have to perform would be their goal task—in fact we can say that the system is in a state that enables them to perform their goal task: an *enabling state.* In the real world, of course, systems are rarely in an enabling state for our goal tasks, and that is when enabling tasks must be performed to modify the state in some way so that it more closely approaches the appropriate enabling state.

The description of an enabling state must include more than a description of the state of the service that is supporting the goal task. In the ATM example, even if the machine knows the user's account number and the amount of cash that they want, the user will not be able to pick the cash up unless they know where to look for it. The actual design makes this very apparent by placing the cash slot directly by their right hand (suitable for 90% of users, at least) and by not covering it up in any way. In doing this the design is attempting to ensure that the user has the appropriate knowledge about performing their goal task—it is anticipating the case of a user who does not know where the money will emerge, and preparing to modify their knowledge. In other words, in an enabling state the user must have certain characteristics (in this case, knowledge of the performance of their goal task). Similarly, the enabling tasks that must be performed to give the machine the knowledge about the amount of money that the user requires also rely upon the user knowing how to provide this information—and the design attempts to provide this knowledge by using large red buttons with amounts of cash written on them. These buttons are carrying out an enabling task themselves: providing the user with the knowledge of how to perform an enabling task. The need for them to perform an enabling task is also provided to them by an illuminated prompt that is extremely salient and so likely to be noticed.

The object of the user's goal task (ie the cash) must also be described in an enabling state, since if the money that they are to pick up has not been counted out and bundled appropriately they will not be able to walk away with it. In the ATM design the machine carries out the enabling tasks of choosing an appropriate distribution of £20, £10 and £5 notes, collecting them from its store of cash and stacking them conveniently to emerge from the cash slot. If it had left them in its store, or presented them from three different slots, the user would have had to perform the enabling tasks of choosing the distribution of denominations, or of collecting them together from different places. As it is, the object has had its characteristics modified by the machine (or possibly instantiated, since it is arguable that the machine is 'creating' the amount of cash from components).

The enabling state, then, consists of a description of the characteristics that the user must have to be able to perform the goal task, a description of the characteristics that the object must have to allow the user to perform the goal task upon it, as well as a description of the state of the service that supports the goal task. This is the three tuple that was described in chapter 1.1:

Enabling State=(User, Service Relationship, Objects)
where:
User = required characteristics of the user, e.g. knowledge, attention
Service Relationship = the interface between the user and the object
Object= required characteristics of the object, e.g. 'file available'.

If the user, the object and the service relationship between them have all of the required characteristics for the performance of a goal task, then the system is in an enabling state. Any mismatches between the state of the system and the enabling state will require enabling tasks to be performed before the goal task can be performed. Each of these enabling tasks will modify the characteristics of the user, the object or the service relationship.

The service relationship is, of course, a crucial part of the enabling state description. The service relationship is similar to the concept of a user-interface common in the design of standalone computers or machines like the ATM. However, for a distributed system of communicating entities sharing goal tasks, the interface between a user and an object becomes more complex, as is described in chapter 1.1, and the term service relationship is therefore preferred.

Chapter 1.1 defined the four basic service relationships that are possible, with the Object being either *represented* or *unrepresented* to the user, and the user's actions upon the object being either *interpreted* or *uninterpreted*. Figure 2 shows how these two dimensions combine to describe the four areas that are of concern in the design of automated systems.

		Goal Task Object	
		unrepresented	*represented*
User's Goal	*uninterpreted*	1. traditional telecommunications	2. virtual reality
Task Action	*interpreted*	3. telepresence	4. traditional human-computer interaction

Figure 2: The domains of the four basic service relationships

The first cell, where the object is unrepresented and the user's task actions are uninterpreted, is the domain of the traditional forms of telecommunication, where the service relationship acts as a channel linking the user and the object (which is typically another person) without building any models of the user's action or the object's characteristics. In describing these service relationships, therefore, the nature and qualities of the channel must be specified so that the relevant characteristics of both the object and the user's task action can be conveyed accurately.

The fourth cell represents the area that has been investigated in conventional human-computer interaction research, where the service relationship must interpret the actions that the user makes in order to effect the desired changes in the object, and must also model the object's characteristics in order to represent them to the user. The user of a database, for example, might give textual search commands that have to be interpreted, and the results of their search (ie a selection from the database) would have to be represented to them by a listing or a set of records. This class of service relationship requires recommendations to be made about the way that the modelling should be carried out, that is, how the user's task actions should be interpreted, and how the object should be represented. The nature of the channels that convey the interpreted actions and the represented object, however, are not so crucial.

In the two cells on the other diagonal of this matrix only one model is constructed by the service relationship. Where the object is unrepresented, so that the user can perceive its characteristics 'directly', yet their actions must be interpreted, the service relationship becomes 'telepresence'—for example, the remote operation of robotic tools that can carry out operations beyond the capabilities of an unaided person, in inaccessible areas or hostile environments.

The remaining cell corresponds to the domain of 'virtual reality', where the user carries out actions that do not need to be interpreted, but where the objects do not necessarily exist and certainly need to be represented in some form by the service relationship. An example might be a flight simulator where the external world and behaviour of the 'aircraft' that is represented to the user is based upon parameters within the simulator's models of how a real aircraft would behave, and not on any actual aircraft responding to the actions carried out by the user in the cockpit of the simulator. Treating service relationship descriptions in this way provides a clear distinction between the 'virtual reality' and 'telepresence' domains, which have quite different requirements and yet which are often confused.

To summarise, an enabling state description includes a description of one of these four classes of service relationship between the user and their goal task object. The field of human-computer interaction research (such as that presented in chapters 2.1, 2.2 and 2.3) deals with the completely modelled cell, and provides information about the way that actions can be interpreted and objects represented. Telecommunications research, embodied within international standards (which will be covered in chapter 3.5), provides us with information about the completely unmodelled cell, dealing with the channels.

4. Criteria of usability

Having described the concept of an enabling state, we can now relate the features of the ATM design that were cited at the end of section 2 to the two criteria defined in chapter 1.1 (and see also Byerley et al, 1993) that must be met if a design is to be 'usable':

- *A usable system ensures that the appropriate enabling states exist for each of the users' goal tasks.*
- *A usable system reduces to a minimum the costs to the user in reaching the appropriate enabling states for their goal tasks.*

The first criterion is straightforward: if the design does not allow an enabling state for a goal task to be reached, then that goal task can, by definition, not be performed, and so the design is not usable for that goal task. In the ATM example that we have been developing in this chapter, the usage context required a rapid cash dispenser, and not a system that could support more time consuming tasks such as querying balances of accounts or printing statements. The enabling states for these goal tasks cannot therefore be reached by the final system, and it would be inappropriate if they could. This criterion makes it clear that a primary stage of any design must be to consider the user's goal tasks that it is to support, and to produce a description of the enabling states that allow these goal tasks to be performed. This necessarily requires descriptions of the intended users of the system, and of the objects that they will be working with (i.e., whose attributes will be affected in some way by their goal tasks), and so makes evident the importance of a thorough understanding of the usage context of any design.

Design advice can also be given on the second criterion, concerning the *costs* for the user which are associated with acquiring an enabling state. In order to provide design advice we need first to consider what is meant by user costs. Long and Dowell (1989) distinguish between *structural* and *behavioural* costs. Behaviours can be thought of as *'what the person does'*, whilst the structures supporting them can be thought of as *'how they are able to do what they do'*.

Structural costs are incurred in developing and maintaining human skills and knowledge, and more particularly in training and educating people and so developing in them the structures which will enable their behaviours. An example of structural human costs might be training a secretary in the layout style of documents and in the operation of the computer by which that style layout can be created. Structural human costs will frequently be borne by an organisation.

Behavioural costs are the resource costs incurred by the user in recruiting human structures to perform work.

Long and Dowell also make a distinction between mental and physical costs associated with both structures and behaviours. Physical structures are neural, biomechanical and physiological. Mental structures are concerned with representational schemes and processes. Physical behavioural costs are, for example, the costs of making keystrokes and attending to a screen display, and may be expressed as physical workload. Mental behavioural costs are, for example, the costs of reasoning and making decisions, and may be expressed as mental workload. Mental behavioural costs are ultimately expressed as physical behavioural costs.

A final distinction is drawn between *cognitive, conative* and *affective* costs. Examples of cognitive structural costs are the costs of developing the knowledge and reasoning abilities of people. Examples of conative structural costs are the costs of developing the stamina and persistence of people. Examples of affective structural costs express the costs of developing in people their patience and care.

Examples of cognitive behavioural costs relate, for example, to the processing of information and its associated physical costs. Conative behavioural costs relate to the repeated mental and physical effort required. Affective behavioural costs relate to the emotional aspects of the mental and physical behaviours required. Behavioural human costs often result in fatigue, stress and frustration and they are costs borne directly by the individual.

The second usability criterion must be satisfied by designing the enabling tasks in such a way that these costs are minimised. This can be done either by having the machine perform an enabling task itself, even if, as in the case of the ATM assuming that the user wants a receipt, this limits the resulting enabling states that can be reached (i.e., the enabling state that would allow the user to take withdraw receipt without a receipt is not reachable in the ATM design). This affects *function allocation,* since it implies that, wherever possible, enabling tasks should be allocated to the machine rather than to the human end-user.

If it is still necessary for the user to perform an enabling task, then this criterion means that it must be designed so the task requires as little cost as possible. In the ATM example, the enabling task the user must perform to define the amount of cash that they want is reduced to a single button press.

The argument that this chapter has presented is that design should proceed in a logically ordered way, centred around an analysis of the context within which the design is to be used and the goal and enabling tasks that it is to support. Two criteria have been presented that provide constraints within which the design process can be carried out. What remains is to provide support to designers in making decisions about how they can meet these criteria. As was described in chapter 2.1, expertise with the design of usable systems can be seen as a 'craft skill', but skills like this are difficult to apply to novel situations, are essentially conservative in the designs that they produce, and are difficult to communicate or transfer between individuals. It is the aim of an applied human factors discipline to provide 'engineering knowledge' in the form of principles that are based upon 'scientific knowledge', so that if the usage context is understood, specific information can be provided for any design decision.

The RACE project GUIDANCE (R1067) has developed a set of specific service integration principles that are targeted to the design stages implied by the arguments presented in the preceding sections. The next chapter, 3.2, will describe the content and application of these principles in detail. In the remainder of this chapter we will present a formalisation of the design method that we have been describing. This method allows the principles to be invoked by providing a structure for the designer to specify in the course of moving from the usage context analysis to the description of goal tasks, enabling states and enabling tasks. At each stage the structure, represented by specification tables, allows the user to record information which can then be used as conditions to match against the 'left hand side' of principles that have been expressed as 'IF...THEN...BECAUSE...' statements. When the designer requires support to complete a slot in one of the specification tables, the 'right hand side' (or consequences) of principles can be examined to find any that give appropriate advice, and their 'left hand sides' can then be considered to identify any usage context attributes that may have been omitted or incorrectly specified at earlier stages of the design.

5. A design method

Various human-computer interaction (HCI) methods for the specification and implementation of user-interfaces have been proposed, but the use of such methods has not spread outside the human factors research community. This lack of progress may be because human factors people have had little interest in integrating their methods with software engineering practice, and those who have not been able to find an HCI method with enough breadth and depth to be useful to system developers. Thus many HCI 'methods' are in fact fragments of methods, or *techniques,* which reach one specific goal without addressing the broader aspects of systems development.

There are many techniques for making specifications for human-computer dialogues and for evaluating user interfaces, but such techniques are almost by definition too narrow in scope and thus do not succeed in covering all phases within the systems development life cycle. Those that really have enough breath, and might be seen as methods, have not shown enough depth and prescriptive recommendations—with a few exceptions, such as the Enabling States Method. This is a layered method for user interface design which has been developed to be appropriate for user interface design, to be independent of any implementation language, to produce an object-oriented specification output, and to be applicable in conjunction with existing design methods.

The output of the method is a set of operational requirements in the form of a specification that can be used as a basis for the software design. The method itself is largely concerned with the 'internal perspective' of use, that is, how the final design will appear to the human end-user, and the design of the interactions between the end-users and the service components that are integrated in the design.

The first stage of the method is to establish the task usage context within which the eventual design will be used. This provides a description of the need for communication, and serves to focus the design process on fulfilling that need. The emphasis is upon the task that is to be performed using the design, the constraints upon its performance, and the acceptable quality and costs of its product. Much of this analysis can be taken from the 'external perspective' of the RACE Project 1077—Usage Reference Model (see chapter 1.1).

The usage context contains a description of the overall task or goal of the system, in terms of the changes in the work domain that are to be brought about by its use. The usability engineer begins the design by breaking down this goal task into a hierarchy of smaller subtasks. For each of the subtasks an enabling state description is then made in terms of the attributes of the user, service relationship and object that are required to allow the subgoal task to be performed.

Following this stage, service integration issues can be considered by examining the task sequences. If two or more subtasks are to be allowed to interleave, for example, then the integration of their enabling states must be designed. This means that the two enabling states must be integrated and an *integrated enabling state* description produced. Thus, service integration issues are treated after the initial enabling states analysis of the separate subtasks, by forming task combinations and examining the resulting sets of service relationships.

Figure 3 shows the method in diagrammatic form. Each of the smaller boxes (labelled Table 1 to Table 7) represents a specification table—the principal notation for representing the design. Other subsidiary notations that can be used as required include message sequence charts and state transition graphs. It is convenient to deal with the specification tables in three groups or layers, as marked in the figure. These layers (discussed below) correspond to the main stages in the design method. The two task layers are organised to describe the goal tasks, enabling tasks, and enabling states on which the goal and enabling tasks operate, and together they produce the operational requirements for the design. The tables within the software structure layer describe in increasing detail the software structures which support the interactive behaviours specified in the goal and enabling tasks.

In general, using the method would involve the designer in working through the layers in order, producing the relevant specification tables. But this is not to say that it is a strict top-down design method. It excludes neither iteration, nor opportunistic specification, nor the use of storyboard-type techniques. Indeed, within each layer, there are no strong procedures concerning the order for producing the tables. It is the logical progression through the method that is important rather than the rigid completion of highly detailed specification tables, although the principles that will be described in chapter 3.2 have been formulated to tie in closely with the parameters of the tables.

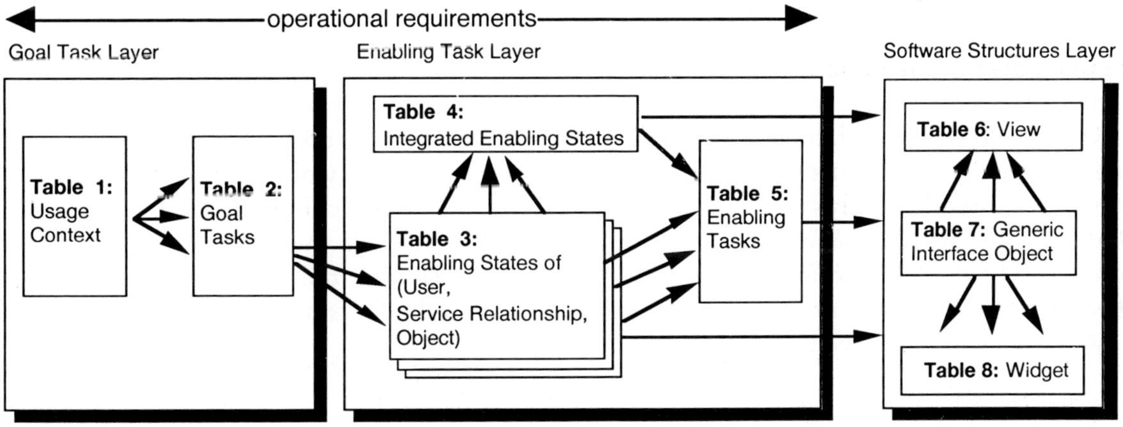

Figure 3: The Enabling States approach to design

The critical points about order in the method are: that the enabling tasks are specified from the enabling states, which are specified from the goal tasks; that the task layers should be specified before the software structure layer; that the software structures are derived from the enabling task layer; and that the final specification (whatever order it was produced in) should express a consistent rationale.

5.1. Analysis of task usage context

The goal of this stage is to describe the need for communication, rather than any recommendations for the services that would allow this communication (the emphasis within this description of the method is upon communication tasks, since the method has been developed within the context of IBC service design, but the method could with little modification be applied within other design contexts). The specification table expresses the aims of the system that is to be designed and the context in which it will operate, e.g. the work to be performed and the characteristics of the users and the their organisation. The usage context description should include the following parameters:

Table 1: Usage Context

Users: how many? simultaneous use? co-operative use?
gender? age range? demographic factors? special abilities? special needs?
technical experience? technical background? attitude to use of system?

Work goal: what is the overall reason for the use of the system?

Work domain: what objects exist to be used, created or modified in reaching the work goal?

Performance parameters: what is the required task quality?
what are the acceptable human and machine resource costs (i.e. time, number of errors, effort, cognitive demands,...)

System: what are the technological constraints upon the design?

Organisation: what sort of organisation will the system be used in (e.g., size, accountability, management styles, permanence, investment potential, etc)

Service Environment: Within what service environment will the system operate? These service environment assumptions are provided at the application level analysis and should include a Reference for User Services diagram.

It is crucial that the usage context is described as accurately and as specifically as is possible, particularly with respect to the overall work goal and the work domain. If the usage context is wrongly analysed, then there is a danger that a perfectly good design may be produced, but one which is targeted to the wrong set of users, to the wrong quality of performance, and so on. However, there is always the possibility to clarify the usage context once the following stages of the method have begun, for they may make it clear that certain aspects of the context have not been considered in sufficient detail, although this does mean the need to check that modifications to the usage context information do not conflict with any intervening design decisions that have been made.

5.2. Goal Task analysis

The aim of this stage is to produce a description of the goal tasks the user wants to carry out. The result of this goal analysis stage of the method is a breakdown of the major goal of the work system that was described in the previous table into a goal task hierarchy. It is important that at this stage the task descriptions remain technology independent, i.e. the task breakdown should, in principle, be implementable on any available technological or service configurations. Decisions concerning technology are taken when specifying the enabling states for each goal task.

The omission of concrete information about how the tasks are currently performed is intentional. This reduces the likelihood of the design simply reproducing existing systems with newer technology. A number of steps have been identified to help produce the task breakdown required at this stage of the enabling states analysis. These steps are described in detail in this section.

The goal tasks that a user wants to carry out need to be described by specifying not only the 'action' of the user, but also the 'object' associated with that action e.g. *retrieve document*, where *retrieve* is the action, and the object of the retrieval is a *document*. Therefore for each goal task description there needs to be a specification of an object and its associated action(s). The first steps show how to describe and decompose the objects and actions to give a breakdown of the major goal task into a goal task hierarchy. The other steps show how to express the sequencing of the different sub tasks within the goal task hierarchy and what the lowest level of the goal task hierarchy description should be.

The highest level in the Goal Task breakdown should be a description of the goal of the overall task being broken down in terms of an object and its associated action(s). This highest level object must then be broken down into its component sub-objects. These sub-objects are in a 'part whole hierarchy' relationship, as shown in figure 4.

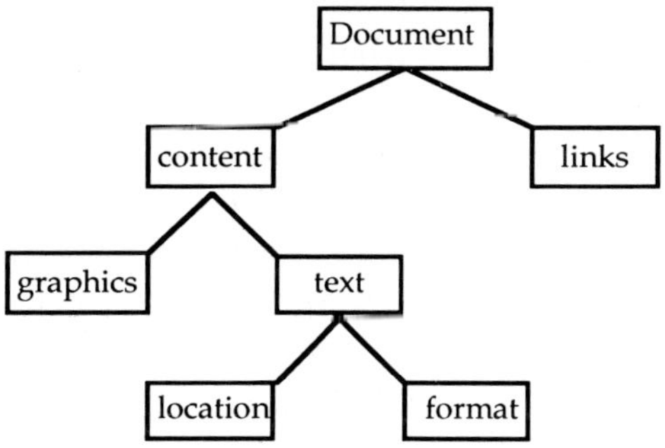

Figure 4: a Part Whole Hierarchy Description of an Object

At each level the attributes of the objects should be described. For example in figure 4 the 'text' might have the attributes of *location* and *format*. Decomposition should continue until the sub-objects can be broken down no further, i.e. until the values specified for the attributes of a sub-object would be physical instantiations of those sub-objects, e.g. text with format y (paragraph) with location x (file z).

Each object/sub-object has an associated action. These actions must also be broken down in line with the breakdown of their associated objects. Actions should be broken down i*n relation to the highest level object*. This should continue until it is no longer possible to break the actions down any further in relation to the main object. Those actions must then be used in relation to the sub-objects, e.g.

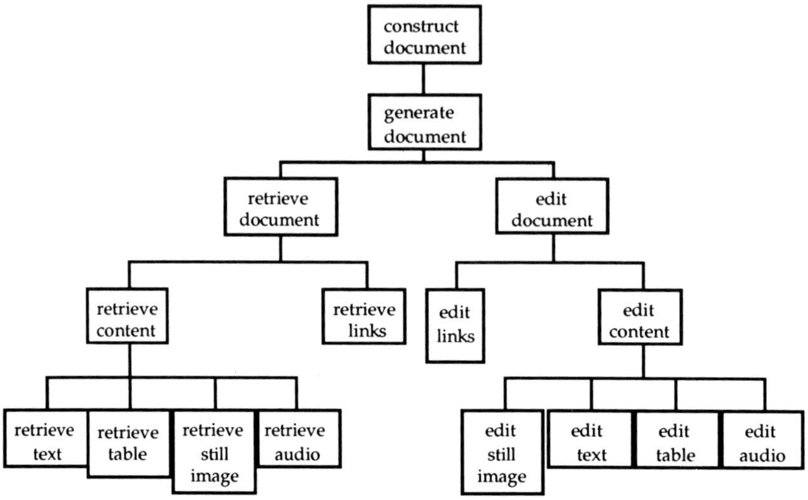

Figure 5: Decomposition of actions with their associated objects

Some diagrammatic notations are recommended for expressing sequencing information. These are:

1) The notation for iteration i.e. where a sub-task must be carried out repeatedly until the required state is met, is a star '*'. (For example, if the task is 'walk', then the sub-task would be 'take a step', which must be repeated.)
2) The notation for selection, i.e. where there is a choice of what to do next, is a circle 'o'
3) There is no added notation for sequencing, which implies that the sequence is from left to right unless otherwise specified.
4) For expressing an ordering different from sequencing from left to right, then an hierarchy construct (Lim 1992) should be used which implies that sub-tasks can be done in any order. The notation for this is an arrow.

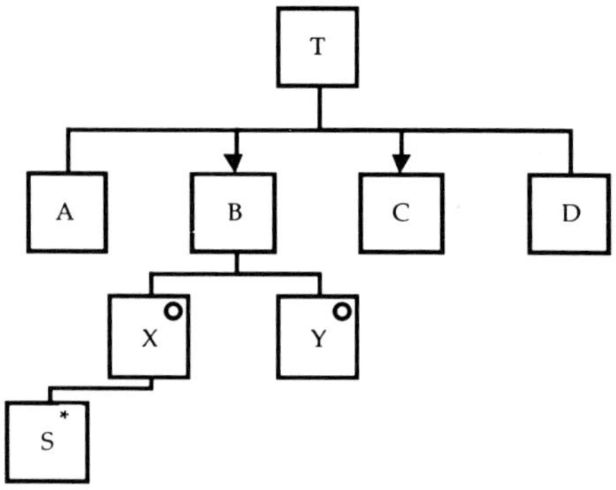

Figure 6 An example notation for expressing sequencing information

For Task 'T', the notation used in figure 6 implies:

Do A, then B and C in any order and then do D. When doing B there is a choice between X and Y. If X is chosen then iterate S. (The choice would be made through constraints that are specified in the task information).

A stopping rule in the goal task breakdown is that *'the breakdown should only be continued to a level that does not require a description of any technology.'* This rule tries to maintain and enforce the idea that the goal task breakdown should be device independent. Each of the goal task sub-object and sub-action pairs can be recorded in the second specification table:

Table 2: Subgoal task: [id number and a short name to reference the table]

Description: a brief textual description of the goal task that this table describes

Sources

a list of the specification tables that refer to this subgoal—these will be other subgoals that are higher in the task decomposition.

Task parameters:

Numerousness:

users: an estimate of the number of human-end users who might be involved in each instance of this subgoal.

other entities: the number of other entities (eg databases) that may be involved in each instance of this subgoal

Temporality:

frequency: how often will this subgoal be performed

duration: how long will the performance of this subgoal take, with respect to the total use of the application

Intertask dependencies:

necessary pre-tasks—list any other subgoals that must have been performed for this subgoal to make sense (although note that they will not be *enabling* tasks)

task interleaving—list any other subgoal tasks that are likely to be performed concurrently with this subgoal

Design aspects:

Functional requirements:

A list of what this subgoal should achieve, ie how it affects the work domain

Interaction requirements:

A list of constraints upon the way the interaction should be performed

References: a list of the tables that this subgoal makes reference to (these may be other subgoal tasks, or enabling states if this subgoal is at the lowest level of the decomposition)

This table shares a number of features in common with subsequent tables, which are used to make it easier to keep track of the large number of tables that will be produced in the course of a specification. These include a brief name and an ID number that identify the table, to make it easy to refer to, and slots for Sources and References. The References slot lists any subsequent tables within the specification that this table requires to be completed—in this case, for a subgoal task, the references might include either a further set of subgoal tasks that it can be decomposed into, or if it can not be decomposed, then the enabling state that is required for it to be performed.

The Sources slot provides an opportunity to discover why this table is required, for it lists all of the tables earlier in the specification that 'reference' this table—as will become apparent at later stages of the specification, this allows redundancy and similarity within the specification to be identified when combinations of tables consistently reappear together as the Sources of other tables. This redundancy can indicate the need for 'class relations' to be considered: where several subgoal tasks have similar decompositions, for example, it may be more economical to describe them each as 'instances' of a superordinate 'parent'. The description of the parent would include information common to each instance, including the references to the further decomposition, and the tables for each instance would only need to record deviations from the description of the parent. This not only simplifies the description of each instance, but it makes the specification of the referenced tables easier, for they will only have one source that needs to be consulted instead of several. There is no need for a separate table structure to allow classes and instances to be specified: any class relations can be recorded in the Sources and References slots.

The task parameters describe aspects of the task that will be relevant to later design decisions, particularly those relating to the integration of multiple subgoal tasks (i.e., integration of services). The Numerousness slots record the number of human-end users and other entities within the work system that will be involved in each instance of this subgoal task—not the number of different entities which may be carrying out this subgoal independently. If they are not interacting with each other, then their simultaneous conduct of identical subgoal tasks is not relevant to the usability of their own task. It is important to remember, however, that 'entities' does not just include the 'actor' or 'user' of a task, but also the 'objects' that they are acting upon: thus several human end-users might be the object of a goal task carried out by another human end-user, and they would all have to be counted here.

The Temporality information records estimates of how often a task is likely to be performed, and how long it will take. This is important because if a task is only likely to be performed rarely it is not so crucial for the usability of the overall system, compared to other tasks. If there are trade-offs to be made between the design of different enabling tasks, whose usability appears to be mutually exclusive, then they can be traced back to the subgoal tasks that they enable to ensure that the more frequently used one takes priority. Intertask dependencies are of crucial importance in the integration of different goal tasks, since they give information that can be used to recognise which enabling states must be coexistent within the system, or at least which states the system may be required to oscillate between frequently. The design of these states must not be made exclusive, and any enabling tasks that change the states from one to another must take high priority for ease of use.

The production of a set of subgoal task tables completes the first layer in the enabling states approach to design. At this point the specification is entirely user-centred, in that it only contains information about what the users intend to do. The next stage begins to introduce information about how these tasks are accomplished, and about the support that must be offered to enable them to be accomplished.

5.3. The Enabling States Analysis

The third stage of the Enabling States approach has two parts. First, each of the subtasks specified in the previous stage is considered, and the state that the work system must be in for each subtask is specified. These states are the enabling states that give the method its name.

Once these states have been identified, in terms of the attributes of all of the users, service relationships and objects which make up the work system, information is taken from the subgoal task descriptions to identify those states that will need to occur simultaneously or in close sequence, perhaps interleaved repeatedly with each other—these are the enabling states for goal tasks that users will want top perform simultaneously or one after the other in rapid alternation. In accordance with the second criterion of usability, that the enabling states should require minimum cost to achieve, it is clearly in the interest of the designer to ensure that these states can be as similar to each other as possible. To simplify the design of these states, then, they are combined where necessary into an integrated enabling state that describes the state of the system when two or more subgoal tasks are to be performed. The specification of these integrated enabling states keeps track of potential conflicts between states and provides a way of resolving mismatches that do occur.

As has been described, an interactive work system consists of entities (or users), working on other entities (or objects), via service relationships. An enabling state describes the state of the work system: what knowledge and processing skills the users must have; the characteristics of the service relationship, and the state that the objects must be in. Each of the subgoal tasks described by a specification table 2 must have an enabling state. Accordingly, the designer must take four steps to describe an enabling state:

1. Describe the states of the *users* for the goal task,
2. Specify the characteristics of the *service relationship* for each goal task
3. Describe the states of the *objects* for each goal task,
4. Consider any interleaving of goal tasks, and then provide an integrated enabling states description for the interleaved tasks.

5.3.1. Step 1: Describe the user for each goal task

The purpose of this step is to describe the *required* user characteristics for a given enabling state. Table 3a gives an example of a user description.

Table 3a: Enabling State—User Description: ID/Name of User

Sources:
A list of the goal tasks that require this enabling state

User Description
A list of the required attributes of the user, including the knowledge that they must have and their capabilities.

User Principles:
A record of any principles that are relevant to this enabling state

References—Enabling Tasks for the User Description
A list of the enabling tasks that are implied by this user description

In all three parts of the enabling state tables, as with the previous specification tables, the specification begins with a list of Sources (in this case, the goal tasks that this user description enables) and ends with a list of References (in this case, Enabling Tasks that will be required to achieve or maintain these user characteristics).

The user description should contain details of any knowledge that the user must have to be able to carry out their goal task—since this is an enabling state for the *goal task*, this should only be knowledge that is related to the work domain, for example, knowledge about the existing and required state of the work objects, and about modifying the object.

If there is a discrepancy between the required characteristics described by the enabling states description and the actual user characteristics when the state is needed (that is, when a subgoal task that it enables is to be performed) then an enabling task will be invoked to modify the user characteristic (these enabling tasks are described in Table 5). For each user characteristic that is described in this table, there should be a corresponding enabling task.

The specification of enabling states is the first point in the Enabling States approach where design principles are applied. The application of these principles will be discussed in detail in chapter 3.2, but briefly they serve to assist the designer in describing the user's characteristics. The slot in this table labelled 'user principles' allows the designer to record any principles that were applied, providing a form of rationale for the design as it is constructed.

5.3.2. *Step 2: Specify the service relationship for each goal task*

Once the designer has described the user the next step is to describe the service relationships for each of the goal tasks. Table 3b gives an example of a service relationship description.

Table 3b: Enabling State—Service Relationship: ID/Name of Service Relationship

Description:
A textual description of the nature of this service relationship and the tasks it enables

Sources:
A list of the goal tasks that require this enabling state

Service Relationship: Are the user's goal task actions interpreted or uninterpreted by the service relationship, and is the object represented or unrepresented?

Description of the User Task Interpretation : If the User's task actions are to be interpreted (ie modelled) by the service relationship, how shall this be done, and what are the constraints and requirements on the performance of the model constructed by the service relationship?

User Task Interpretation Principles: a record of any principles that were used to derive the User Task Interpretation description

Description of the Object Representation: If the Object is to be represented (ie modelled) by the service relationship, how shall this be done, and what are the constraints and requirements on the performance of the model constructed by the service relationship?

Object Representation Principles: a record of any principles that were used to derive the object representation description

Service Attributes: If the user's task actions are uninterpreted, or if the object is unrepresented, then information must be provided about the channels of communication that are used to convey the user's task actions to the object, or the object's attributes to the user. If the service relationship is both interpreted and represented, this section can be omitted.

Involved Entities: which entities are linked by this relationship? eg user, workstation, recipient

Communication Type: The type of communication involved in this relationship (eg Human-Human Conversation; Machine-Machine Distribution)

Channel characteristics:

Information Content: A description of content of the information that is to be communicated by this relationship, (eg head-and-shoulders view with sound)

Service Components: What service components (eg audio, video) are involved in this relationship, and whether they are mandatory or optional.

Synchronisation requirements: a description of any synchronisation required between service components and the temporal tolerations.

Format Preservation: What degree of loss is tolerable in the information communicated by this service relationship (ie what are the constraints upon compression)

Quality of Audio Service Component: Constraints upon the quality of any audio components (eg telephone quality, CD/HiFi quality)

Quality of Video Service Component: Constraints upon the quality of any audio components (eg frame rate, colour depth, etc)

Quality of Service: What attributes of the service relationship are most important in ensuring the acceptability of the service relationship to the user?

Service Attribute Principles: A record of any principles used in completing any of the Service Attribute information

References: Enabling Tasks for the Service Relationship
A list of the enabling tasks implied by this service relationship description.

The service relationship description should specify whether the user's goal task actions are interpreted or uninterpreted (i.e., does the service relationship construct a model of the user's actions, and interpret them to carry out actions upon the object, or does operate upon the object in direct correspondence to the user's actions) and also whether the object is represented or unrepresented (again, does the service relationship construct a model of the object which it presents to the user, or does it convey the attributes of the object directly to the user). If either of these aspects of the communication are modelled, then the designers must specify how the modelling is to be carried out (e.g., what aspects of the user's task actions must be picked up, and with what granularity must they be communicated to the object; what attributes of the object must be communicated to the user, and how accurately?), and what constraints apply to the performance of the model. This sort of information is the subject of the human-computer interaction research reported in Section 2 of this book.

The next step is to describe the service attributes. As the table shows, these are a set of requirements that begin to specify the communication facilities that the service relationship must offer to the user. These are only relevant, however, if the service relationship is conveying the user's task actions to the object without interpreting them, or if it is conveying the object attributes to the user without representing them. In these cases the characteristics of the communication channels that convey this information can influence usability (as will be described in the chapter on usability standards, chapter 3.5). If the task actions or the object are modelled by the service relationship, then it is the nature of the model that is of most relevance, and the service attributes do not need to describe that side of the service relationship. Because of the detailed goal task breakdown, these attributes can be derived from the functional and operational requirements in the source subgoal tasks tables. Design principles can be applied at this point, and any that are used can be recorded in the slot for 'Service Attribute Principles'.

5.3.3. Step 3—Describe the object for each goal task

Table 3c, which describes the state that an object must be in for a goal task to be performed, is similar in content to the user description recorded in table 3a. As with the User description, any mismatch between the specified characteristics and the actual characteristics will require an enabling task to be performed, and so for each characteristics listed there should be a reference to an enabling task. Principles can be applied to derive these characteristics, and any that are used recorded in the slot for 'Object Principles'.

Table 3c: Enabling State—Object Description: ID/Name of Object

Sources:
A list of the goal tasks that require this enabling state

Object Description
A list of the required attributes of the object, including any knowledge it must have and any actions that can be performed upon it (ie changes that can be made to it).

Object Principles:
A record of any principles that are relevant to this enabling state

References: Enabling Tasks for the Object Description
A list of the enabling tasks that are implied by this object description

5.3.4. Step 4: Integrated Enabling States

If two or more subgoal tasks are to be allowed to occur simultaneously or to interleave then the integration of their enabling states must be designed. This means that the two enabling states must be combined and a third *integrated* enabling state description produced. This description will allow any modifications to the individual enabling states descriptions to be highlighted and any influences on software design to be considered. Table 3 gives an example of a integrated enabling states description.

Table 3: Integrated Enabling State: Name of Integrated Enabling State

Description: A textual description of the tasks that are enabled by this state

Sources:
The enabling states that are integrated by this table

Description of the User :
Any user characteristics that are required for this state and that are not already listed in the source enabling states. If there are any contradictions in the source enabling states, they must be resolved here.

Interactions and Modifications to Individual Service Relationships:
A textual description of the number and nature of service relationships being integrated in this state. A description of the changes that will need to be made to any of the relationships while this state exists, including recommendations for simultaneity of user task actions and the simultaneity of representation of task objects.

Description of the Objects
Any object characteristics that are required for this state and that are not already listed in the source enabling states. If there are any contradictions in the source enabling states, they must be resolved here.

References: Enabling Tasks for the Integrated Service Relationship:
A list of the enabling tasks implied by this integrated enabling state that are not already referenced by the individual states (nb these may be novel enabling tasks).

The integrated enabling state description contains a list of Sources, which are the enabling states that it integrates, and a list of the additional characteristics for the users, service relationships and objects involved in these states. Each of these characteristics may require further enabling tasks to be specified, and these are listed in the References slot.

5.4. Analysis of the Enabling Tasks

In describing the users, the service relationship and the objects the designer will have identified a number of enabling tasks which have to be carried out in order for users to reach the appropriate enabling states for their goal tasks. Specification Table 4 of the design method describes these enabling tasks and applies principles which help the designer meet the second usability criterion, to reduce to a minimum the costs to the user in reaching the appropriate enabling states for their goal tasks. The principles applied at this stage are concerned with user costs and function allocation (see chapter 3.2).

Table 4: Enabling Task number and title of enabling task

Description: a brief textual description of the role of this task

Class Relationships: Is this task a parent of other, more specific tasks, that inherit its attributes? Does it in turn have a more generic parent task from which it inherits attributes?

Sources:

Enabling States: a list of the enabling states that might require this task to be performed

Usage Aspects

Allocation: should this task be performed by the user, or can the machine carry it out?

Allocation principles: any principles that were used to derive the allocation of this task

Cost principles: any principles that affect the cost of performing of this task

Flow of Work: what other tasks (enabling or goal) will tend to occur before or after this task? (This can be represented by a figure or graph)

Design Aspects

Functional Requirements

A list of the constraints upon what this task should do.

Operational Requirements

A list of the constraints upon how this task should be performed.

References:

A list of the software structures implied by this task (eg GIOs)

The specification tables for these tasks contain a brief description of the task, information about the sources and references of the table, and a design aspects section that lists the functional and operational requirements for the task. The functional requirements describe what the results of the task should be. The information needed to complete them is derived from the enabling states specifications. The operational requirements describe constraints upon the way in which the task can be performed, and these would ideally be prescribed by principles. The subgoal task whose enabling state is being created or maintained by the enabling state in question would supply the task parameters to fulfil the conditions of the principles which should be applied. Additionally, the enabling task tables contain a section for the usage aspects which must be considered, and these would also be derived from principle prescriptions. The description of the usage aspects replace the section for task parameters in the subgoal task tables.

These enabling tasks replace the information which was omitted from the previous layer: that describing the way in which the subgoal tasks are currently performed. Because the concepts which are used to describe them are taken solely from the analysis of the work domain, however, this redescription is now in terms of the users' conception of their overall goal. Basing design decisions on the users' conceptions of the work domain ensures that the users will be able to interpret the actions and states of the machines within the work system.

As with goal tasks, there may be an interleaving between enabling tasks, and the designer may refer to integration principles for assistance in the specification of the respective GIOs. As with subgoal tasks, it is likely that the enabling tasks will form groups that can be economically described by using class relationships. A number of enabling tasks that all operate upon the characteristics of the same entity or service relationship, for example, may share a great deal of their individual specifications in common. In this case they would be better specified by one superordinate enabling task and several briefer tables that list the deviations or additional aspects of each 'instance' enabling task. These relationships can be recorded in the Sources and References of the individual tables.

5.5. The Software Structure Layer

The position after the completion of the first five tables in the Enabling States approach is thus a complete description of the tasks which must be performed by the work system, including those tasks which are not directly related to the overall goal and which may vary from design to design. The next step is to begin the specification of the work system software.

The final layer in the Enabling States approach produces specifications of the appearance of the interface objects required to satisfy the operational requirements described in the previous layer, together with the actions that they respond to, and their behaviour. This is done largely by collecting together the information specified in the earlier tables. Where this information is not sufficient to identify one particular design option or feature of a GIO, then either that feature is not directly relevant to the usability of the system or no definitive advice can be derived from the user requirements and principles. In this case the GIO specification table will reflect the design decisions made by the software implementation team.

In the specification method GIOs which are to be used in reaching a particular enabling state are clustered into a *View* for that state, together with the objects which must be present to provide the service relationship defined in that enabling state. All interface objects that are likely to be used together (or in close sequence) and the consequent interaction behaviours are gathered together. These tables (table 5 within the method) can be used to ensure that certain elementary properties that are central to usability have actually been adhered to within the design of elements of an interface that appear together (these properties, such as *consistency, coherence* and *salience* will be described further in chapter 3.2).

Table 6: GIO: A name for the Generic Interface Object and an ID number

Description: a textual description of the GIO, together with a sketch or diagram of its representation

Toolbox ref: if an accepted toolbox of software structures is being used, this slot can be used to indicate an appropriate element

Sources: a list of any enabling states, enabling tasks or goal tasks that reference this GIO.

Design aspects:

- **interface model:** is the design of the GIO intended to be a metaphorical representation or model of some aspect of the work domain or other 'real-world' objects and actions that the user is expected to have prior experience of? If so, what aspects of this model are the most crucial and why?
- **design principles:** any principles that have been applied to influence the choice of interface model
- **design discrepancies:** are there any ways in which the GIO performs differently to the way that the interface model might lead the user to expect?
- **design rationale:** why has this particular design been specified in this manner?

Properties: This section consists of a set of properties that must be specified to describe the required appearance of the GIO. The list below gives a few examples of such properties.

- **name: layout of parts:**
  - **value:** eg how are the components of the GIO to be related to each other, spatially or temporally
  - **controlled by:** can this be modified by the machine, or by the human end-user
- **name: position**
  - **value:** eg where should this GIO appear in relation to other GIOs
  - **controlled by:**
- **name: size**
  - **value:** eg any constraints upon size that might not otherwise be apparent
  - **controlled by:**
- **name: border-width**
  - **value:** eg the style of border that should be used for a screen representation (to ensure consistency with or discriminability from other GIOs)
  - **controlled by:**
- **name: background-colour**
  - **value:** eg the style of border that should be used for a screen representation (to ensure consistency with or discriminability from other GIOs, or to make the GIO salient or easily ignorable)
  - **controlled by:**

Interaction procedures: A list of atomic actions that the GIO will respond to. The list below gives an example

- **name: clicking on background**
  - **preconditions:** what attributes must the GIO have for this interaction to be possible
  - **actor:** who can perform this interaction, the human end-user or the machine (eg other software structures)
  - **event:** what event signals the occurrence of this interaction (ie what does the actor do)
  - **effects:** what should the GIO do in response to the event

References:

a list of the widgets that comprise this GIO, and of the view(s) to which it belongs

The Views, whose table is very similar to the GIO table and so is not reproduced here, also have Design Aspects which may differ from their constituent GIOs (the view may follow a metaphor such as 'a desktop', for example, while individual GIOs may be representative of work domain objects). Views can also have Properties and Interaction Procedures, since they need not fill up the whole display (if they are to be shown on a computer screen) and may need to be positioned with respect to other Views, and the user may be able to interact directly with a View rather than through one of its GIOs (e.g., clicking on the background to deselect GIOs). Views can also directly reference Widgets, such as a 'close-box' within a window.

Widgets are described in table 7 of the method, and since these are also very similar to the GIO table, they will not be reproduced here either (a completed example of a View, a GIO and a Widget table can be found in chapter 3.3). Widgets do not have Design Aspects, but they do have Properties and Interaction Procedures. In fact, most of the interaction procedures to be specified within a design will be recorded on the Widget tables, since Views and GIOs are built out of them, and Widgets are the lowest element of the software structure specification layer. As with the GIO, it is possible that a toolbox of software structures may be used by the design team, so that the code for each widget does not need to be written from scratch, and most of the basic interaction procedures will be extant.

The software structures layer has not been described in as much detail as the tables in the two task layers because, while it is important that the individual interaction procedures and the appearance of the elements of the interface objects are specified, it is not so important that they follow the precise schema recommended here. It is a matter of convenience whether a scrollbar, for example, should be considered as a GIO that consists of several Widgets, each of which can be interacted with separately and which require individual specification, or whether it suffices to describe it as a Widget itself, and to include all of the interaction procedures on one table—this may seem economical if all scroll bars are to be exactly the same in appearance, size, location and performance, but if not it may be better to describe separate GIOs each of which reference a small number of Widgets, many of which can be reused between the GIOs as required.

The Widget table forms the final level of this specification method, and so does not have a reference slot to further tables. The software structure tables (Views, GIOs and Widgets) could be used in isolation to drive the software implementation, but as will be reported in chapter 3.3, it is better to supply the whole specification so that the rationale for various design decisions can be seen, and the relationship between various enabling states and goal tasks can clarify any ambiguities or omissions in the lower level tables.

6. An automated Design Support Package

Although the specification tables described in the previous section provide a well structured format for prompting and recording design information, for recruiting the advice and support of design principles, and are comprehensively cross-referenced through the Sources and References slots, anything more complex than a trivial design example inevitably results in a large number of tables that can be hard to keep track of. Primarily with the intention of providing support for the collation of the tables and the navigation through the specification sequence the method was developed in conjunction with an automated Design Support Package (DSP). Other main purposes of the DSP were to automate the supply of design information, advice or other assistance of value in improving user-interface designs via the design principles that were being codified; and to support decision making about particular designs both during the initial development and during the subsequent iterations when the designs are modified.

The DSP as a decision support environment could theoretically provide support at several levels:

- as design principles,
- as links to reports, books and other materials presenting the most relevant knowledge in e.g. cognitive psychology (in a simple but informative way),
- as advice and examples of good designs.

Regarding establishing links to cognitive psychology the kind of support provided is of course dependent upon the knowledge and interest of the designers, as well as the values of their company regarding the utility of psychological skills for its progress in business. The essential question is here not to support the designer in answering questions of how to design a specific user-interface but to simply identifying the general kinds of support that might be needed. In fact, such kinds of support are essential for the designers' own development of skill in the field. These kinds of support include:

- references to the most relevant state-of-art literature in cognitive psychology (and related studies in human factors);
- some of this support available online within the design support package,
- online mini-tutorials on specific aspects of the specification and design of user interfaces,
- ancillary online information of previous products/systems/ services the company has developed or worked on as well as names of persons (including designers) within or outside the company which might be able to offer specialist advice the type of problem involved
- information on current in-house courses in the area of interest and public courses related to the problem of design.

The aim of principles and advice is of course to give direct support in solving specific design questions. In certain situations, a designer may be familiar with the problem but wish to consult a specific principle to refresh their memory about some particular issue. In other instances, they may want a more in-depth exploration of the design problem. So the individual designer or design team may want to pursue a particular design problem to whatever seems appropriate in the circumstances.

The type of support provided by the DSP divides into three main areas: support for using the enabling states method itself (The Design Environment), support for accessing the principles (The Principles Stack) and support for learning how to use the enabling states method (The Tutorial). These three strands were the key areas identified as requiring immediate support as a result of an analysis of designer's needs and requirements, which preceded DSP development.

Using the method entails the construction of a task model comprising a number of different elements (e.g. main goal, subgoals, enabling states, enabling tasks, generic interface objects etc.), and the subsequent filling in of specification tables corresponding to each of these elements. A key function of the Design Environment aspect of the DSP therefore is to provide support for task model construction and table generation. Having constructed a task model and generated the appropriate tables users also require navigational facilities to enable them to move easily between different levels of the tables and between the tables and the task model representation. Support for navigation is thus also a key function for the Design Environment.

These key functions have been implemented in Lisp running on a Macintosh IIcx with a 21" (A3) colour monitor. This design environment allows users to create goal and subgoal icons on a model building canvas (window) in order to construct a graphical model of the task. As each element is added to the model, so a corresponding (empty) specification table is generated which may be subsequently filled in by the user (see figure 7—it should be noted that the DSP screens shown in figures 7 and 8 were originally colour images of A3 size, and that some quality has been lost in the transition to smaller black and white images). The user may indicate on the task diagram whether or not the corresponding specification table has been completed, and this is shown by a tick on the task diagram. A state transition diagram editor may be invoked from each of the specification tables to allow corresponding state transition diagrams to be created.

A number of methods of navigation have been provided within the Design Environment. The task diagram itself acts as the first means of navigation: clicking on any of the elements within the diagram will automatically bring up the appropriate specification table. Within the tables themselves are 'Source' and 'Reference' slots which show a particular table's position within the overall task structure. Clicking on the entries in these slots will bring up the appropriate tables and so this also acts as a method of navigation. A third means of navigation is through an 'Index' window that provides an ordered list of all the task elements created so far and users may click on appropriate elements to go to corresponding tables.

A primary function of the Design Environment is to provide easy and timely access to the usability principles. To this end a mechanism for accessing the Principles Stack was designed to allow the principles to be browsed and applied at any stage of the design process, from directly within the Design Environment. In a similar vein, users may want to know more about the method, how to use it, how to use the DSP and how to apply the principles. This information is provided by the Tutorial which can be similarly accessed directly from within the Design Environment.

The Principles Stack, the second major constituent of the DSP, has been designed in order to present the principles to designers in manner both quick and easy to understand. The stack may be accessed from either within the Design Environment section of the DSP, or independently as a separate principles repository. Two main access mechanisms to the Principles have been designed. Firstly, illustrations of principles concerned with specific application areas such as service integration, CSCW and hypermedia can be accessed; this is the 'Application First' route. Alternatively, designers may first wish to gain an overview of the generic design principles available before narrowing the scope to specific examples areas; this is termed the 'Principle First' route. A third section has also been incorporated into the Principles Stack to illustrate how it could be customised to provide direct access to specific principles for a specific application, namely Multimedia Multiauthor Document Production (MMMADP).

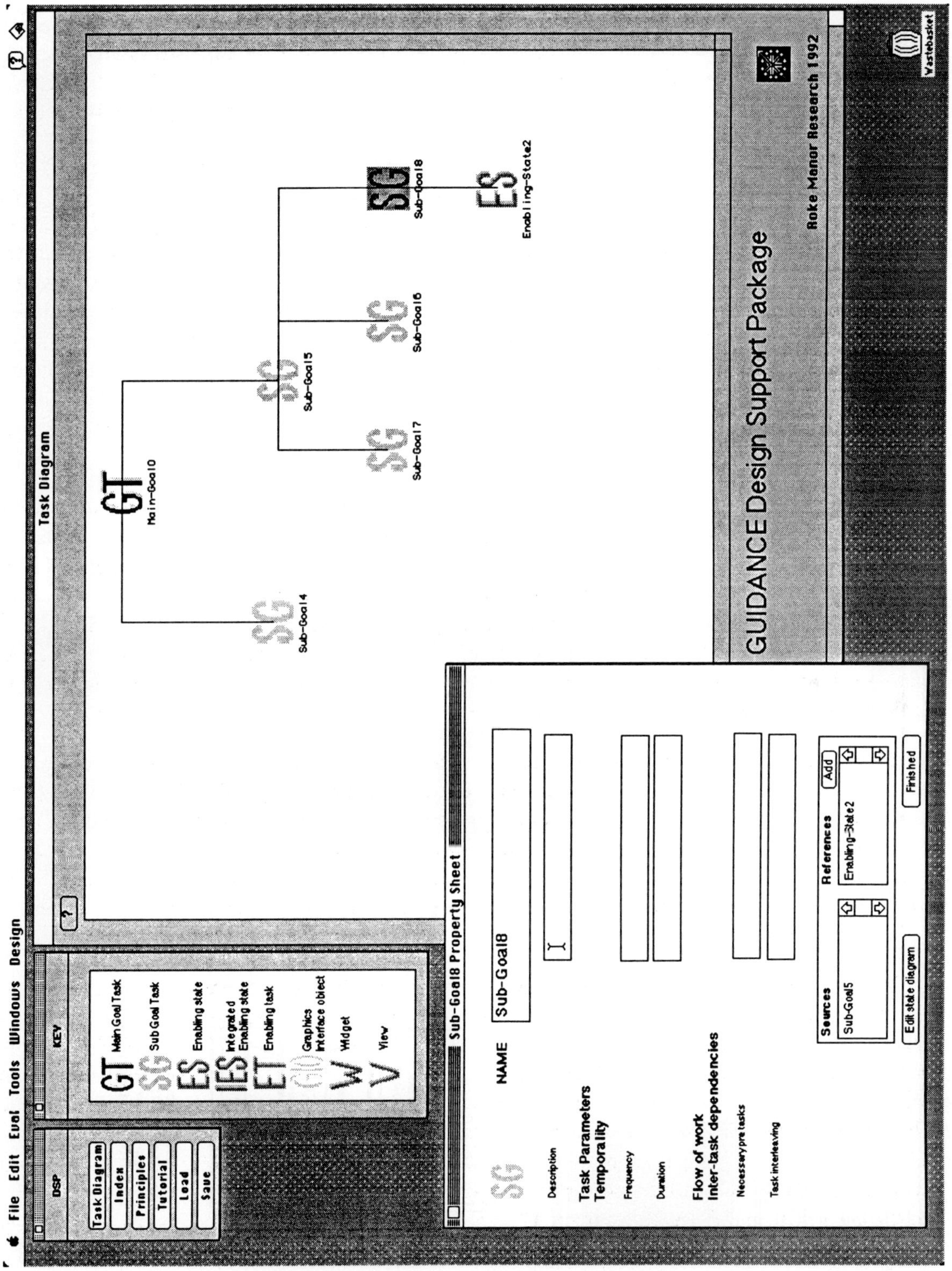

Figure 7: The DSP 'Design Environment'

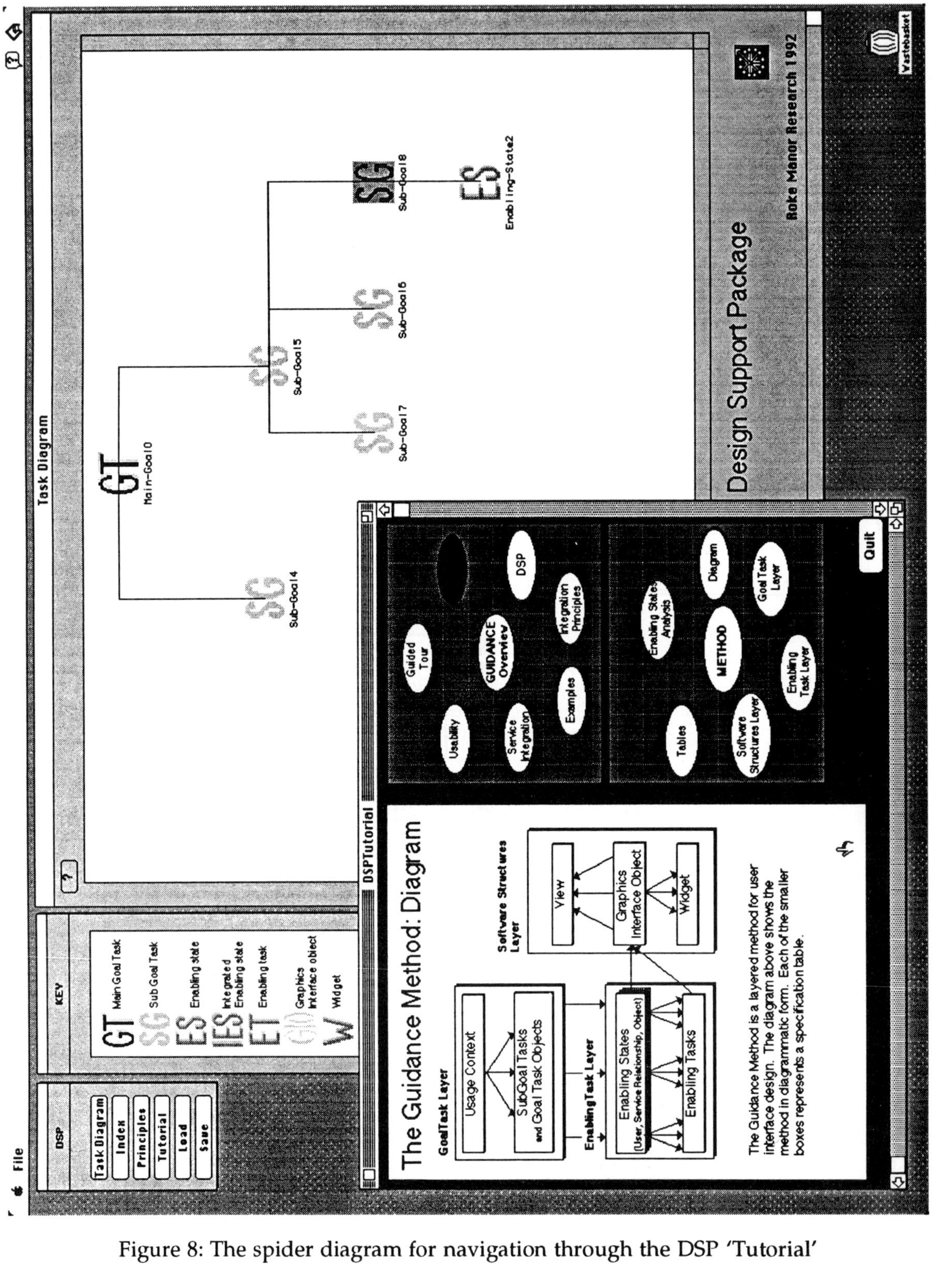

Figure 8: The spider diagram for navigation through the DSP 'Tutorial'

The third major role of the DSP is to introduce the design methodology to the designer. As part of this, it incorporates a Tutorial on how to use the method and how to use the DSP to support this activity. For this it was necessary to design an overview to introduce the package, explain what it does and how to use it, and to provide a detailed explanation of the method in non-specialist terminology.

In order to provide the desired level of support, the Tutorial has been designed to enable the user to reach anticipated goals by using an interactive, multimedia on-line guide. The user can search the Tutorial by means of overview diagrams which guide access to the various sections contained within, as listed below:

- **Guided Tour**—to guide the first-time user through the use of the Tutorial package
- **Method**—to explain the use of the Method, Method diagram, task analysis, and production of tables
- **DSP**—to explain the use of the Designer Support Package
- **Integration Principles**—to give examples of the use of Integration Principles
- **Service Integration**—to explain the meaning of Service Integration, and how to use the Service Relationship Diagram
- **Usability**—to explain the importance of Usability issues
- **Examples**—to give examples of the use of the method in producing specifications for different applications

It was recognised early on that ease of navigation through the vast amount of detailed information held within the Tutorial was an essential requirement. To meet this need much effort during the Tutorial design phase was directed at developing the 'Spider Diagram' navigation system illustrated in figure 8.

The Tutorial system was designed to provide domain specific information consistently on the left hand side of the screen, and navigational aids on the right (with the exception of full size illustration screens). Simply clicking on the required Spider Diagram button facilitates navigation to the desired location. The diagrams continuously update to reflect the user's current position within the stack, whilst always providing a means of navigation back to previously visited destinations.

Wherever possible detailed examples, taken from the MMMADP Case Study, have been provided within the Tutorial to accompany theoretical explanations. Implementation has made considerable use of the facilities offered by hypermedia technology. Non-sequential links are possible to any section within the Tutorial, and using simple mouse clicks, users can progress through layers of explanation until they reach the required level of detail.

7. Comparison with other methods

The purpose of this section of the chapter is to gain an appreciation of the various methodologies and tools currently available to aid software engineers in the overall task of system design. This will be achieved by reviewing both traditional software design methods, and the emerging HCI specific methodologies and supporting software tools. This process will enable the enabling states method to be put into the context of practical design activities. This review discusses current design practice in terms of structured design methodologies, and how these are deficient from a human factors point of view.

The various methods and HCI tools developed to compensate for these deficiencies are described in sections 7.2 to 7.5, and section 7.6 provides an investigation of how the user centred enabling states method can be integrated with one of the structured design methodologies commonly in use today. As the review of the various methods and tools is necessarily brief, a bibliography section is included with this chapter to provide appropriate references should the reader wish to investigate further.

7.1. Current Design Practice

As computer systems have evolved from their early role of basic calculators through to the complex multi functional machines of today, so the software required to drive them has become equally complex. It was soon obvious that as the software became more intricate, so the approaches used to design that software needed to be more structured and methodical. In response to this need, a number of methodologies were developed which sought to impose a structure on the software development process to make it more of a disciplined engineering process and less of an art. These approaches became known collectively as structured design methodologies. In general, structured design methodologies provide some combination of the following elements (from Bott 1988) :

- A description of a set of documents or other products to be produced
- A set of standards prescribing the format of these documents
- A set of procedures for preparation of the products
- A set of quality assurance procedures for verification and validation of the products
- A collection of software tools to support these activities

A variety of different structured methodologies are now in common use within the software engineering arena, e.g. Structured Systems Analysis and Design Method (SSADM) Cutts (1987), Jackson System Development (JSD) Jackson (1983), Structured Design—Yourdon and Constantine (1979), Real Time Structured Analysis and Structured Design (RTSASD) Ward & Mellor (1985). However, it is not within the scope of this report to describe these methods in any detail; the bibliography provides a list of where more in depth reviews may be found. It has been suggested recently (e.g. by Walsh et. al., 1989) that the integration of human factors techniques to software engineering methods would be facilitated by techniques which use structured design management. Such structured analysis and design methods would:

- provide a support for human factors knowledge in terms of appropriateness, timing, scope. context and granularity,
- stimulate the development of an almost complete, general method for human factors design, and
- facilitate the uptake of human factors input for designer.

This approach has, however, been less successful than expected. While the newly developed structured design methods made headway in turning the general software process from an art to a structured engineering process, the more specific software skill of user interface design did not make a similar transition. The new methods in general did not provide techniques for the specific design of the user interface. As a result, decisions about the user interface were left until the implementation phase. The user interface was then 'bolted on' to match the needs of the system rather than the needs of the user. If subsequent user testing revealed that the interface was inadequate, little could be done to improve it as it was too late in the design cycle and therefore too costly to change. Identifying the reasons why structured methods are deficient in this area has provided a popular topic for investigation. In a review of HCI integration with structured methods, Anderson (1988) lists a number of deficiencies of typical structured methods, as summarised below:

1. Structured methods do not provide formalised procedures for eliciting and expressing the operational requirements for a system. Information about user roles, instrumental goals and skills needs to be available at the earliest stage.
2. Structured methods do not support the allocation of function between user and computer. The tendency is to automate those functions which can be automated, and leave the rest for the user.
3. Structured methods provide no formal methods for assessing whether automated functions and the way they in which they are implemented will form a set of coherent tasks for the user.
4. Some structured methods do not support an object oriented approach to design, but this concept has been found to be of use in both system design and user interface design.
5. The notation of structured design methods is not readily assimilable to end users or project managers. In the absence of rapid prototyping this means that frequently, an inadequate design can be released.
6. Dialogue design is highly iterative in response to changes in system requirements and in attempting to define a 'quality' interface. Structured methods are not good at recognising this iterative process.
7. Structured methods provide only limited techniques for specifying the human-computer dialogue, e.g. Augmented Transition Networks.
8. There is no physical dialogue design stage and no reference to human-computer guidelines or style kits. Very often, screen design is left to common sense.

These deficiencies often result in design processes where the interests of the potential user population are placed outside the scope of the system design process. The outcome is that only the information which remains within the design process, which can be used for UI design, is lists of machine generated events and users' responses to these. In response to the lack of available structured methods to meet the needs of the user, a collection of HCI methods have evolved which specifically address the user issue. A subset of the more well developed and commonly used methods & tools is reviewed in the following section of this chapter.

The purpose of this review is to gain an appreciation of those methods and tools on the market which deal with the same elements of the design cycle as the enabling states method. This review is therefore limited to those methods and tools which support the early stages of the design activity, i.e. requirements capture, definition and analysis. Three core HCI activities are performed at the requirements definition stage, namely user requirements analysis, task analysis and usability specification. These activities provide ways for the design team to ascertain the characteristics of the people who will use the system, the tasks that the users want the system to support, and the ways in which the system will actually be used. A number of HCI methods have been developed to support each of these activities, as briefly described below (for details refer to the DTI HCI Service's Usability Now: Tool & Methods Handbook).

7.2. Methods for User Requirements

Traditional methods for requirements analysis tend to concentrate on the purely functional aspects of the system and the various hardware requirements deemed necessary to meet those functions. Information concerning what the intended users actually do and how they achieve it is often overlooked. The various HCI methods detailed below are all attempts at formalising and structuring the collection of user oriented information to aid in the production of a requirements set which meet the needs of the users, not the system.

7.2.1. IT-USE

ESPRIT project 2144 Information Technology Uptake Support Environment (IT-USE) aims to help users recognise, understand and actively manage the range of human and organisational problems at all stages of the design cycle. To this end, a multimedia product is under development which will meet the three objectives of: Creating an awareness of the importance of human and organisational aspects of IT systems; Providing in-depth understanding of the constituent human and organisational aspects i.e. ergonomic factors, work design, training and uptake plan; and providing tools and templates as practical support to system implementers in addressing human and organisational aspects of specific situations. The tool is intended to be a full multimedia system utilising text, graphics, animation, video and sound as a means of presentation, and will require little or no training. Currently, the system is in a demonstration status using a case study for illustrative purposes. Upgrade to complete working status is planned, but not yet fully implemented.

7.2.2. HUFIT

Project HUFIT is the ESPRIT initiative's focal point for Human Factors and Ergonomics in the Office Systems Programme. The project aims are to provide the European IT industry with the methods and techniques to produce world class products; products which have flexible, usable and effective user interfaces, and products which closely match the tasks, characteristics and needs of the users.

One way in which HUFIT aims to achieve this is through the provision of the 'HUFIT Toolset' to integrate the human factors contribution to the product design and development cycle. The toolset comprises a variety of techniques which prompt designers to consider the user perspective at various stages within the design cycle.

A User Requirements Toolset has been developed within the project to be used specifically at the requirements definition and analysis stage of the life cycle. The purpose of the toolset is to encourage greater consideration of users and tasks in product planning for the development of user requirements specifications. The User Requirements Toolset is not a set of tools in the software sense, but rather a collection of paper based activities and techniques. Four primary techniques have been developed as described below :

1. User Mapping—to identify all user groups for the planned product, and to examine the benefits and costs for those groups of using the product.
2. User Characteristics—to describe the characteristics of the user groups and to derive the product requirements resulting from them.
3. Task Characteristics—to encourage designers to consider various task characteristics and their implications for product design.
4. User Requirements Summary—to bring together and correlate the output of the previous tools.

7.2.3. User Cost-Benefit Assessment

This is a form of user analysis where positive gains and major losses to each user group of a system are examined in detail. It is primarily concerned with organisational issues such as job design, job security, personnel policies etc, and how these perceived factors will be affected by the introduction of a new system. The method is performed as a paper based exercise, with the aid of a set of checklists.

7.2.4. Contextual Enquiry

Using contextual enquiry to talk with users in their natural working environment provides information on how to better support the work of individuals, teams and organisations. The information gained helps to form an understanding of the nature of the users work, and can be used as a basis for the effective design of the software (or hardware) system. One disadvantage is that the method appears to be both time consuming and resource intensive especially when collection and analysis of video data is involved.

7.2.5. MoD/DTI Guidelines

Although the Guidelines were originally designed for the military arena, the techniques are in fact applicable to a range of areas. The Guidelines exist as a set of 6 paper based documents (Human Factors Guidelines for the Design of Computer Based Systems) providing a structured methodology for incorporating human factors into all aspects of the design cycle. The third document deals specifically with human factors in requirements specification, and is intended to be used by non human factors specialists.

7.3. Methods for Task Analysis

Task analysis is a fundamental human factors activity. Task analysis is used to collect and assimilate detailed information from the users about the precise nature of the operations of a system in certain set of circumstances. When applied at the initial stages of the design cycle, it plays an essential role in ensuring that the final system supports those tasks that the users wish to perform rather than simply those tasks which are most easily supported by the system. Numerous methods for task analysis have been developed within the human factors community and can be usefully divided into three categories; macro techniques, micro techniques and those techniques which fall naturally between the two.

7.3.1. Macro Techniques

Task analysis at a macro level is concerned primarily with the organisational, social and environmental aspects of a system. The ETHICS method (Effective Technical and Human Implementation of Computer-based Systems, Mumford 1983) explicitly involves users in the analysis process and concentrates on the exploration of users' goals, values, sources of job satisfaction in addition to traditional information flows and key tasks. It provides the design team with a means of trading off social and technical features of the design to ensure an optimised socio-technical outcome.

Functional Analysis of Office Requirements (FAOR, Schafer 1988) is a second macro technique which concentrates on using a multiperspective approach to allow the office to be perceived from a number of different viewpoints. FOAR utilises the Soft Systems Methodology (Checkland 1988) approach to guide the analyst and assist in the choice of perspectives. The disadvantage of both of these methods is that they require considerable training before they may be used by non human factors specialists.

Open Systems Task Analysis (Harker & Eason 1985) can also be regarded as a macro technique of task analysis, and involves describing functions and roles before attempting to define information requirements for IT systems. While OSTA is seen as a reliable vehicle for involving users at the requirements definition stage, it still requires some training for non human factors experts.

7.3.2. Micro Techniques

Micro level task analysis is concerned with cognitive modelling whereby the cognitive knowledge and physical actions required for users to perform their tasks is modelled. These methods involve the hierarchical decomposition of small, discrete tasks into smaller cognitive units. Numerous different methods have been developed at this micro level of analysis with a plethora of similar acronyms :

- CCT—Cognitive Complexity Theory (Kieras & Polson 1985)
- CTA—Cognitive Task Analysis (Barnard 1987)
- CLG—Command Language Grammar (Moran 1978)
- GOMS—Goals, Operators, Methods and Selection rules (Card et al 1983)
- KAT—Knowledge Analysis of Tasks (Johnson & Johnson 1989)
- PTR—Personalised Task Representation (Gregory 1979)
- TAG—Task Action Grammar (Payne 1985)
- TAKD—Task Analysis for Knowledge Descriptions (Johnson et al 1985)

All of these paper based methods require considerable training before they may be used. There are currently no publicly available software tools to aid in this process, although moves have been made to develop in house support for some of these methods by a number of organisations e.g. Sun based support for PTR Diagrams at HUSAT; support for KAT, implemented in KEE, a knowledge engineering tool, running on SUN 3/160s at QMC, University of London; and an expert system tool to support CTA developed by the Esprit BRA3066 Amodeus project, at the Medical Research Council Applied Psychology Unit (May, Barnard & Blandford, 1993).

7.3.3. *Intermediate Task Analysis Techniques*

Lying in-between the high level organisational analysis of the macro techniques and the fine grain analysis of the micro techniques lies Hierarchical Task Analysis (HTA).Using this method, tasks are broken down into operations and suboperations in a hierarchical fashion. Plans are drawn up to indicate how each stage of the operations are to be carried out. The analysis may be represented in both tabular and graphical format. This technique has been widely adopted in industry, particularly for use in training needs analyses. HTA is a relatively simple technique, requiring a minimum of training (in comparison to other TA methods).

7.4. Methods for Usability Specification

Specifying the usability requirements of a given system at an early stage in the design cycle provides a measure of acceptable performance of the system for specific types of users carrying out specific types of tasks. Two such methods for specifying usability are the System Attribute Specification approach, and an explicit statement of usability as advocated by the International Standards Organisation. Both approaches are described below.

7.4.1. *System Attribute Specification (SAS) Tables*

When usability is quantitatively specified in advance of implementation, it is known as usability engineering. The objectives of usability engineering are to give usability an operational definition through the use of metrics, and to set planned levels for usability attributes. Usability attributes in this sense relate to learnability, throughput, flexibility and attitude.

One method for specifying the usability requirements is through the use of system attribute specification tables as proposed by Bennett al (1989) and Whiteside et al (1987). Using these tables it is possible to specify which usability attribute is being measured; what users will be performing what tasks under which circumstances; what metric will be used to gauge performance; which method will be used to collect the information and finally what criterion will be used as a mark of success. It is intended that this method be used by the design team and checklists are available to help them in the construction of the SAS tables.

7.4.2. ISO 9241/11 Usability Statements

Currently, there is limited help offered by the various HCI standards for the requirements definition and analysis phase of the design cycle. ISO 9241 (Ergonomic Requirements for Office Work with Visual Display Terminals) does provide some general advice on the design process in part 10—Dialogue Principles. ISO 9241 Part 11 also offers some support for usability specification by the provision of a framework for communicating usability and functionality requirements. This framework takes the form of a 'Usability Statement' in which the usability requirements for a proposed system may be stated in terms of its intended users, uses and environments for use. The statement may be used by designers seeking to establish a systematic basis for the consideration of usability in the design process, or by consumers seeking a suitable product or work system. This standard is beginning to have more significance in the light of EC Directive 90/270/EEC on the minimum safety and health requirements for work with display screen equipment. The relationship between international standards and usability will be discussed in more detail in chapter 3.5

7.5. HCI Tools

When considering the term 'tools' in a strict sense to mean software, rather than paper based, support there appears to be little available to support HCI methods and techniques. Some CASE tools are emerging which provide for the inclusion of limited HCI activities and these are discussed below.

Computer Aided Software Engineering (CASE) tools are now well utilised within the software engineering fraternity. The majority of CASE tools are intended to support the Requirements Capture and Analysis phase of the design cycle. They provide the software engineer with a set of diagramming facilities working around a central data dictionary. The information processing requirements of an organisation can then be analysed from which an appropriate design may be derived. Frequently, CASE tools support a specific structured design methodology e.g. TEAMWORK (Cadre Technologies Inc) supports the standard Yourdon method, with a real time extension to support RTSASD (Real Time Systems Analysis and System Design).

While CASE tools are acknowledged as being powerful tools for the capture and analysis of formal information system characteristics, they seldom provide support for the involvement of users in the early stages of design, or for the capture of specific user requirements. More recently, an awareness of the need to incorporate user input early on in design has emerged, and a few CASE tools are now appearing which provide for such user involvement. The provision of graphical facilities for the design of user interfaces within CASE tools allows managers and the potential end users of the system to visualise and discuss the design, and ascertain whether or not it will meet their requirements.

7.5.1. Software Through Pictures

Software Through Pictures (Interactive Development Environments, IDE) is one of the rare breeds of CASE tools that does support an approach to requirements capture that involves potential users of the system. The tool provides a number of editors to handle design data in various formats e.g. Dataflow Diagram Editor, Entity Relationship Editor, Control Flow Editor. In addition to these, it also provides the Transition Diagram Editor, a facility for prototyping the interface.

This prototyping facility is incorporated within the User Software Engineering approach advocated by IDE. The Transition Diagram Editor is a graphical front-end to the RAPID/USE language. A textual user interface is specified with a hierarchical state transition network of nodes and arcs. The diagram is compiled to an executable prototype giving prospective system users the opportunity to work directly with the interface. One obvious drawback of this tool is that it is restricted to the production of purely alphanumeric type interfaces.

7.5.2. Teamwork

Cadre Technologies Inc's Teamwork is a set of fully integrated tools which support different viewpoint of system requirements evaluation, based upon well proven and widely promoted methods. Each tool is a multiwindowed multiuser analysis module, all of which provide automatic documentation facilities. They share a single project database containing all project information, from data dictionary to the text and graphic notes required for tracking design decisions. Examples of the tools are Teamwork /SA—for structured analysis through functional decomposition; Teamwork/RT—the real time extension to basic Yourdon notation; Teamwork/IM—an environment for information modelling; Teamwork/ADA—for the graphical automation of Ada software design.

A recent addition to the Teamwork suite of tools is a prototyping facility which has been developed by Instrumatic UK Ltd. ShortCut allows aspects of the system not represented by the standard model building notation to be modelled. For example, system interactive requirements can be verified and users can confirm whether the prototype conforms to their ideas for the look, feel and behaviour of the end system.

ShortCut allows the designer to create a number of views of the intended system, using a graphic interface design tool. This has libraries of graphics icons and symbols which are user configurable, which allows the construction of a graphical representation of the final system. These representations are tied to the functionality described in Teamwork's data flow diagrams, which ShortCut compiles to provide the drive behind the graphics during actual prototype testing and verification.

7.5.3. DESIGN/1

DESIGN/1 belongs to the larger set of FOUNDATION CASE tools available from Andersen Consulting. In addition to a range of CASE facilities, such as a central data repository and various data representation editors, DESIGN/1 also provides a screen/report design facility to encourage user involvement in the system analysis process. Screen/report images may be created using elements stored in the design repository, and then manipulated using the screen or report painter.

A simple prototyping facility is also provided which allows limited navigation of the application to be simulated. DESIGN/1 supports the design of OSF/Motif, OS/2 Presentation Manager and MS Windows 3.0 user interfaces.

7.5.4. Excelerator

Excelerator from Intersolv plc has been developed to provide specific support for communication with users in addition to the usual CASE facilities. It allows for the creation of user interface designs and report prototypes for testing with potential end users. The diagrams are created using a graphical interface which allows objects to be exploded to give more detail e.g. a computer icon can be exploded to illustrate a prototype data entry screen. In addition to the interface editors and the usual data representation editors, there is also a facility for the creation of high level presentations for end users and managers.

7.6. Using Enabling States with RTSASD

If the Enabling States Method is to have any impact on current design practice, it is essential that the methods advocated are actually usable by the design team within the context of their own methods and techniques. There are many different types of design methods currently in use, as described above, and it is beyond the scope of this review to ascertain how the enabling states method relates to all these methods. However, it is useful to consider a case study of how it can be tailored to meet one particular method. This section of the report is concerned with how the enabling states method can be used within the context of the Real Time Structured Analysis and Structured Design (RTSASD) methodology.

RTSASD emerged in the 1980s as a real time extension to the structured analysis and structured design approaches of the 1970s. In essence, it allowed for the dynamic behaviour of the system to be specified in addition to its functionality. A full and detailed description of the RTSASD method may be found in Ward & Mellor (1985). For the purposes of this report it is sufficient to consider the main elements of a RTSASD design (shown in figure 9).

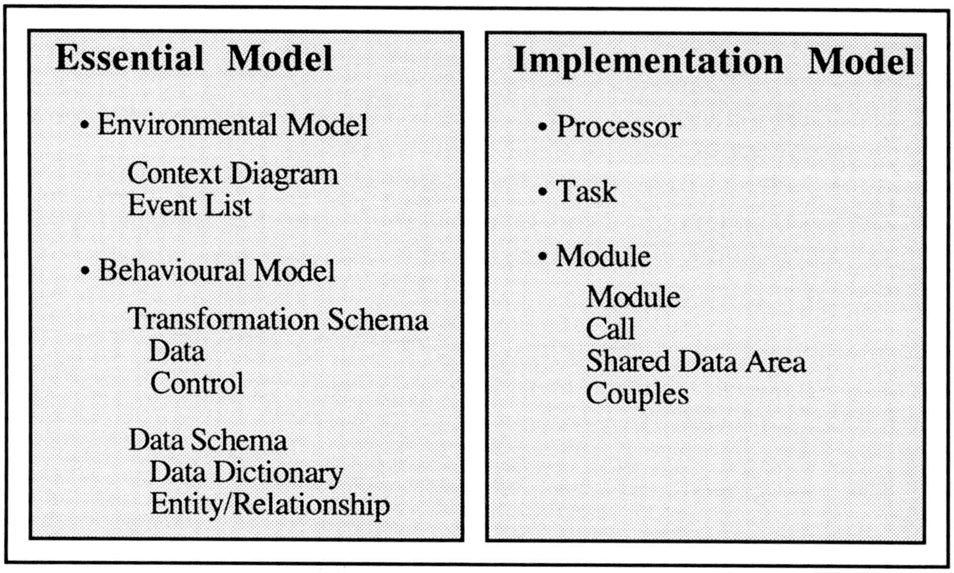

Figure 9. RTSASD Elements

The Environmental Model is a description of the environment in which the system operates. It has two parts : the Context Diagram, a description of the boundary that separates the system from its environment; and the Event List, a description of the external events to which the system must respond. The Behavioural Model is the description of the behaviour required of the system in response to events in the environment. This also has two parts : the Transformation Schema, a graphical representation (data flow diagrams together with control flow diagrams) of the transformations the system makes in response to events; and the Data Schema, a description of the information the system must have in order to respond. Finally, the Implementation Model is primarily concerned with the descriptions of the various allocations to Processors, Tasks and Modules.

One major drawback of RTSASD is that the Essential model phase of design is primarily concerned with requirements *analysis* rather than the earlier requirements *capture* stage. A second, and more fundamental drawback of RTSASD is that the design of the user interface (or man machine interface MMI) aspect of the system is almost completely ignored until the implementation phase. Any reference to the MMI within the behaviour model data flow diagrams is likely to be at a very high level e.g. 'operator input', where input could imply anything from a single digit/keystroke entry to large scale data configuration inputs of many megabytes. This 'operator input' is then left to be further defined only at the implementation phase. Unfortunately at this point the design and implementation of the MMI is often as large a task as the rest of the system design, but with only a fraction of the resources allocated to it.

Discussion with experienced RTSASD designers has led to a number of suggestions on how the enabling states method can be incorporated into the RTSASD method. It was acknowledged that the output of the enabling states method i.e. object and action specifications, message sequence charts, state transition diagrams and storyboards were all useful, but at different stages of the RTSASD process. Figure 10 illustrates one way in which the enabling states method can be used in conjunction with RTSASD.

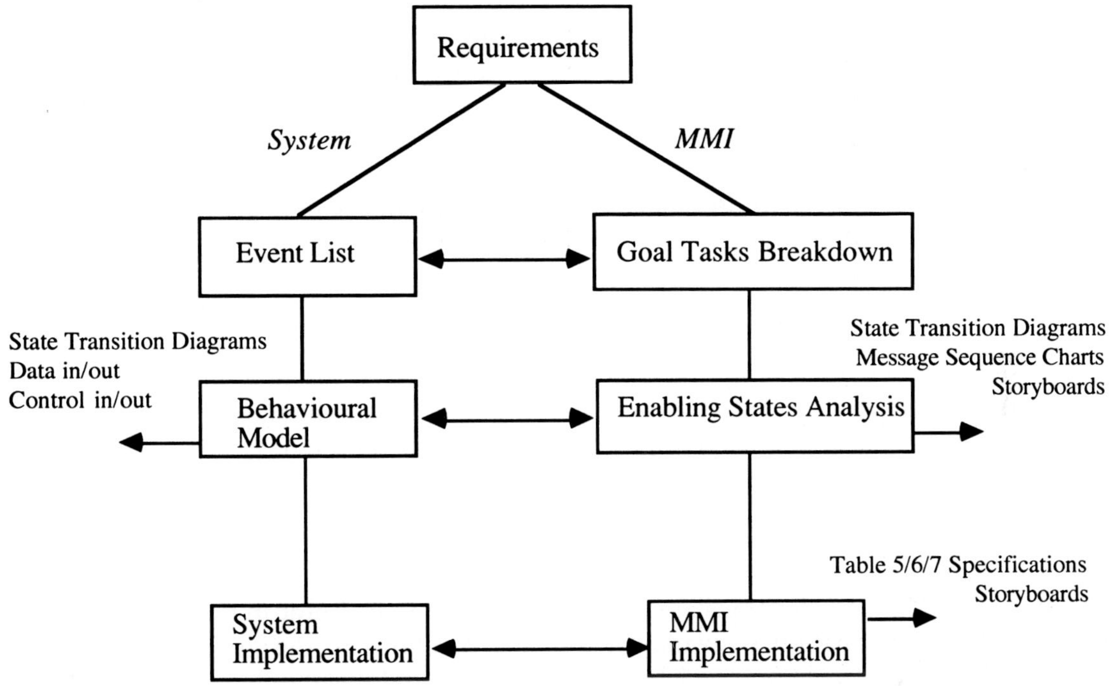

Figure 10. RTSASD/enabling states method integration

On receipt of the initial requirements statement two complementary strands of activity may be pursued. On the RTSASD strand, an event list is drawn up, which gives brief descriptions of the inputs to and outputs from the system. A behavioural model is also created which provides state transition diagrams and the data and controls in and out. Within the enabling states strand the corresponding events are the goal state and goal task definitions, and the enabling states analysis. Output from these two stages include storyboards and state transition diagrams, both of which can usefully feed in to the behavioural model of the RTSASD strand. The design approach described thus far has dealt solely with what the system has to do. Once we consider how we want the system to achieve its tasks we are moving into the implementation modelling phase. It is at this stage that the descriptions of the interface objects and their behaviour, as specified in the specification tables 5, 6 and 7 can feed into the RTSASD implementation strand (see figure 10).

Using this dual pronged approach, it is possible to see how the enabling states method can be usefully integrated into a real time design. The merit in this approach is in its modularity. Thus while separate support tools may be employed for both strands (i.e. CASE tools for RTSASD (e.g. TEAMWORK) and the Design Support Package (DSP) for the enabling states strand) the output of the enabling states specification can be readily incorporated into the RTSASD strand without any major transformation.

8. Conclusion

This investigation of current design methods has illustrated the scarcity of mechanisms facilitating relevant and timely human factors input. Various HCI methods and tools are emerging to fill this void, but are rather fragmented in their approach. Most methods reviewed seem to offer assistance for discreet stages of the design cycle, and need to be supplemented by, or used in conjunction with, other methods and tools. The tools reviewed within this chapter do offer the advantage of being integrated with existing and well used CASE tools, which will obviously enhance their uptake and continued use. However, the level of support offered by the tools appears to be limited solely to prototyping activities.

It can be concluded from these findings that there is still much scope for the development of further methods, and particularly tools to support the software design activity, be it the design of basic standalone user interfaces or the design of complex interfaces for fully networked telecoms services. Thus the user centred enabling states method and tool support, as viewed in the context of existing design & HCI methods, can offer a valid and significant contribution to this field.

The specification of human-machine interfaces must be approached in a controlled and sequenced manner so that each design decision is based upon well-founded information and not on assumptions, that the decisions are taken at an appropriate point in the sequence so that the appropriate information is available, that the solutions to design problems that are considered are drawn from the complete space of options, and that design advice is formulated in a comprehensible manner and its content is relevant to the appropriate design stage. The enabling states approach to design, coupled with the principles that will be described in the next chapter, provide a conceptual framework to specification that has these characteristics.

The enabling states method is a formalisation of the process designers need to follow if they are to specify a design that is centred on supporting a user's goals. It is not intended to be used in place of or in opposition to existing design practice—the case study using the enabling states method in conjunction with RTSASD shows that the method can successfully complement existing design methods. It provides structured and empirically justified information that designers currently have to try to obtain for themselves as they go through the design on an ad hoc basis. Where 'craft skills' are currently used to bridge gaps in design, chapter 3.2 will show how principled design advice can be codified and presented in conjunction with the enabling states method. Chapter 3.3 will describe the experience of implementers who have used the enabling states method, from both the human-factors and the software engineers' points of view.

References

Anderson, J. M. (1988) The Integration of HCI Principles in Structured System Design Principles. In *Conference proceedings Milcomp 88, Military Computers, Graphics and Software* 27th-29th September, London.

Barnard, P.J. (1987) Cognitive Resources and the Learning of Human Computer Dialogues. In J. M. Caroll (ed) *Interfacing Thought: Cognitive Aspects of Human Computer Interaction*. MIT Press, Cambridge, Mass.

Bennett. J., Butler, K. & Whiteside, J. (1989) Usability Engineering. Tutorial presented at *CHI '89*, Austin, Texas.

Bott, M. F. (1988) Systems Analysis and Design Methods and Integrated Development Environments : The State of The Art and the Future. In E D Megaw (Ed) *Contemporary Ergonomics 1988*. London. Taylor & Francis.

Byerley P., Denley I., Melchior E-M., Voigt U-B., Floruß S., and Hermann S., (1993). A unified Method for the design of Telecommunication Services. *Ergonomics. Special Issue on Telecommunication.*

Card, S. K., Moran, T. P. & Newell, A. (1983) *The Psychology of Human Computer Interaction.* Lawrence Erlbaum Associates, Hillsdale, New Jersey.

Checkland, P. B. (1988) Soft Systems Methodology : An Overview. In E D Megaw (Ed) *Contemporary Ergonomics 1988*. London. Taylor & Francis.

Cutts, G. (1987) *Structured Systems Analysis and System Specification* Yourdon Press. New Jersey.

DTI HCI Service (1991) *Usability Now HCI Tools & Methods Handbook* Department of Trade & Industry.

Gregory, R. (1979) *Personalised Task Representation* (AMTE9E) Report TM79103, Copyright Controller HMSO.

Harker, S. & Eason, K. (1985) Task Analysis and the definition of user needs. In G. Johannsen, G. Mancini and L. Martensson (eds) *Analysis, Design and Evaluation of Man-Machine Systems*, Pergammon Press

ISO 9241 *Ergonomic Requirements for Office Work with Visual Display Terminals*

Jackson, M.A. (1983) *System Development* Prentice Hall, New Jersey.

Johnson, P. & Johnson, H. (1989) Knowledge Analysis of Tasks : Task Analysis and Specification for Human Computer Systems. In A. Downton (ed) *Engineering the Human Computer Interface*. McGraw Hill, London.

Johnson, P., Diaper, D. & Long. J, B. (1985) Tasks, Skills and Knowledge : Task Analysis for Knowledge Based Description. In B. Shackel (ed) *INTERACT '84, Human Computer Interaction*. Elsevier.

Kieras, D. & Polson, P.G. (1985) An Approach to the Formal Analysis of User Complexity,. *International Journal of Man Machine Studies, 30*, pp 47—62.

Lim, K.Y., Long, J.B., Silcock, N. (1992) Integrating human factors with the Jackson System Development method: an illustrated overview. *Ergonomics, 35*, 1135-1162.

Long, J.B., Dowell, J. (1989). Conceptions of the discipline of HCI: Craft, applied science and engineering. In A. Sutcliffe & L. Macaulay (Eds.), *People and computers V (pp.* 9-32). Cambridge: Cambridge University Press.

May, J., Barnard, P.J. & Blandford, A (1993) Using structural descriptions of interfaces to automate the modelling of user cognition, User Modelling and User Adapted Interfaces, 3 (in press)

Moran,T.P. (1978) *Introduction to the Command Language Grammar* Report No 55L-78.3 Xerox Corp, Palo Alto, California.

Mumford, E. (1983) *Designing Human Systems for New Technology.* Manchester Business School Publications, Manchester.

Payne, S. J. (1985) Task-Action Grammars. In B. Shackel (ed) *INTERACT '84,* Human Computer Interaction. Elsevier.

Schafer, G. (1988) *Functional Analysis of Office Requirements—A Multiperspective Approach'* John Wiley. Chichester.

Walsh, P., Lim, K.Y., Long, J.B. (1989) Jackson System Development and the design of user interface software, *Ergonomics, 32,* 1483-1498.

Ward, P.T. & Mellor, S.J. (1985) *Structured Development for Real-Time Systems* Yourdon Press. New Jersey.

Whiteside, J., Bennett, J. & Holtzblatt, K. (1987) Usability Engineering : Our Experience and Evolution. In M. Helander (ed) *The Handbook of Human Computer Interaction.* North Holland Press.

Yourdon, E. & Constantine, L.L. (1979) *Structured Design* Prentice Hall, New Jersey.

Computers, Communication and Usability:
Design Issues, Research and Methods for Integrated Services
P.F. Byerley, P.J. Barnard and J. May (Editors)

Chapter 3.2

Usability Principles for Service Design

Ian Denley[1], Leif R. Hedman[2], Becky Hill[1], Andy Whitefield[1],
Kiri Papadopoulos[2], Anne Clarke[3], Nick Hine[4]

[1]Ergonomics Unit, University College London, 26 Bedford Way, London, UK, WC1H 0AP

[2]Telia Research, Hjälmaregatan 3, Box 85. S-20120, Malmö, Sweden.

[3]HUSAT Research Institute, The Elms, Elms Grove, Loughborough. Leics, UK, LE11 1RG

[4]The Microcentre, Department of Mathematics and Computer Science, University of Dundee, Dundee, UK, DD1 4HN

1. Introduction

Integrated Broadband Communication (IBC) systems will merge computing, networking and telecommunication technologies to deliver a large number of services to users at a single terminal. The GUIDANCE project (see acknowledgements) addressed the general problem of how to ensure that the services provided to the end-user within an IBC system can be used efficiently and effectively; and more specifically, the problem of how might a number of services required by an end-user be integrated to maximise their ease of use?

This chapter presents part of the project's solution to the problem of service integration—the notion of service integration principles for designers. The project has developed a service design method which takes a designer through a set of design decisions, at a number of which prescriptive advice in the form of service integration principles is offered. The problem of service integration is described in chapter 1.2; an introduction to the design method as a whole can be found in chapter 3.1, and the use of the method in the design of a prototype system can be found in chapter 3.3. This chapter, however, concentrates on the service integration principles. There are two related issues to be addressed when considering design principles. The first is concerned with the Human Computer Interaction (HCI) knowledge which they embody and how it is expressed, i.e. with 'what do we know?' The second is concerned with their practical application by designers, i.e. with 'what should we do?' These two issues are discussed in sections 2 and 3 below. In section 4 a case history in multimedia conferencing is used to illustrate the application of the principles in service design.

2. The content and expression of principles

The term *principle* has been used in a number of ways in the HCI literature, a fact which often leads to confusion. Long & Dowell (1989) have identified three different approaches to HCI: the *craft, applied science,* and *engineering* approaches. Each approach can be seen as utilising a form of principle, although these differ in many ways, including their derivation and expression.

The first type of principle is the *heuristic* or rule of thumb. This form of principle is utilised within the *craft* approach to HCI. The knowledge embodied in an heuristic is acquired by practice and example, and is neither explicit nor formal. An example of successful application of such an approach may be found in Bornat & Thimbleby (1989) who describe the development of the text editor 'ded'. Numerous heuristics were applied and tested and those found to be beneficial included simplicity of operation, coherence, and modelessness. The successful application of such heuristics, however, depends primarily upon the expertise, guided intuition and accumulated experience of the practitioner.

The second type of principle is the *guideline*. This form of principle is utilised within the *applied science* approach to HCI. The knowledge embodied in a guideline is scientific knowledge acquired through the hypothesising and testing of theories and models. In contrast to heuristics, scientific knowledge is explicit and formal, operational, testable and generalisable. The applied science approach may be exemplified by the work of Gardiner & Christie (1987) or Smith & Mosier (1986). As one of over two hundred guidelines offered by Gardiner and Christie, the following may be taken as an example: 'In the design of tasks, task elements may be combined in the following ways—*simple + simple, complex + simple,* but not *complex + complex.*' Guidelines, however, are poorly integrated into system development, do not adequately describe behaviour, and carry no guarantee of effectiveness. Moreover, science is explanatory and predictive whereas design is prescriptive. Also, the successful application of science-based guidelines is often in fact craft-based, being dependent upon experience.

The third type of principle is the *engineering principle*. An engineering conception of HCI, as favoured by Long & Dowell (1989), would seek to develop a codified body of engineering principles. These do not yet exist; they would have the positive properties of guidelines, but in addition would be prescriptive in form and with a known probability or guarantee. They would be prescriptive in that they would explicitly and systematically describe the conditions under which specified system structures and behaviours are recommended; they would have a guarantee which relates to the consequences of following the prescription in terms of the resulting system performance.

The work on the GUIDANCE project has aimed to develop one known form of principle (the guideline) towards another, relatively unknown, form (the engineering principle). That is, it has tried to formalise the expression of guidelines: in a prescriptive manner; in line with a conception for HCI intended to support the development of engineering principles; and with some indication of an associated guarantee. The term principle in the remainder of this chapter should, unless specified otherwise, be taken to mean a principle of this type that addresses service integration.

2.1 Parameterising the Principles

The basic parameters that constitute the conditions for a principle are *users* and *machines* (i.e. the basic components of an IBC interactive worksystem {IIWS}), *tasks* (i.e. what the IIWS does), and *performance* (i.e. how well the IIWS does it).

The prescription embodied in a principle specifies the structures and behaviours of the system which will enable specified classes of IIWS to achieve desired performance on the task. Provided that the conditions are met, then application of the principle should give some guarantee or assurance that the IIWS will achieve desired performance. But given the current incomplete state of HCI knowledge, such guarantees cannot assume the same levels of certainty as is enjoyed by principles advanced by the physical sciences. This means that IBC designers must diagnose the effectiveness of the design (based on the application of the principles) through iterative evaluation of the design.

It is to be expected that there will be frequent cases of only partial conformity between a principle's parameter values and those faced by the designer. For this reason, the rationale underlying a principle is expressed, enabling the designer to make a judgement concerning the impact of partial conformity on the principle's prescription.

In summary, then, the full expression of a principle contains the following elements:

If $X = fn(T,U,M,P_d)$
then Y
because $R(Y = fn(X))$
with guarantee p>threshold

where

X is the triggering attribute set
T is the task attribute set
U is the user attribute set
M is the machine (user service) attribute set
P_d is the desired task performance
Y is the prescription
R is the rationale

2.2 Levels of Generality

Principles can be classed along a dimension of generality of expression. For example, a generic principle might be applied across a range of IBC work domains, whilst a specific principle might only be applied to a particular work domain. Principles with varying degrees of generality between these two extremes are also possible. For the purposes of this chapter principles are classified into four levels of generality with increasingly detailed design parameters and design prescriptions:

- Generic Principles—a basic set of twelve high level usability principles intended to be applicable to all IBC work domains
- Generic Principles: Integration Issues—the twelve basic usability principles as applied to the issues of service integration, work structures integration (computer supported cooperative work {CSCW}) and interactive media integration.
- Generic Principles: Application Level Services—the basic set of principles as applied to distribution, retrieval and conferencing services and their logical combinations when these services are integrated. These principles have at least two of their parameter slots filled (task and performance) and also provide a rationale for their prescriptions.
- Specific Principles— these principles can only be applied to a particular work domain. This chapter describes specific principles for the domain of multimedia multiauthor document production (MMMADP) work. Examples of specific MMMADP principles are given in section 4. These principles are fully parameterised and are embedded in a design method for end-user operational requirements.

The principles described here, then, offer improvements upon current guidelines and heuristics by deriving and making explicit their scope, content and expression. However, only the lowest levels of generality can fully utilise the format described in section 2.1. Additionally, empirical work was performed by the project to elaborate, refine or reject proposed principles. The empirical work enabled tighter specification of the conditions and rationale for particular prescriptions of IIWS behaviour, which will support the designer in reasoning about the principles' applicability and relevance to the system under design.

Each of these four levels of generality are discussed in the following sections.

2.2.1 A Set of Generic Principles

Esgate et al (1990) carried out a survey and review of Human Factors (HF) design advice reported in the literature and identified over 300 design recommendations (heuristics, guidelines or possible engineering principles) which had some relevance to the design of IBC systems. This large set of principles was reduced using criteria of applicability to service design and service integration and completeness of parameterisation to identify twelve classes of service integration principle. This basic set is listed below with a brief description of each.

Compatibility: Users should be able to apply knowledge they have gained from outside the system.

Coherence: The user should be able to generalise their experience between system components.

Simplicity: The number and complexity of necessary actions should be reduced to a minimum.

Redundancy: The user should be provided with summary information in several ways.

Salience: Critical information should be presented to the user in a sufficiently intrusive way.

Reversibility: The user should be able to restore pre-existing states of the system.

Transparency: Information about tasks which can be performed, as well as about the states of the system, should be stated unambiguously and clearly.

Completeness: Information about tasks which can be performed, as well as about states of the system, should be complete in such a way that all options are presented simultaneously.

Controllability: The user should be able to control the actions of the system.

Flexibility: The user should be able to choose the modalities of their task performance with respect to the input media available and the experience they have.

Feedback: The user should be informed about the consequences of their actions.

Support orientation: If the information to be presented is too complex or covers more than is possible to present at one time, the user should be helped to find the relevant information by giving them support in orientation.

2.2.2 Generic Principles: Integration Issues

This section gives more detailed descriptions of the set of generic principles described above and provides specific annotations to the issues of Service Integration, Work Structures Integration (Computer Supported Cooperative Work—CSCW) and Interactive Multimedia Integration.

1. Compatibility

Description

Users should be able to apply knowledge they have gained from outside the system.

Rationale:

This will help the user in learning to work with the system and reduce the amount of new information to be interpreted and remembered (e.g. cognitive load, memory load).

Detailed description:

Service Integration:

There are two alternative ways of understanding compatibility with respect to the integration of services:

1. Integrated services should exploit abstract external knowledge.

2. Integrated services should exploit specific external knowledge.
For example, if telephone and fax are integrated, the first alternative would suggest that the user use the same external knowledge for both components of the integrated service (this abstract knowledge would be akin to "communication use"). The second alternative would suggest that users should utilise their knowledge of existing audiotelephony for the telephone component of the service, and their existing fax knowledge for the fax component of the service.

Work Structures Integration:
The design model of any CSCW application should exploit common domain knowledge of the team members. Such common domain knowledge creates a shared perspective between the team members which facilitates distributed team-work.

Interactive Multimedia Integration:
The input and output modalities should, as default, be in the "natural" medium, known from other applications outside the system.

2. Coherence

Description
The user should be able to generalise their experience between system components.

Rationale
This will help the user in learning to work with unfamiliar system components and reduces memory load.

Detailed Description

Service Integration:
The range of system states across the different services that can be assumed in response to a group or sequence of actions should form a meaningful set.

Work Structures Integration:
All team members should have a common understanding of the application and be able to transfer knowledge between components of the application. For example, the commands, labels or names used should have the same meaning across the application and across team members.

Interactive Multimedia Integration:
The same media should be available in all components of the application and should be used with the same set of commands in each of these components.

3. Simplicity

Description
The number and complexity of necessary user actions should be reduced to a minimum.

Rationale
Thus, cognitive costs (e.g. memory load) and physical costs (e.g. keystrokes) are minimised.

Detailed Description

Service Integration: As the integration of a variety of services may cause more workload (memory load) than working with the services on separate basis, it is important that the actions to be performed are easy to carry out and remember.

Work Structures Integration: As CSCW may cause higher workload than working alone the number and complexity of actions required to carry out cooperative tasks (e.g. communication and joint editing) should be reduced.

Interactive Multimedia Integration: The use of different media at the same time may lead to higher workload. This can be offset by reducing the complexity of actions necessary to invoke and control different media.

4. Redundancy

Description
The user should be provided with summary information in several ways.

Rationale
This will reduce the memory load placed on the user.

Detailed description

Service Integration: With IBC, users may have access to a large number and variety of services, and should, therefore, be provided with summary information about the state of these integrated services. For example: which service is available, which service is already in use etc.

Work Structures Integration: As cooperative work might increase workload it is necessary to provide the user with summary information about their own work state and the work state of other users or team members.

Interactive Multimedia Integration: Multimedia can be used to present information in a number of ways (e.g. redundancy of combined visual and audio signals). However, media redundancy can also confuse or irritate the user, especially if the user is to be expected to be engaged in more than one task at the same time.

5. Salience

Description

Critical information should be presented in a sufficiently intrusive way.

Rationale

This draws attention to the most critical part of the system at a given moment, and helps users avoid missing important information.

Detailed Description:

Service Integration: Users must be made aware of the consequences of using different services. For example, users must be given clear and unambiguous information concerning the financial costs being incurred through service use, or their loss of privacy when using a shared application.

Work Structures Integration: Cooperative work requires an awareness of the actions of other team members, the consequences of these actions, and the effect of one's own actions on other people. System behaviour must be sufficiently intrusive to support such awareness amongst users.

Interactive Multimedia Integration: The use of contrasting media can be used to attract attention. However, the medium used for attracting attention has to be carefully selected with respect to the media already presented on screen. (e.g. attracting attention when a video is playing).

6. Reversibility

Description

The user should be able to restore pre-existing states of the system.

Rationale

This allows users to change their mind without being forced to perform an excessive number of enabling tasks.

Detailed Description

Service Integration: Set-up, close-down and configuration of services must allow users to change service components without incurring excessive costs.

Work Structures Integration: Cooperative work can result in changes to the work of other team members. Users must be able to return work to former states. This return must be possible regardless of when the changes occurred.

Interactive Multimedia Integration: Users should be able to undo all media conversions i.e. to undo the conversion from text to speech back to text.

7. Transparency

Description

Information about tasks which can be performed, as well as about the states of the system, should be stated unambiguously and clearly.

Rationale

This should support the user in being able to predict the consequences of their planned actions. No trial and error behaviour is necessary.

Detailed Description

Service Integration: Transparency is particularly important if a great variety of services are offered to the user. Selecting inappropriate services will be costly both in terms of financial costs and decreased task quality.

Work Structures Integration: 1. The application (including commands, dialogues etc.) should be unambiguous for all members of the team. 2. Organisational influences, the existing hierarchy and its consequences and the aim of the cooperative work in this context should be clear to all team members.

Interactive Multimedia Integration: The medium which is used as a default for presenting information should enable the user to gain a clear and unambiguous understanding.

8. Completeness

Description

Information about tasks which can be performed, as well as about states of the system, should be complete in such a way that all options are presented simultaneously.

Rationale

This should reduce the memory load of the user and thus give them the opportunity to decide between all potential options and task actions.

Detailed Description

Service Integration: All available services should be offered.

Work Structures Integration: Members of a team should be informed about the work of other team members.

Interactive Multimedia Integration: Information presentation should exploit different media as appropriate to task performance.

9. Controllability

Description

The user should be able to control the actions of the system.

Rationale

This should give the user the locus of control and thus enable them to perform their actions as intended.

Detailed Description

Service Integration: The user should be able to control the set-up and configuration of services.

Work Structures Integration: Team members should be able to control what is being presented about their own work and their actions in order to maintain privacy.

Interactive Multimedia Integration: User control over the actions of the system should not be compromised by the presentation media used.

10. Flexibility

Description

The user should be able to choose the modalities of their task performance with respect to the input media available and the experience they have.

Rationale

This will enable the users to perform the tasks in the most appropriate and efficient way, and encourages the development of specific user strategies.

Detailed Description

Service Integration: The integration of services should not lead to a reduction in the flexible use of services. The user should not be forced to use a specific (e.g. expensive) services for a specific tasks.

Work Structures Integration: 1. Team members should be able to select their own work strategies without compromising the groups' work as a whole. For example, team members may wish use different text editors. 2. The application should itself reflect changes in the work team or the organisation of the work/task.

Interactive Multimedia Integration: The user should be able to use alternative input media e.g. speech, gesture etc. Users should also be able to specify the media used to present information and make real-time changes to this specification.

11. Feedback

Description

The user should be informed about the consequences of their actions.

Rationale

This enables the user to understand the actions and states of the system and allows them to avoid errors and plan their next actions.

Detailed Description

Service Integration: Users should be informed about the state of the services they are using (e.g. quality of transmission, costs etc.).

Work Structures Integration. Team members should be informed about relevant changes of the other team members or relevant actions of the other team members.

Interactive Multimedia Integration: Feedback should be given in the appropriate medium for the task.

12. Support orientation

Description

If the information to be presented is too complex or covers more than is possible to present at one time, the user should be helped to find the relevant information by giving them support in orientation.

Rationale

This enables the user to find the relevant information and helps them to perform even complex tasks.

Detailed Description

Service Integration: Users should be able to navigate though all available services.

Work Structures Integration: Team members should be able to locate desired information regardless of whether they are working on their own objects or the objects of other team members.

Interactive Multimedia Integration: Navigation tools should be provided to support the user in locating themselves (and the desired information) in multimedia documents.

2.2.3 Generic Principles: Application Level Services

Chapter 1.1 identifies three possible Application Level service types, namely:

- Distribution (one or more send services)
- Retrieval (one or more retrieval services)
- Conferencing (one or more conversational services)

Distribution occurs when the source of the information has control and sets up the call without negotiation with the sink. Communication is thus unidirectional.

Retrieval occurs when the source of information is held by one communicating entity, but the controller of the call is a different entity who wants the information. Communication is, therefore, required in both directions, and for both directions the initiator is the sink. No simultaneity is possible. The send from the sink is the signal to the source for information, to which the source responds immediately with a send, thus both sends are contained within one call set up by the sink.

Conferencing occurs when the possibility exists for sends to occur bidirectionally. Neither communicating entity has control, and either could have set up the call, and both must agree to partake.

These three types of tasks (illustrated in figure 1) are discussed in the following sections. There are a number of usability issues associated with each type of task. Broadly speaking, however, for each type there is a dominant usability issue determined by the nature of the services on offer, thus:

- Distribution: Feedback/ message confirmation issues
- Retrieval: Authentication
- Conferencing: User to user negotiation procedures

The three basic Application Level service types are considered in sections 2.2.3.1 to 2.2.3.3. Each section highlights a number of usability issues associated with each service type and gives example generic application level principles for one of the issues. The usability issues and the principles were identified from the data, results and Human Factors experience of a number of RACE projects. The data on which they are based was presented at a RACE Usability Consolidation Workshop held in London in October 1992 (see acknowledgements).

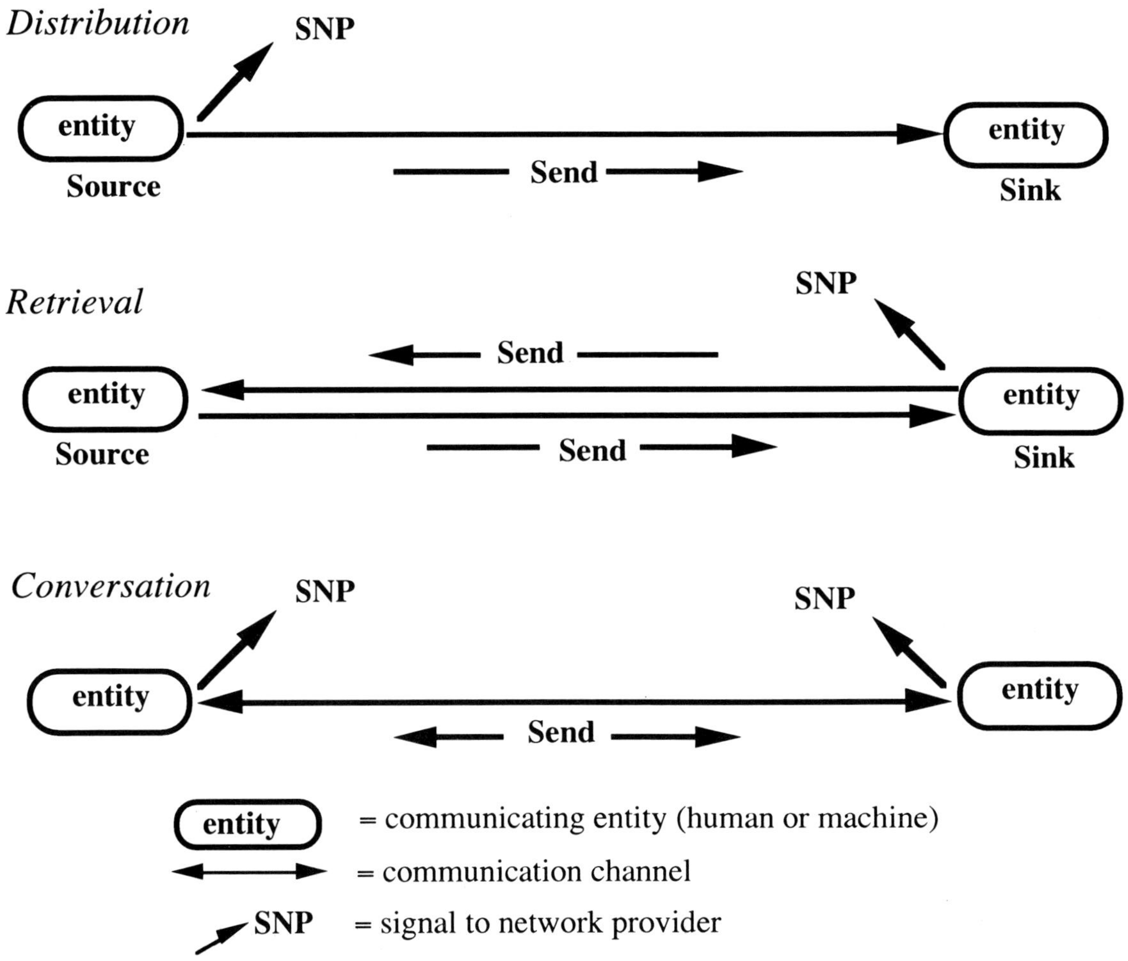

Figure 1: The three service types

2.2.3.1 Distribution

There are a number of usability issues associated with distribution. These are highlighted below, and generic application level principles are then described for the major issue associated with distribution i.e. message confirmation.

1. Message Confirmation
Information on the status and progress of a distribution task

2. Addressing
User procedures for message addressing

3. Filtering
The filtering of unwanted messages (e.g. junk mail etc.)

4. Priority
Specifying the priority of a distribution task

5. Message Abort
Aborting a distribution task

6. Ordering
Information concerning the ordering of messages on the sink

7. Security/Privacy
Access rights to messages

8. Reliability
The reliability of a distribution task e.g. alarm systems

9. Service resource availability
Information concerning availability/capabilities of sink prior to connection

10. Defaults
Default configurations for distribution tasks

11. Cost feedback
Information on the costs of distribution tasks

Generic Principles:

Application Level Services Associated with Message Confirmation in Distribution Services

The most important generic application level principles associated with the issue of message confirmation are **feedback, controllability** and **salience.**

Feedback can be provided to users at a number of stages within the distribution task. These stages, and the direction of feedback to different users are shown in figure 2. Feedback can be provided to *senders* during the writing of a message (e.g. the status of draft or stored messages); when a message is sent from the user's terminal (e.g. about the validity of the message address); during the progress of the message through the network (e.g. network locations where failure occurred); the sender can also be offered feedback about when the message arrives, and whether it has been read. *Receivers* are provided with feedback about the presence of incoming messages and whether they have been read and discarded.

```
sender                              receiver
        <------------   compose
        <------------   send
        <------------   network
        <------------   receive  ------------>
        <------------   read      ------------>
                        delete    ------------>
```

Figure 2: Stages of Feedback within a Distribution Task

Feedback

IF the task is information distribution
AND desired performance is that all messages should arrive in the form they were sent within a specified timespan
THEN distinct classes of feedback should be specified to senders concerning the progress of the information distribution
BECAUSE: distinct reasons for message failure will allow the user to discern the point in the process where failure occurred and then take appropriate remedial actions.

These desirable classes of feedback can be decomposed as follows:

IF the task is information distribution
AND desired performance is that all messages should arrive in the form they were sent within a specified timespan
THEN senders should be given the following feedback concerning the progress of the information distribution:

1) Before the message is sent:
   - that the message will be delivered before a stated time
   - that the receiving terminal and network capabilities are adequate to deliver the message.
2) When the message is sent:
   - the message has been sent
3) During message transmission:
   - if the message cannot be delivered by the stated time or with the capabilities desired at the start, or due to receiver filtering.
4) When the message has been received
   - that the message has been received, the time it arrived and the quality level
5) When the message has been read
   - that the message has been read and the time.

BECAUSE: distinct reasons for message failure will allow the user to discern the point in the process where failure occurred and then take appropriate remedial actions.

Controllability

IF the task is information distribution
AND desired performance is that all messages should arrive in the form they were sent within a specified timespan
THEN the sender should be able to interrogate the recipients' system to check that inappropriate message bandwidths or media types are not sent
BECAUSE: knowledge of the recipients terminal characteristics will allow senders to control and modify their own actions.

IF the task is information distribution
AND desired performance is that all messages should arrive in the form they were sent within a specified timespan
THEN
• users should be able to specify desired classes of feedback
• users should be able to specify media format of the feedback
BECAUSE: users may have preferred strategies for identifying message failure and taking appropriate remedial actions.

Salience

IF the **task** is information distribution
AND **desired performance** is that all messages should arrive in the form they were sent within a specified timespan
THEN
• users should be able to specify priority of intrusiveness of feedback
• feedback on messaging errors should be sufficiently intrusive to attract users attention and may be related to priority of message (if specified by user).
BECAUSE: users may have preferred strategies for identifying message failure and taking appropriate remedial actions.

IF the **task** is distribution of emergency signals
AND **desired performance** is to have a zero failure rate in sending an alarm, where failure in this instance is attributable to sender uncertainty concerning the progress of such alarm messages
THEN feedback on emergency alarms should always include:
• sent confirmation (where the user should be made aware that the alarm message will override any other communication that is taking place)
• received confirmation
• acted upon confirmation
BECAUSE: users need to be reassured that all stages of the alarm process have progressed as desired.

2.2.3.2 Retrieval

There are a number of usability issues associated with retrieval. These are highlighted below, and generic application level principles are then described for the major issue associated with retrieval i.e. authentication.

1. **Authentication**
   Confirmation of user access rights and privileges to the information to be retrieved
2. **Specification of information**
   User specification of classes of information to be retrieved
3. **Presentation of information (Indicating desired processing rate by sink)**
   The presentation and representation of retrieved information to users
4. **Service resource availability**
   The service resource capabilities of the source and the sink with respect to the information to be retrieved
5. **Security**
   Information concerning different classes of user access

6. Confirmation of connection

Feedback on the progress of the retrieval task

7. Access rights/ filtering (at sink)

Informing the users of their own and the required access rights

8. Costs

Determining the costs of a retrieval task

Generic Principles:

Application Level Services Associated with Authentication in Retrieval Services

The most important generic application level principles associated with the issue of authentication in retrieval services are **flexibility** and **simplicity**.

Flexibility

IF the task is information retrieval
AND desired performance is that required access procedures should support users of varying capabilities
THEN:
- authentication should not be based on entry time
- authentication should not be based on one media i.e. users should have a choice of different media so that users who cannot use a particular media can still retrieve information
- macros or alternative ways of access should be allowed

BECAUSE: these will reduce access failures and reduce user's affective costs.

Simplicity

IF the task is information retrieval
AND desired performance is that required access procedures should support users of varying capabilities
THEN:
- the sink must automatically identify itself so that the user does not have to enter their name/address (once the user is recognised by the server)
- the sink's address should be defaulted as the receiving address for the information
- there should be an option for forwarding the information

BECAUSE: these will reduce access failures and reduce user's affective costs.

2.2.3.3 Conferencing

There are a number of usability issues associated with conferencing. These are highlighted below, and generic application level principles are then described for the major issue associated with conferencing i.e. call control/user negotiation.

1. Call Control

User or third party control of conferencing tasks

2. Call set-up and close down

The initiation and ending of conferencing tasks

3. Authentication
Caller identification/ lock-out of unwanted callers/ database browsing and access to information

4. Addressing
User procedures for identifying receivers e.g. work group visibility

5. Call ownership/ Call hand over
The control of a conversational task with respect to billing and negotiation

6. Collaborative manipulation of information
Collaborative work and the control of information

7. Costs
Information concerning financial costs of conferencing. e.g. on-line/ default billing

8. Service Dropping/Alteration and Fallback
Changing of service components/media during conferencing e.g. video to audio

9. Feedback
Feedback concerning call attributes

Generic Principles:

Application Level Services Associated with Call Control in Conferencing Services

The issue of call control has a number of generic application level principles which can be applied in the design of conferencing services. Call control can be managed by the users through negotiation or alternatively through an intermediary or third party (for conferences, say).

a) User Control

The principles of concern here are: **controllability, feedback and salience.**

Controllability

IF the task is conferencing
AND desired performance is to ensure that a user's conversational tasks are not disrupted by other users or third parties
THEN:
- users should be to promote demote calls in queue
- users should be able to interrupt current conferencing to respond to queued calls
- users should be able to resume held call
- users should be able to alter their own call interaction—visibility to others and privacy if required
- users (especially People with Special Needs—PSN) should be able to alter media separate from call control
- users should not be able to alter their view to other parties when not currently on the floor—i.e. zooming on self not permitted etc.
- audio should be off until speech concerning conversational task is initiated (e.g. off-line privacy possible)

- during multipoint conferencing with single video component (i.e. participants can only see one video image but can hear all participants) and automatic video-switching is provided it must be supplemented by a switching method under voluntary user control.
- for users of a multi-party conversational service it should be possible (without the use of a third party):
- for a new user to join the call
- for a user to leave the call
- to split the call into two distinct calls
- to transfer ownership of the call to another user

BECAUSE: in cooperative workgroups where there are potentially a very large number of interactions between users and systems, users should be able to plan their actions to minimise disturbance in line with their task requirements and reduce their cognitive and affective costs (frustration).

Feedback

IF the task is conferencing
AND desired performance is to ensure that a user's conversational tasks are appropriate with respect to intended audience
THEN:
- users should be informed of how they are received by other parties—i.e. notification of speaker broadcast etc.

BECAUSE: users should always be aware of their own visibility to others in cooperative workgroups.

Salience

IF the task is conferencing
AND desired performance is to ensure that users can rationalise and control the flow of incoming calls
THEN: • users should be able to identify incoming calls during current connection
- users should be able to identify calling queue and identities of callers

BECAUSE: this will allow users to plan their actions in line with task demands

b) Third Party Control

The principles of concern here are: **controllability, salience and flexibility**

Controllability

IF the task is multi-point conferencing
AND desired performance is to ensure that previously negotiated and agreed conference organisational objectives are met for all participating users
THEN:
- the third party should be able to set-up procedures with service provider for how the conference should take place—all users with/without self-view, flexible facilities
- the third party should be able to promote demote calls in queue
- the third party should be able to interrupt current conferencing to respond to queued calls
- the third party should be able to resume held call
- the third party should be able to control a peripheral device and be aware of control actions and outcomes

BECAUSE: in cooperative workgroups where there are potentially a very large number of interactions between users and systems third party control may be necessary to meet organisational requirements (Note, this will inevitably reduce end-user control).

Salience

IF the task is multi-point conferencing
AND desired performance is to ensure that previously negotiated and agreed conference organisational objectives are met for all participating users

THEN:
- the third party should be able to identify incoming calls during current connection
- the third party should be able to identify calling queue and identities of callers
- the third party should be able to indicate request for floor
- the third party should be able to identify other requests for floor in a queue
- users should know the identity of the third party, and be notified of the third party action prior to its execution
- the third party should not be able de-allocate all the service components.

BECAUSE: this will allow both first and third party users to plan their actions in line with task demands and conference requirements.

Flexibility

IF the task is multi-point conferencing
AND desired performance is to ensure that previously negotiated and agreed conference organisational objectives are met for all participating users
THEN:
- the third party should be able to accept general control (chair), and control should be negotiable and not necessarily default to user who sets up the call. This has to be balanced by cost implications which also need to be negotiated if users have different aims for the conference
- the third party should be able to negotiate chair of group, and therefore, nominate floor holder, remove floor, take floor

BECAUSE: this will allow both first and third party users to plan their actions in line with task demands and conference requirements.

2.2.3.4 Integrating Basic Services

As chapter 1.1 describes, these three basic service types can also be carried out concurrently i.e. there are four possible combinations to consider. The four combinations are:

- Distribution and Retrieval
- Distribution and Conferencing
- Retrieval and Conferencing
- Conferencing, Distribution and Retrieval

The following are some examples of a number of problems that occur during the integration of the three basic services. The examples are restricted to the major issues that we have identified above. i.e. message confirmation (distribution), authentication (retrieval) and call control (conferencing).

For instance, feedback on the progress of information distribution may interfere with authentication procedures for information retrieval. Different media (either user specified or machine controlled) may be used to increase the *salience* between the two tasks, but the user should be able easily to *control* which service becomes primary depending on the priority of each task. For example, users may *only* be prepared to accept disturbance from feedback concerning the failure of a high priority message to reach its destination and *only* in this situation may be prepared to switch tasks in order to take remedial action.

Also, because both distribution and retrieval services use feedback in different instances the user must be easily able to distinguish between the different types of information being provided and be able to act upon this information i.e. *support for orientation* must be provided. For example, during distribution feedback will be provided on the progress of the messages sent, whilst during retrieval feedback will also be provided on the progress and the costs of the retrieval. Feedback, then, needs to be coherent within service types but also salient between them in order to support appropriate user actions. Similarly, message confirmation from distribution tasks may be particularly disruptive when combined with conferencing tasks where the number of real time messages, dialogues and conference interactions may be very high. As for distribution and retrieval different media (either user specified or machine controlled) may be used to increase the *salience* between the two tasks, but the user should be easily able to *control* which service becomes primary depending on the priority of each task. Again, the user must be easily able to distinguish between the different types of information being provided between the two services and must be able to act upon this information i.e. *support for orientation* must be provided.

It was also recommended above that authentication procedures for retrieval tasks should be flexible i.e. that the users should be able to input relevant data in any order and in any media. If retrieval is combined with conferencing then users should be able to work concurrently using different media for different tasks. For example, authentication could be performed using text whilst conducting an audio/video conference. However, using two services and two media concurrently will lead to decrements in task performance due to media interference as well as workload considerations. As a further example, in retrieval tasks the user may be asked to identify themselves to the source in order to gain access to restricted information. Similarly, in conferencing tasks the receiver of a private or confidential call may also be asked to identify themselves before call connection. In this instance, there may be possible confusion between the user identification procedures necessary to receive secure messages and those necessary to receive private calls. The user, then, needs to be given clear indication of the requirements for each service i.e. authentication requirements should be *salient*.

In sum, then, it is clear that by combining service types, complex interactions occur which require extra design decisions to be taken which are critical to the concurrent use of such integrated services. When all three service types are integrated the design problems become compounded by the increased complexity of such task and service combinations.

2.2.4 Specific Principles

As described in section 2.2.2, principles have been developed at different levels of generality and expression. The GUIDANCE project used, as a vehicle for its research, systems to support multimedia multiauthor document production (MMMADP) work. Consequently, the specific principles developed by the project were restricted to tasks related to multimedia document production, users who are authors (multiple and distributed) and machines running integrated services to support document preparation.

The specific application chosen was the production of multimedia holiday documents. This allowed us to collaborate with another RACE project which was implementing holiday information databases on prototype multimedia terminals (see Jerke et al 1990).

Over 150 specific principles were developed and tested by the project, each of which is fully parameterised in line with the format described in section 2.1. Figure 3 gives an example of the relationship between these parameters and the design prescription with its associated rationale. Considerably more examples are given in section 4.

Design principles expressed in this way can help designers to meet the two usability criteria described in chapters 1.1 and 3.1 (and see also Byerley et al, 1993). The relationship between the two usability criteria and the principles is discussed in section 3. The principles need to be applied during the design process. A design method has been developed by GUIDANCE to make the application of principles and the achievement of the two criteria easier; this method, has already been described in detail in chapter 3.1. Section 4 of this chapter provides an illustration of the use of this method using example specific principles (see also chapter 3.3).

IF	the **task** is to gain an overview of all members of the workgroup, objects and functions available to plan and carry out actions
AND	the **users** are a limited number of co-operatively working multi-media authors and editors creating/modifying objects and exchanging messages
AND	the **service environment** supports graphical representation of information
AND	**performance** should be not slower or more error prone than if working in an open plan office
THEN	APPLY PRINCIPLES OF: a) **COMPATIBILITY** b) **SIMPLICITY**
	a) **COMPATIBILITY**
THEN	**Layout:** use an open plan office model: (all members of the workgroup, objects and tools represented at the same time)
	Interactions: similar interactions/actions possible as in an open plan office
BECAUSE:	the users can recruit existing knowledge about open plan offices to plan and carry out actions
	b) **SIMPLICITY**
THEN	**Layout:** distinct and easily understood display for objects, persons and functions available
BECAUSE:	an unambiguous overview should reduce user memory load

Figure 3: Examples of Specific MMMADP Principles

3. The application of principles by designers

The second section of this chapter dealt with the issue of the content and expression of principles. The second issue for principles is how to overcome the problems of applying current HCI guidelines and heuristics to system design. For example, one reason why current guidelines and heuristics have only a modest impact on design is because they fail to meet designers' hopes for more integrated, structured and formal tools (Hammond et al 1983). They also fail to influence designers because designers either may not be aware that HCI information is pertinent to their problem or they may not be able to find the information even if it is available (Mosier & Smith 1986).

However, the potential benefits of guidelines and (by implication) principles should not be underestimated. They are an accessible medium which can provide a useful adjunct to designers' intuitions, and can act as a springboard for ideas or as a checklist for omissions (Gould & Lewis, 1985). Principles can certainly contribute to the design process (particularly if they offer improvements to current guidelines and heuristics), but their role must be seen as an adjunct to other design techniques. This approach has been taken in the GUIDANCE project where principles are an integral part of the design method developed on the project (see chapter 3.1). The design method is a layered method for user interface design, and was developed to be independent of any implementation language, to produce an object-oriented specification output, and to incorporate the principles and the task analysis developed on the project. It is a method in that it contains an explicit set of notations for representing the design (the principal notations are specification tables), and some (often implicit) procedures for creating and transforming the representations. (See Lim et al, 1990, and Walsh et al, 1989, for discussions of design methods with reference to user interface design.) The design method is also supported by a software tool (see Terrins-Rudge et al, 1992) which assists designers in selecting appropriate principles associated with a given part of the design method or a design parameter of interest.

Both the conditions and the prescriptions of the principles are closely tied to components of the method. For example, the conditions reflect possible values specified in the specification tables of the method. As the designer fills in a specification table a number of values for a particular parameter may be suggested. The selection of these parameter values influences the presentation of appropriate principles in support of necessary design decisions. For instance, if the designer specifies an audio/video editing task then relevant principles for such a task may be accessed if required. These principles' prescriptions would provide recommendations for the completion of particular 'slots' in the specification tables, and would reflect the design decisions required for those tables. The prescriptions applying to functional software design are expressed in terms of conceptual/semantic design (e.g. how domain concepts are represented by the system), whereas the prescriptions applying to physical software design are expressed in terms of lower level design decisions such as dialogues, interactions, layouts and device considerations (see example principles in figure 3).

The application of principles to a target design, then, is supported by a close linking of the service integration principles with the design method, and by supporting the designer in accessing and using the principles and the method by providing a software tool. The remainder of section 3 discusses some of the concepts of the design method and their implications for the application of principles

3.1 Two criteria of usability

The analysis of usability requirements for IBC begins with the concept of a network of users (humans or machines) connected by communication channels. This network as a whole has the function of supporting the achievement of an application goal. Figure 4 shows an example network description for group authoring of multimedia documents. This example will be used throughout this chapter to illustrate the use of the Enabling States approach (see chapter 3.1) and design principles. Each communication channel illustrated in figure 4 can represent one or more services, and each service may involve more than one service provider.

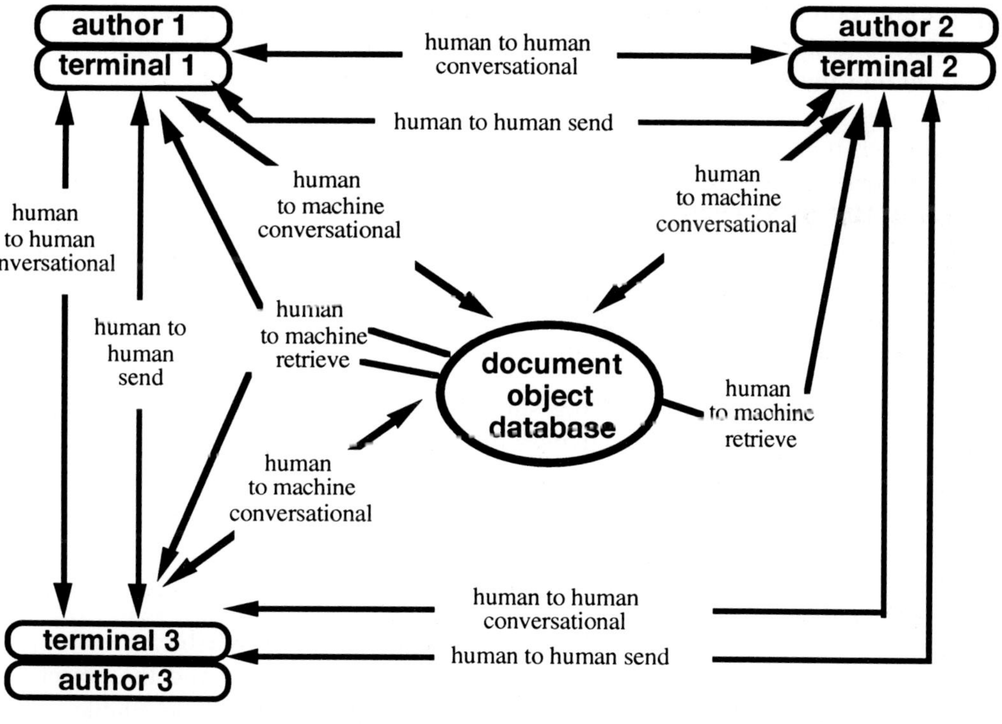

Figure 4: An Example Network for Group Authoring of Multimedia Documents

The Enabling States approach takes the perspective of the human user within such a network i.e. it takes the internal perspective described in chapter 1.1 (see also Byerley et al, 1993). The human user will have certain ***goal tasks*** which they must accomplish in co-operation with other users (humans or machines), with which they must interact via communication services. To achieve these goals, the user will require particular combinations of services.

These combinations will vary between different tasks, and also between different instances of the same task.

The project has proposed a distinction between two types of task that must be performed. There are tasks by which the users intend to move closer to their goal (*goal tasks*), and there are also tasks (*enabling tasks*) which create or modify the system to bring it into a state which enables the user to execute their goal tasks. In IBC, these enabling tasks include accessing and modifying services, by interacting with service providers. Chapter 3.1 describes these types of tasks in more detail, and proposes two criteria for a usable system which state that a usable system:

- ensures that the appropriate enabling states exist for each of the users' goal tasks
- reduces to a minimum the costs to the user in reaching the appropriate enabling states for their goal tasks

The method assists designers of IBC systems in meeting these criteria by explicitly representing goal tasks and enabling tasks and providing design principles which are applied at appropriate points in the design method. The relationship between the two usability criteria and the design principles is discussed in sections 3.2 and 3.3, and the application of principles within the method is illustrated in section 4.

3.2 Meeting the First Usability Criterion

A usable system ensures that the appropriate enabling states exist for each of the users' goal tasks.

As described in chapter 3.1, an enabling state describes the preconditions for a goal task. A goal task is conducted by a particular user on a particular object. The task may require the support of a service in order to be executed. Therefore, an enabling state description consists of a triplet:

Enabling State=(User, Service Relationship, Object)
where:
User = required characteristics of the user, e.g. knowledge, attention;
Service Relationship = the interface between the user and the object
Object = required characteristics of the object, e.g. 'file available';.

Design principles which aim to assist the designer in meeting this first criterion will therefore need to provide recommendations concerning all three components of an enabling state.

For instance, the enabling state will describe the state of the user that is required for them to execute their goal task. If there is a mismatch between the *required* user characteristics described by the enabling states description and the *actual* user characteristics then a principle can be applied which suggests a solution to this problem. For example, a principle might recommend help pages or training to provide the user with the knowledge necessary to execute their task.

Similarly, a principle may provide recommendations about an *object* of a goal task. An obvious example in telecommunications is that a user who wishes to talk to another person needs that person (the object of their task) to be alert and receptive. This state of alertness could be created by an enabling task conducted by the system, such as making their phone ring, or could be conducted by the user saying "excuse me". The design recommendation in this case would concern that object state which is conducive to the goal task to be performed.

The *service relationship* is a crucial part of the enabling state description. The service relationship is similar to the concept of a user interface used in the design of stand alone computers. However, for a distributed system of communicating entities sharing tasks, the interface between a user and an object becomes more complex, as described in chapters 1.1 and 3.1, and the term service relationship is therefore preferred.

Four basic service relationships were proposed in chapters 1.1 and 3.1, where users *actions* may be *interpreted* or *uninterpreted*, and the *object* of the users' goal tasks may be *represented* or *unrepresented*.

Design principles can provide recommendations about service attributes (e.g. media and media quality), about the representation of objects and about the interpretation of tasks. Figure 5 summarises the types of principles applied for the four different service relationships.

User's Goal Task Actions		*Goal Task Object*: *unrepresented*	*Goal Task Object*: *represented*
	uninterpreted	1. Service Attribute Principles (task and object).	2. Object Representation Principles Service Attribute Principles (task only).
	interpreted	3. User Action Interpretation Principles Service Attribute Principles (object only).	4. User Action Interpretation Principles Object Representation Principles

Figure 5: Principles applied to the four basic service relationships

In case 1, principles provide recommendations about the service attributes of the service in question. These may concern, for example, the characteristics of the channel and might include the video and voice qualities needed to enable a conversation.

In case 2, principles provide recommendations concerning the mapping of the object to the model. For example, the representation of a goal object should be *compatible* with the user's experience, and so a compatibility principle might recommend that a text file be represented as a small page of text. Additionally, because the object is represented no service attributes are necessary for the object but may be provided, however, for the actions.

In case 3, principles provide recommendations concerning the mapping of the user's actions to the model. For example, a *flexibility* principle might recommend that users should be able to perform tasks in different ways, for example with different input media. Additionally, because the task action is interpreted no service attributes are necessary for the action but may be provided, however, for the object.

In case 4, principles provide recommendations concerning the mapping of the user's actions to the model and the mapping of the object to the model. There will be no prescriptions concerning service attributes.

Meeting the first usability criterion, then, involves the use by the designer of a number of different types of principles. Further examples of all these principles are given in section 4.

3.3 Meeting the Second Usability Criterion

A usable system reduces to a minimum the costs to the user in reaching the appropriate enabling states for their goal tasks.

Design advice can also be given on the second criterion given in chapter 1.1, concerning the *costs* for the user which are associated with acquiring an enabling state. In order to provide design advice we need first to consider what is meant by user costs. Chapter 3.1 has considered this issue and proposed that the following types of user costs can be identified: structural and behavioural; mental and physical; and cognitive, conative and affective. This section considers some theoretical and practical issues in applying the notion of user costs within design principles, and illustrates these with respect to those user costs associated with attention and stimulus processing.

Many of the issues addressed in classifying user costs can be related to the area of selective attention. Because attentional resources appear to be limited (Kahneman,1973; Schneider and Schiffrin, 1977), attending to a goal task or enabling task necessarily limits the degree to which other tasks can be attended to. There is a considerable amount of data on the consequences of the focusing of attention, but the processes by which we control attention, be they overt or covert, have not received much study until recently. Future studies in usability would therefore review what is known in psychology on the control of attention. They might begin by identifying the body of literature that bears on the processes by which the focus of attention is chosen:

- What do we know about users' ability to control the focus of their attention?
- What do we know about the extent to which the unattended is monitored and how attention is captured away from its focus?
- What do we know about individual differences in the control of attention?
- To what extent is the control of attention specific to modality, situations and motivation?
- To what extent is knowledge about the control of attention related to work in Engineering Psychology and HCI, and in particular to the concept of mental workload?

Any technique purporting to design for, and to assess, the usability of a system should address these issues. Unfortunately, no single usability index is capable of providing all of the relevant information. For this reason, an assessment battery employing a wide variety of indices is clearly in order. Recent conceptualisations of the complex interaction between a user and a task have stressed the multidimensional nature of this interaction (Kinsbourne & Hicks, 1978; Laughery, Drews & Archer, 1986; Navon & Gopher, 1979; Polson & Freidman, 1988; Sanders, 1979; Wickens, 1980, 1984a). The most effective analysis of such a system should, therefore, also be multidimensional in nature.

Models of human performance in complex multi-tasking situations can be divided into three general categories:

1) *Energetical models* (Hockey, 1986)
emphasise the role of internal and external sources of stress.
2) *Cognitive/Capacity models* (Wickens, 1984a; Rasmussen, 1990)
focus upon the specific information processing demands that the interface/task places upon the user.
3) *Cross-talk/Competition models* (Senders, 1983)
deal with problems arising between input channels and output control devices.

For the most part, these three general models have been presented as conflicting descriptions of the types of constraints which impact upon user performance and user costs in human-machine systems. However, it is clear that all of these considerations are likely to be important when assessing the usability of a given system.

Moreover, the model of user costs presented in chapter 3.1 can be related to Rasmussen's SKR model of human performance (1986, 1990), where skill, rule and knowledge based behaviour describes decreasing levels of familiarity with a given task (e.g. enabling task) as well as the environment in general.

At Rasmussen's skill-based level, behaviour is seen as controlled by pre-programmed and automatic sensorimotor patterns acquired by training and developing skills and knowledge. Familiar tasks, where decisions are taken based on stored rules, occur at the rule-based level.

In solving novel tasks, actions are planned on-line, using mental processes (e.g. identification, decision, planning) and structures, and are assumed to be executed at a conscious and knowledge-based level.

In relating the breakdown of user costs described in chapter 3.1 with the findings concerning automatic and controlled behaviours, it can be assumed that much of the physical and mental user costs are either automatic and effortless or controlled and effortful depending upon the user's degree of learning and familiarity associated with acquiring a specific enabling state. The behavioural costs for the user are reduced to a minimum in reaching the appropriate enabling states for their goal tasks if the user can use their existing knowledge in a skill-based fashion.

Mental cognitive costs can also be expressed as mental workload. There is, however, no general agreement on the definition of this concept. But in its broadest sense it can be thought of as the proportion of information processing capacity expended in meeting demands during task performance, in this context the performance of enabling and goal tasks. In fact, general mental work load can be used to characterise the user-task interaction, here called IWS behaviour. This implies that mental workload is a function of task demands, user's and system's capacity and the user's goal-setting.

The amount of available mental capacity or resources determines the quality of user performance according to theoretical models claiming that information processes are resource dependent (Kahneman, 1973; Navon and Gopher, 1979). The type of processing which requires considerable attentional control (Schneider and Shiffrin, 1977) and is under subjective control is also sensitive to motivational factors (Navon, 1984). In the terminology of chapter 3.1 affective physical and mental costs relate to the emotional aspects of cognitive physical and mental costs e.g. during reasoning and making decisions (and therefore to the amount of information processing involved in mental workload).

Thus, a principle (such as feedback, say) that effects motivational processes will have an influence on resource dependent and controlled information processing. The amount of mental effort incurred by (or invested into) a given enabling task for their goal tasks will therefore determine the degree of controlled processing that the given user is engaged in.

As mentioned previously, the user's goal-setting also regulates the amount of mental resources invested into a series of enabling and subgoal-tasks leading to the desired goal task, and thus the degree of mental workload. It can be suggested that some feedback principle recommendations (e.g. desired performance) might limit the user's freedom with respect to goal-setting and force them to cope more actively with changing enabling and goal task demands.

In summary, then, the breakdown of user costs described in chapter 3.1 can be interpreted with respect to the twelve generic usability design principles. For example, each of the twelve generic principles can be interpreted with respect to mental cognitive costs. An example of mental cognitive costs is the cognitive workload for the user associated with presentation of multiple sources of complex information. A *compatibility* principle recommends that such information should be structured and presented to the user in such a way that the user can use their existing knowledge from outside the system to locate appropriate information and establish its relationship to other information sources. Similarly, a *salience* principle recommends that critical information should be clearly identifiable, whilst *reversibility* suggests that users should be able to reverse their action thus bringing the worksystem back to its former state.

Another aspect to reducing user costs concerns the decision of who (or what) carries out the enabling task. For this reason, there are also principles which give advice about the function allocation between user and machine. One overall aim is the system property of *simplicity,* which says that the actions that the user has to perform should be reduced to a minimum. Other important function allocation principles are *controllability* which recommends that the user should be allocated control over critical system functions, and *flexibility* which recognises that there is not always a single best way for operating a worksystem and recommends that where performance varies with organisational contingencies there should be flexible allocation such that a function can be performed either by the machine, the user or by both.

3.4 Principles and Service Integration

So far we have discussed how design principles provide design recommendations that assist the IBC designer in meeting the two criteria for a usable system. That is, the principles provide prescriptive advice concerning the triplet of users, service relationships and objects which comprise the enabling states description, and also provide prescriptive advice concerning the acquisition of an enabling state, via the enabling tasks.

Whitefield et al (1992) and Denley et al (1992) have suggested that lack of integration between services may seriously affect the uptake of IBC systems, and that for end-users the problems caused by such lack of integration between services may make the usability of integrated systems even worse than that of existing independent services.

A major aim of the design principles proposed in this chapter is to help the designer overcome the problems of service integration.

Integration principles address the interactions or interleaving of the end-user's goal tasks. For instance, taking a medical example, a radiologist seeking expert advice from a remote expert may wish to retrieve patient information including x-rays, stored as image or text files in a remote database and send these to the remote expert whilst continuing to communicate using a multimedia conferencing service. Obviously, there is an interleaving between the goal tasks communicating with a colleague and retrieving and sending files, which may cause service integration problems for the end-user. For instance, the usability of these services will be reduced if there are inconsistent screen layout conventions (e.g. different positions for common buttons in screens or windows for different services); or if there is confusion between operating procedures (e.g. similar database services may require very different access commands depending upon where they are physically stored).

Service integration principles assist the designer in solving such integration problems between services. Figure 6 summarises the different types of principles described in this chapter and shows the relationship of each type of principle to the Enabling States approach. Section 4 illustrates where and how the different types of principles shown in figure 6 are used in the IBC system design process.

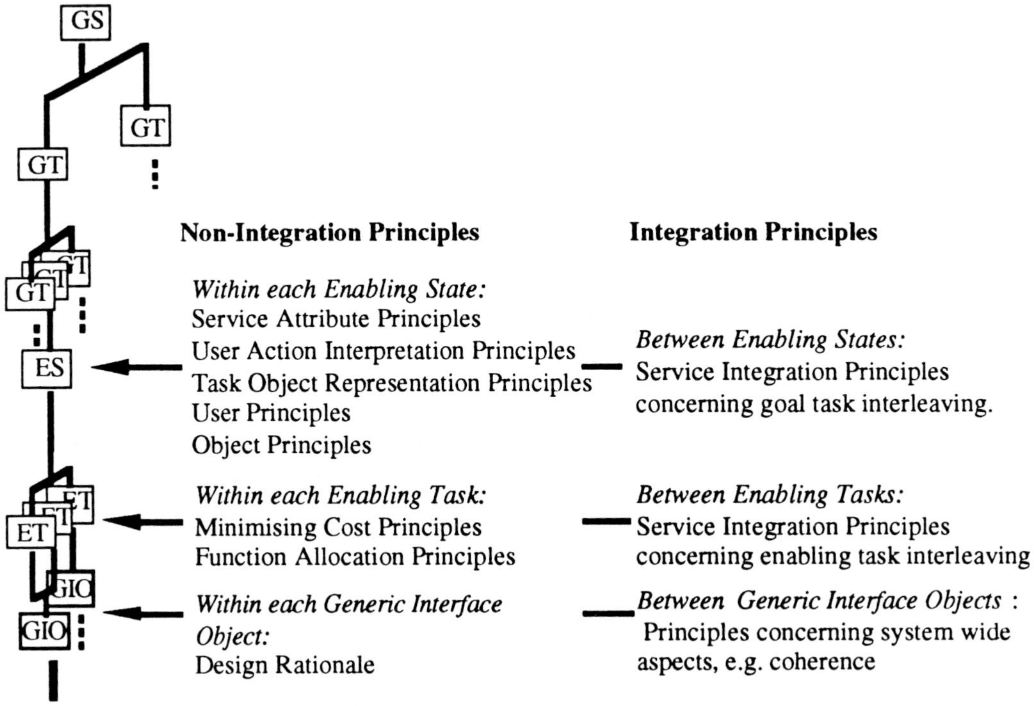

Figure 6: The identification of service integration principles within the Enabling States approach

4. Service integration principles in design: A simplified case history in multimedia conferencing

Conferencing can be described as a human to human conversation task to which human to machine conversation tasks may be added (Melchior et al 1992). In this context, two broad categories of multimedia conferencing can be distinguished: videoconferencing and videotelephony. Videotelephony is a relatively low quality and low cost version of multimedia conferencing with mainly point to point communication between two users. The service components are audio (speech) and moving image (motion video). Additional human to machine conversational tasks may be required from time to time including, for instance, sharing a view of a document to be discussed between the users. Videoconferencing is the higher quality version of multimedia conferencing which involves multi-point communication between more than two users. In this scenario we are concerned only with point to point videotelephony.

Videotelephony, as a subset of multimedia conferencing, consists of three high level goal tasks: human to human conversation, human to machine conversation and human to machine retrieval. The scenario used here concerns two authors collaborating in the editing and linking of information to form a multimedia document. In particular their task is to discuss changes to a video clip which forms part of the multimedia document. A third author is also part of the workgroup but is not involved in this particular task scenario.

The task network which can support the tasks described in this scenario, also called a *Reference for User Services* (RUS) for this application was shown earlier in figure 4 (for further details see Byerley and Bruins, 1992; Byerley et al, 1993). This network describes the communicating entities and the channels that link them. The communicating entities involved, namely the authors and the document object database, are connected by an internal network.

For human to human conversation tasks a point to point connection is required which means that only two users can be involved in any telephony conversation. Both audio and moving image service components are required. The audio service component must be mandatory whilst the moving image service component should be optional. Audio is needed for the transmission of speech, moving image for the transmission of the users' faces.

For human to machine conversation tasks the connection required is multipoint since all users have access to the database.

For human to machine retrieval tasks a multipoint connection is also required. The information content retrieved from the database may be still images, moving images, text documents etc. The information transfer is from the database to the user.

The specific scenario used here takes a snapshot of the group authoring task to illustrate how the interleaving of tasks affects the design through the application of principles. Imagine that two remote authors wish to discuss and edit a video document that is stored on the document object database. This scenario involves three types of generic tasks which, of course, may interleave:

- Human to human conversation: the authors discuss proposed changes to the video clip
- Human to machine retrieve: the authors retrieve the video clip from the database
- Human to machine conversation:
    - i) the authors search the database for the desired video clip,
    - ii) the authors use a remote video editor

This illustration works through the design of the services to support *two* of these tasks, and demonstrates the importance of the principles in the design. The two tasks considered here are:

- Human to human conversation: the authors discuss proposed changes to the video clip
- Human to machine conversation where the authors use a remote video editor

This following illustration is a *highly simplified* example, which although based on a real design, is used for illustrative purposes only. The approach has been described in full in chapter 3.1, and will not be described in detail here. The method is summarised in figure 7. The stages in the design have been faithfully reproduced, but the content of each of the specification tables has been abridged. The purpose of this illustration is primarily to show how design principles influence the design of services.

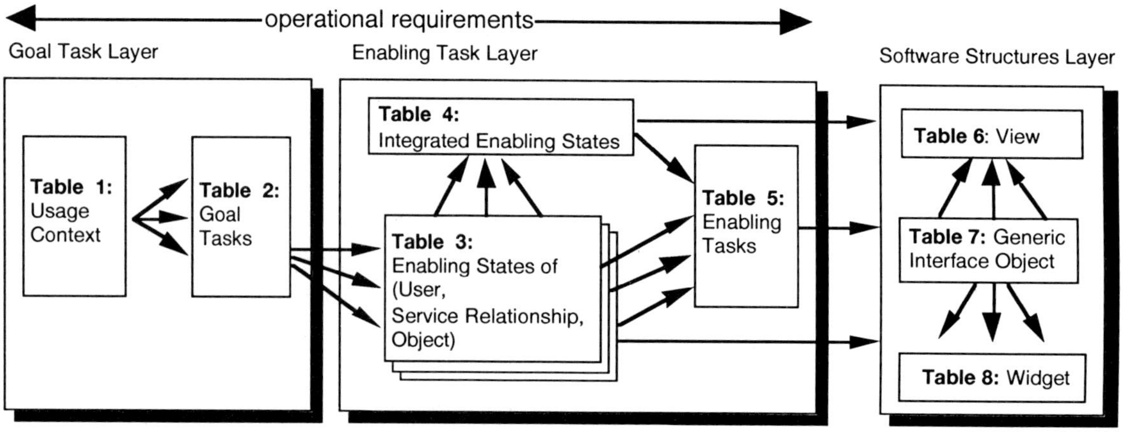

Figure 7: The Enabling States approach to design

4.1 Stage 1: The Usage Context

The goal of this stage is to describe the need for communication rather than the recommendations for the services that would allow this communication. The specification table expresses the aims of the system to be designed and the context in which it will operate, e.g. the work to be performed and the characteristics of the users and the their organisation. Specification Table 1 gives an abbreviated usage context description.

Specification Table 1: An Abridged Usage Context Description

Users: more than one, simultaneous, either gender, adult, at least average intelligence, sharing European cultural norms, working in the area of holiday industry and/or document production, experienced in the use of interactive computer work systems.

Work goal: to have a multimedia document which allows holiday makers to select a suitable holiday.

Work domain : transformations of control information (e.g. comments, plans),transformations of document elements to form a complete document (i.e. assembly and (some) editing of pre-existing elements into a document, but no creation of those elements).

Performance parameters: task quality, human and machine resource costs (i.e. time, number of errors, effort, cognitive demands,...)

System: could include all possible IBC services, in any combinations, operating in standard office environments.

Organisation: a large travel agency or tour operator, physically distributed, with democratic management style, and hierarchic management (one person has final responsibility for the document).

Service Environment: multimedia videotelephony, and a multimedia document database. These service environment assumptions are provided at the application level analysis and should include the Reference for User Services diagram shown in figure 4 above.

4.2 Stage 2: The Goal Task Analysis

The aim of this stage is to produce a description of the goal tasks the user wants to carry out. The result of this goal analysis stage of the Enabling States approach is a breakdown of the major goal of the work system (i.e. human end–users, and other communicating entities such as machines, databases, terminals etc) into a set of subgoals, and a description of the objects that are transformed or created in performing these goals. It is important that at this stage the task descriptions remain technology independent, i.e. the task breakdown should, in principle, be implementable on a number of technology and service configurations. Decisions concerning technology are taken when specifying the enabling states for each subgoal–task.

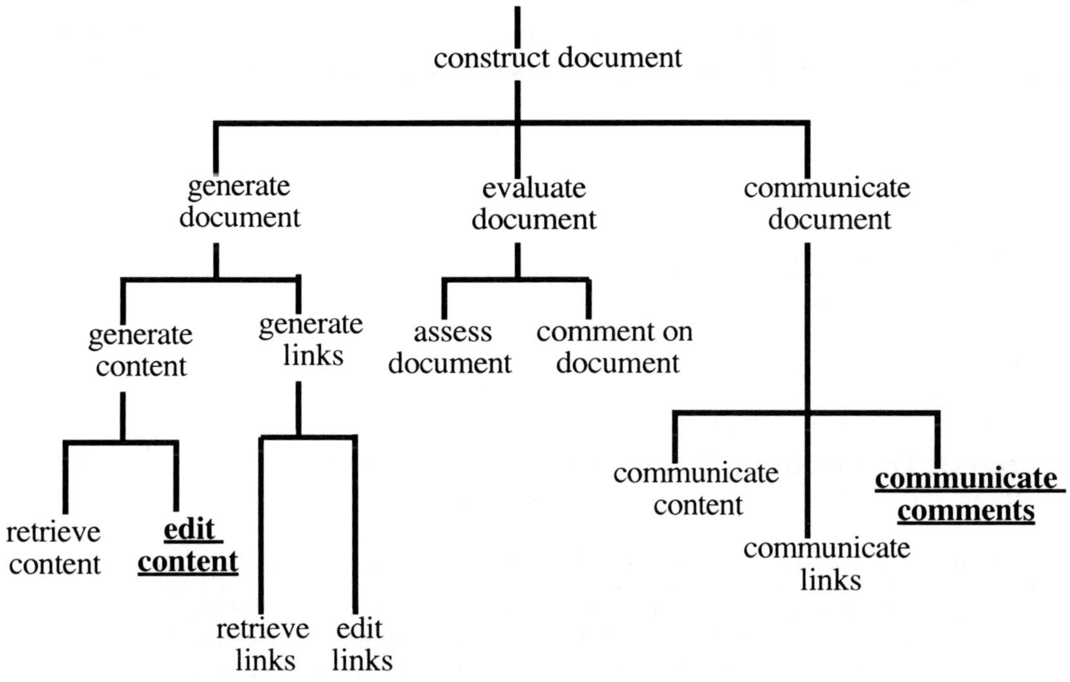

Figure 8: Goal Hierarchy for high level goal tasks 'Construct Document'.

There is no subgoal–task *Create Link/Content* shown in figure 8 because, for the sake of simplicity, we have made the assumption that the authors are not concerned with creating the document data (e.g. filming the video clips), but merely in their collation and editing. Each of the highlighted subgoal–task above is described with a specification table. These tables contain the information that describes what should be achieved by the task, with any constraints that might affect the way in which it is done. Specification Tables 2a and b give example specification tables for two subgoal–tasks.

***Specification Table 2a*:** An Example for the Subgoal-Task *Communicate Comments*

Description: The user wishes to communicate with another author about part of the document from the document object database

Sources: Communicate Document

Task Parameters

Temporality:

frequency: Many times per day

duration: Average call time 10 minutes (Range 3 minutes to 1 hour).

Flow of work

Intertask Dependencies:

necessary pre-tasks: None

task interleaving: This task may be concurrent with human-machine conversational services (e.g. edit content)

References: None

***Specification Table 2b*:** An Example for the Subgoal-Task *Edit Contents*

Description: The user wishes to make changes to part of the document from the document object database

Sources: Generate Document

Task Parameters

Temporality:

frequency: Many times per day

duration: Average editing time 40 minutes (Range 15 minutes to 3 hours).

Flow of work

Intertask Dependencies:

necessary pre-tasks: The user must have retrieved the object from the database

task Interleaving: This task may be concurrent with human to human conversation services (e.g. communicate comments)

References: None

4.3 Stage 3: The Enabling States Analysis

An enabling state describes the state of the work system: what knowledge and processing skills the *Users* must have; the characteristics of the *service relationship*, and the state that the *objects* must be in. Each of two subgoal-tasks highlighted above must have an enabling state. This is the first point where design principles can be applied (figure 9).

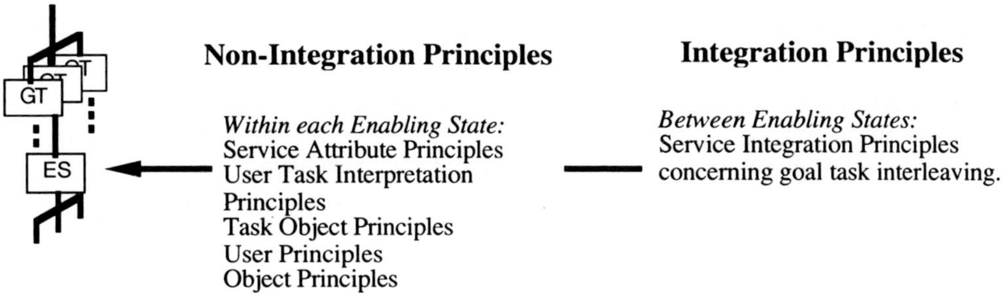

Figure 9: Application of design principles to enabling state specification

There are 3 steps for the designer to describe an enabling state:

1) Describe the states of the *users* for the subgoal-task, and apply any user principles that are appropriate
2) Specify the characteristics of the *service relationship* for each subgoal-task, and apply any Service Attribute principles, User Action Interpretation principles or Task Object Representation principles that are appropriate
3) Describe the states of the *objects* for each subgoal-task, and apply any object principles that are appropriate

Following this, the designer must consider any interleaving of subgoal-tasks, apply any service integration principles that are appropriate, and then provide an integrated enabling states description for the interleaved tasks (see section 4.4).

Step 1: Describe the User for Each Subgoal -Task

The purpose of this step is to describe the *required* user characteristics for a given enabling state. If there is a discrepancy between the required characteristics described by the enabling states description and the *actual* user characteristics then a principle can be applied which suggest a solution to this problem. Example user descriptions and user principles are given below in Specification Tables 3a-i and 3a-ii. This step also describes the required enabling tasks that produce the enabling state. These enabling tasks are then described in Specification Tables 5a and b (see section 4.5).

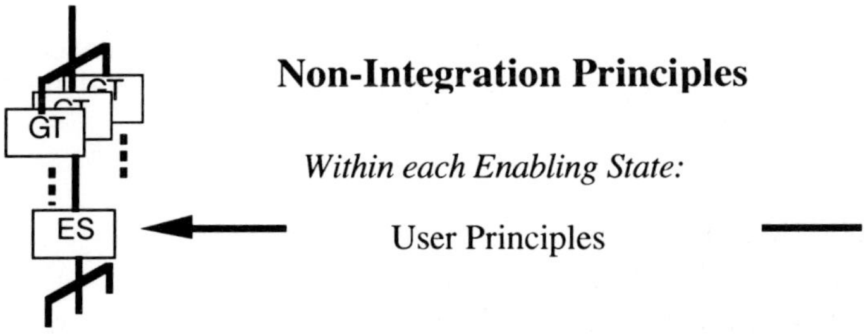

Figure 10. Application of design principles to user state specification

Specification Table 3a-i: Example User Description for the Enabling State *Comments Communicable*

User Description

The user has to know how to start and end communication
The user has to know how the recipient will receive the image of the user
The user has to be able to change between services
The user has to be able to determine and control quality of the communication
The user has to be aware of incoming calls

User Principles:

Compatibility :

IF the task is to communicate with other persons in the workgroup
AND users already have knowledge of team working in an open plan office
AND the service environment supports graphical display of objects
AND the performance should minimise users' affective costs, e.g. frustration
THEN Apply principle of compatibility:
The user should be made aware of all members of the workgroup, their presence/absence and their relationships
BECAUSE (rationale) this will allow users to exploit existing knowledge about office work, and the work environment with respect to the availability of persons and the actions which may be performed on these persons.

Enabling Tasks for the User Description

Start/end call
Obtain/adjust selfview
Change service
Control quality of the communication
Call alert
Allow hold, resume and forward call

References: Enabling Tasks and Goal Actions, which are included in the Generic Interface Objects (GIOs)

Specification Table 3a-ii: An example User Description for the Enabling State *Contents Editable*

User Description

The user has to know how to start and end editing
The user has to know what the recipient can see during editing
The user has to be able to 'hand-over' editing services, etc.

User Principles:

Compatibility :

IF the task is to edit existing document objects
AND users need to have knowledge of existing objects, and already have knowledge of team working in an open plan office
AND the service environment supports graphical display of objects
AND the performance should minimise users' affective costs, e.g. frustration
THEN Apply principle of compatibility:
The user should be made aware of document objects and their relationship to existing office organisational structures
BECAUSE (rationale) this will allow users to exploit existing knowledge about office work, and the work environment with respect to the availability of objects and the actions which may be performed on them.

Enabling Tasks for the User

Start/end editing tool
Make editing actions visible/not visible
Hand-over edits
Give the user knowledge of how to perform the intended actions
Alert user to system feedback

References: Enabling Tasks and Goal Actions, which are included in the GIOs

Step 2—Specify the characteristics of the service relationship for each subgoal-task

Once the IBC designer has described the user the next step is to describe the service relationships for each of the subgoal-tasks identified in the goal task breakdown. The application of principles will contribute to the designer's overall rationale for the design of these enabling states, and the principles themselves are described with respect to this rationale in the service relationship description. Specification Tables 3b-i and 3b-ii illustrate the service relationship tables for the two subgoal-tasks of edit content and communicate comments.

When specifying the service relationship the designer will also access non-integration principles concerning service attributes, task action representations and task object representations within each enabling state:

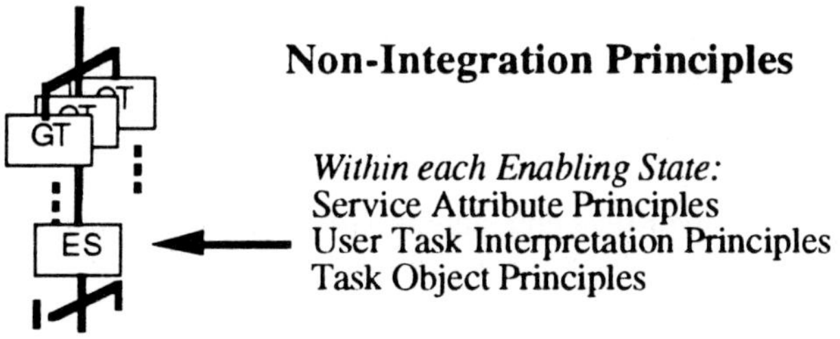

Figure 11: Application of design principles to service relationship state specification

The service relationship of a particular subgoal-task determines what principles should be applied, and also determines what parts of the specification table the designer must complete. In this illustration the following service relationships apply:

• **Communicate Comments** = Uninterpreted Task/Unrepresented Object

In this instance the service relationship description requires consideration of only the service attributes of the system, and only *Service Attribute Principles; User Principles and Object Principles* may be applied, (see figure 6). The **service environment** fed forward from the application level analysis is a real time, point to point, videotelephony service

• **Edit Content** = Interpreted Task/Represented Object

The service relationship description in this instance requires consideration of the objects and their representations, the tasks and their representations and the service attributes of the system to be designed. The principles that can be applied are: *User Task Interpretation Principles; Task Object Principles, User Principles* and *Object Principles,* (see figure 6). The **service environment** fed forward from the application level analysis is a multimedia document object database and multimedia editing tools.

<u>*Specification Table 3b-i*</u> An Example Service Relationship Description for the Enabling State 1: *Comments Communicable*

Description: The user can communicate with a recipient directly via a face to face videophone call.

Source: Communicate comments

Service Relationship: uninterpreted/unrepresented

Service Attributes:

Involved Entities: user, workstation, recipient

Communication Type: Human-Human Conversation

Channel characteristics:

Information Content: Head and shoulders view of the participants and their voices are transmitted

Service Components: Audio and video service components are required. The audio component must be mandatory, the video service component may be optional

Synchronisation requirements: Lip synchronisation is required for the combination of audio and video service components

Format Preservation: Not required for audio and video service components, compression acceptable

Quality of Audio Service Component: Low quality. i.e. comparable to standard telephone

Quality of Video Service Component: Standard video quality (4:3 aspect ratio, 625 lines, 120 000 pixel, 20-40 MBit/s peak bit rate) comparable with NTSC, PAL or SECAM standard resolution TV quality

Quality of Service: Adequate quality of loudness, brightness, contrast.

Service Attribute Principles:

IF audio and moving images are provided simultaneously (videotelephony conversation)
THEN audio transfer rate is <64 KBit/s and standard moving image quality is required.
IF audio only is provided
THEN quality is always voice (3.1—7 kHz)

Enabling Tasks for the Service Relationship

Establish synchronous transmission
Establish service compression
Establish low quality audio
Establish standard quality video
Determine quality of service

References: Enabling Tasks and Goal Actions, which are included in the GIOs

<u>*Specification Table 3b*</u>*-ii* An Example Service Relationship Description for the Enabling State 2: *Contents Editable*

Description: The user can make changes to a document object using remote editing tools

Source: Edit contents

Description of the Object representation: Document objects represented as icons

Object Representation Principles:

Transparency:

IF the *task* is to distinguish between different types of document objects
AND *users* wish to work on objects
AND the *service environment* supports graphical display of objects
AND the *performance* should minimise users' cognitive costs and be error free
THEN *Apply principle of transparency:*
represent the nature (media type) of each object by a distinct and meaningful symbol
BECAUSE *(rationale)* clearly distinguishable object types will reduce user retrieval errors. For

example:  = video object,  = Text object, etc

Coherence: Objects of similar classes should be represented with the same icon.
Feedback: The state of the object should be recognisable, e.g. in use
Completeness: All existing objects should be displayed

Description of the Task representation: indication of an object (or part of an object) to work on is done by selection, this is done by using the mouse (movements and actions represented by arrow on the screen), which is placed on the object, by clicking the object is selected

User Task Interpretation Principles:

Simplicity :
IF the task is to edit document objects
and users are multimedia authors
and the service environment supports editing using remote tools by pointing at objects or the contents of objects with minimal typing of arguments
and the performance should be quick and error free
THEN Apply principle of simplicity:
allow the pointer to move in all directions and change the shape of the pointer according to possible actions
BECAUSE (rationale) this will reduce the number and complexity of actions required, and intended actions will be easy to perform and remember

Reversibility: Deselection should be possible by clicking on the background, and edits should be undoable
Feedback: The state of the object should be recognisable
Completeness: All possible user actions represented

Enabling Tasks for the Service Relationship
Select object
Present object
Place object at convenient place on screen

References: Enabling Tasks and Goal Actions, which are included in the GIOs

Step 3—Describe The Object for Each Subgoal -Task

In common with the user description discussed above, the purpose of this step is to describe the *required* object characteristics for a given enabling state. If there is a discrepancy between the required characteristics described by the enabling states description and the *actual* object characteristics then a principle can be applied which suggest a solution to this problem. Example object descriptions and object principles are given below in Specification Tables 3c-i and 3c-ii. This step also describes the required enabling tasks that produce the enabling state. These enabling tasks are then described in Specification Table 5a and 5b (see section 4.5).

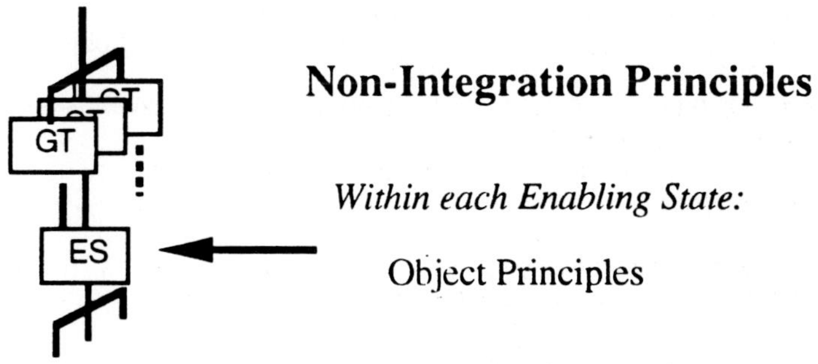

Figure 12: Application of design principles to object state specification

Specification Table 3c-i: An Object Description for the Enabling State 1: *Comments Communicable*

Object (Recipient)
The recipient has to be identified
The recipient has to be informed about call
The recipient has to accept the call

Object Principles:
Salience :
IF the task is to communicate with other persons in the workgroup
AND users have no special needs
AND the service environment supports audio output
AND the performance should minimise costs of unattended calls
THEN Apply principles of salience and redundancy:
use simultaneous audio and visual signals to alert recipient to incoming call
BECAUSE (rationale) combined audio and visual signals (e.g. flashing and ringing alarms) are most effective in gaining attention.

Enabling Tasks for the Object
Identify recipient
Call alert
Accept call

References: Enabling Tasks and Goal Actions, which are included in the GIOs

Specification Table 3c-ii: An Object Description for the Enabling State 2: *Contents Editable*

Object (Document)
The object has to be indicated
The object's availability (non-availability) has to be indicated
Object attributes should be changeable
Object version should be changeable

Object Principles:
Controllability :
IF the task is to edit existing document objects
AND users can specify attributes of existing objects, and already have knowledge of team working in an open plan office
AND the service environment supports graphical display of objects
AND the performance should minimise users' affective costs, e.g. frustration, annoyance
THEN Apply principle of controllability:
only allow users to change essential attributes (e.g. privacy) of objects which they have created or have been given control of by another author
BECAUSE (rationale) editing of objects for users engaged in collaborative tasks should be a negotiated or managed process.

Enabling Tasks for the Object
Indicate object
Indicate availability of object e.g. busy dialogue
Change object attributes, e.g. privacy
Version control

References: Enabling Tasks and Goal Actions, which are included in the GIOs

4.4 Stage 4: The Interleaving of subgoal-tasks

Whilst describing an enabling state designers will need to consider what types of technology are available and appropriate for the subgoal-tasks identified. This information is available from the application level description (i.e. from the customer's perspective). This information becomes particularly pertinent when there is potential interaction between goal tasks, and hence between services. The service environments for the two subgoal-tasks of communicate comments and edit contents are:

- a real time, point to point, videotelephony service
- a multimedia document object database and multimedia editing tools

With the two subgoal-tasks of communicate comments and edit contents there are two interleaving relationships important to our simplified example:

- Users want to discuss the editing of a video object whilst carrying out these editing changes in real time (i.e. interleaving between editing and communication tasks).
- Both users want to edit the same video object (i.e. interleaving between editing tasks).

Applying Service Integration Principles

For each of these two possible subgoal-task relationships there are service integration principles that influence how the IBC designer describes their enabling states. The information for the service environment descriptions included in these principles is drawn from the Reference for User Services description described in section 4.1.

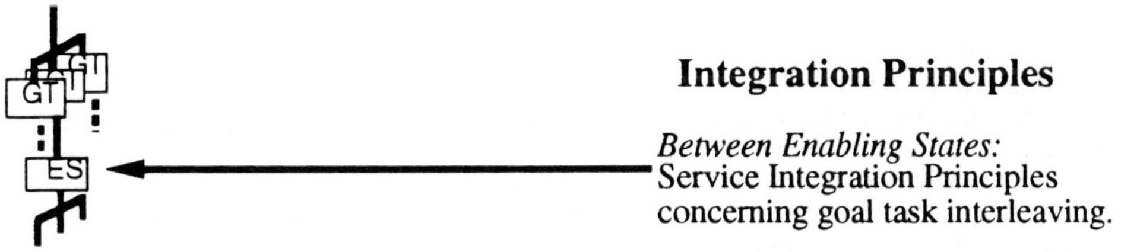

Figure 13: Application of design principles to enabling state interleaving

Examples of these principles are given below:

1. Human to Human Conversation and Human to Machine Conversation

IF	the *task* involves human to human conversation <u>and</u> human to machine conversation performed congruently
AND	there are at least two *users*
AND	the *service environment* supports audio and video communication and includes a multimedia object database

AND the *performance* on both tasks should be similar to that for each independent task (i.e. no decrement in performance nor increase in costs to the user should occur through multi-tasking)
THEN *Apply principles of simplicity and controllability :*
ensure that the number and complexity of actions is kept to a minimum and that users can control which service becomes primary
BECAUSE *(rationale)* interference between tasks should be minimised and the locus of control should remain with the user (i.e. the machine should not adopt unpredictable states).

IF the *task* involves human to human conversation and human to machine conversation performed congruently
AND there are at least two *users*
AND the *service environment* supports audio and video communication and includes a multimedia editors
AND the *performance* depends on both tasks being carried out simultaneously
THEN *Apply principle of salience:*
ensure that the primary service component does not interfere with the secondary service component (e.g. mute a non-primary video service component or mute a non-primary audio component).
BECAUSE *(rationale)* machine should assist users in focusing on primary tasks and allow use other input/output modalities for secondary tasks (e.g. when editing a video document users can utilise a free audio channel for conversational tasks).

2. Human to Machine Conversation and Human to Machine Conversation

IF the *task* involves human to machine conversation and human to machine conversation (e.g. joint editing)
AND there are at least two *users*
AND the *service environment* includes a multimedia object database and multimedia editors
AND the *performance* should be quick and error free
THEN *Apply principle of controllability :*
allow only one user to edit the object, and allow the editing user to specify whether his/her actions will be visible on the workstation of the second user
BECAUSE *(rationale)* users engaged in collaborative tasks should be able to control what is being presented to other users to ensure privacy

IF the *task* involves human to machine conversation and human to machine conversation (e.g. joint editing)
AND there are at least two *users*
AND the *service environment* includes a multimedia object database and multimedia editors
AND the *performance* should be quick and error free
THEN *Apply principle of controllability :*
allow only one user to edit the object at a time and allow this user to manage the handover of editing capabilities to the other user
BECAUSE *(rationale)* users engaged in collaborative tasks should be able to control their own actions and those of their workstation

Integration principles, then, provide usability recommendations to the designer which will influence the design of the enabling states for the two subgoal-tasks of communication and editing which have been identified as potentially interleaving. Next the designer must design an integrated enabling state to describe the interleaved tasks.

If two or more subgoal–tasks are to be allowed to interleave then the integration of their enabling states must be designed. This means that the two enabling states must be combined and a third *integrated* enabling state description produced. This description will allow any modifications to the individual enabling states descriptions to be highlighted and any influences on software design to be considered. Specification Table 4 gives an example of an integrated enabling states description.

Specification Table 4: An Example of an Integrated Enabling States Description for *Communicate Using Videotelephony and Edit Document*

Description: The user can communicate with a recipient directly via a face to face videophone call and retrieve and jointly edit multimedia documents simultaneously

Sources: Comments Communicable and Contents Editable (Enabling States 1 and 2)

Description of the User

See Enabling States 1 and 2

Integrated Service Relationship :

Interactions and Modifications to Individual Service Relationships: There are two service relationships in this enabling states description. Multimedia objects are presented simultaneously with the presentation of video images of the recipient. As simultaneous, multiple video outputs will distract the users from their primary tasks one of the video outputs is set to pause. Simultaneous audio conversation is permitted.

Simultaneous joint editing is not permitted, but co-workers can hand over control of the editing tool. Hand-over should be negotiated and controlled by user currently using the tool. The visibility of user actions to the co-worker is allowed, but should be controlled by user currently using the tool.

Description of the Objects

There are two objects: People and Document objects. See Enabling States 1 and 2.

Enabling Tasks for the Integrated Service Relationship

Inform users of machine actions (e.g. video freeze)
Allow user override of machine actions if desired
Make editing actions visible/invisible to co-worker
Hand over editing to co-worker

References: Enabling Tasks and Goal Actions, which are included in the GIOs

4.5 Stage 5: The Enabling Task Analysis

In describing the users, the service relationship and the objects the designer will have identified a number of enabling tasks which have to be carried out in order for users to reach the appropriate enabling states for their goal tasks. Stage 5 of the Enabling States approach describes these enabling tasks and applies principles which help the designer meet the second usability criterion and reduce to a minimum the costs to the user in reaching the appropriate enabling states for their goal tasks. The principles applied at this stage are concerned with user costs and function allocation (see section 3.2)

There may be an interleaving between enabling tasks, and as with goal tasks, the designer may refer to integration principles for assistance in the specification of the respective Generic Interface Objects (GIOs). For example, in this simple case history there is a potential interleaving between the enabling tasks of 'adjust self view' and 'place object at convenient place on screen'. Cognitive costs (e.g. workload) to the user because of this potential interleaving can be alleviated by allocating to the machine the task of placing the object conveniently so that there are no overlaps between the videophone window and the window containing the object.

Examples of two of the enabling tasks identified during the enabling states analysis are given below in Specification Tables 5a and 5b. 'Adjust self-view' is an example of an enabling task allocated to both the user and the machine, and 'identify recipient' is an example of an enabling task allocated only to the machine.

Specification Table 5a An Example of an Enabling Task Description for *Adjust Self View*

Description: This enabling task relates to the user description of the Enabling State *Comments Communicable* and allows the user to see themselves as the recipient will see them during face to face communication.

Class Relations: None

Usage Aspects:

Allocation: user and machine

Allocation Principles:

Simplicity:

IF the task is to adjust the camera for self view
AND users have no special needs
AND the service environment supports head and shoulders video
AND the performance should minimise users' cognitive and affective costs, e.g. workload and frustration
THEN Apply principle of simplicity:
providing default settings and make as many adjustments as possible automatically (e.g. focal length)
BECAUSE (rationale) this will minimise the number of user actions necessary

Cost Principles:

Simplicity and compatibility:

IF the task is to adjust the camera for self view
AND users have no special needs
AND the service environment supports head and shoulders video
AND the performance should minimise users' cognitive and affective costs, e.g. workload and frustration
THEN Apply principle of simplicity and compatibility:
keep the camera static and allows users to adjust their own position, e.g. in a similar manner to using a mirror
BECAUSE (rationale) reverse control of camera actions demands skill and would be time consuming

Design Aspects

Functional Requirements

The user should be able to perform a call without checking self view if this is desired.

The user should be able to perform a self view at any time during a conversation without terminating the conversation

Operational Requirements

The machine should be able to maintain quality of image output throughout a call without requiring the user to make adjustments (e.g. automatic focusing)

References: GIO Self View Dialogue

Specification Table 5b An Example of an Enabling Task Description for Identify Recipient

Description: This enabling task relates to the user description of the Enabling State *Comments Communicable* and supports automatic dialling of recipients for face to face communication.

Class Relations: None

Usage Aspects:

Allocation: machine

Allocation Principles:

Simplicity:

IF the task is to identify the recipient of a call
AND users are limited in number and have knowledge of other users and their roles

AND the service environment has knowledge concerning the availability of recipients and their locations
AND the performance should minimise users' cognitive and physical costs
THEN Apply principle of simplicity:
design the user interface to support the rapid selection of the limited set of user numbers (addresses) without the use of a dialling keypad
BECAUSE (rationale) the restricted number of communicants makes number dialling unnecessary

Design Aspects

Functional Requirements

The user should be able to be perform a call by clicking on a person icon and the Real Time Communication Service (RTCS) icon in any order.

Operational Requirements

Machine dialogues should confirm status of call and service attributes

References: GIOs Person Icons, Real Time Communication System (RTCS) Icon, Call Confirmation Dialogue

5. Software design using the enabling states approach

The completion of the design stages so far results in a complete description of the tasks which must be performed by the work system, including those tasks which are not directly related to the overall goal and which may vary from design to design. The next step is to begin the specification of the work system software.

The final layer in the Enabling States approach produces specifications of the appearance of the interface objects required to satisfy the operational requirements described in the previous layer, together with the actions that they respond to, and their behaviour.

Figure 15 shows the how the design principles used in the specification of enabling state and enabling tasks influence the design of Generic Interface Objects (GIOs), and shows the relevant parameters of the GIO specification table.

GIOs which are used within different services may interact if these services are used simultaneously, and service integration principles, therefore, provide information about task interleaving and the usability of different services which may affect GIO design. Similarly, enabling task integration principles serve the same function with respect to the enabling tasks the user has to perform on a given set of GIOs when using simultaneous services. User action interpretation principles help the designer in representing user actions on a given GIO and are applied to the 'event' slot in the GIO specification table. For example, principles might recommend different cursor shapes to indicate appropriate user actions. Task object representation principles and object principles influence the design of the 'effect' slot which describes how objects are represented. This may concern, for example, icon and dialogue design. User principles, function allocation principles and cost principles apply to the 'actor' slot in the specification table and make recommendations concerning which communicating entity should perform a particular action and why. Additionally, the software implementation team may use software libraries to ensure consistency between GIO components such as widgets.

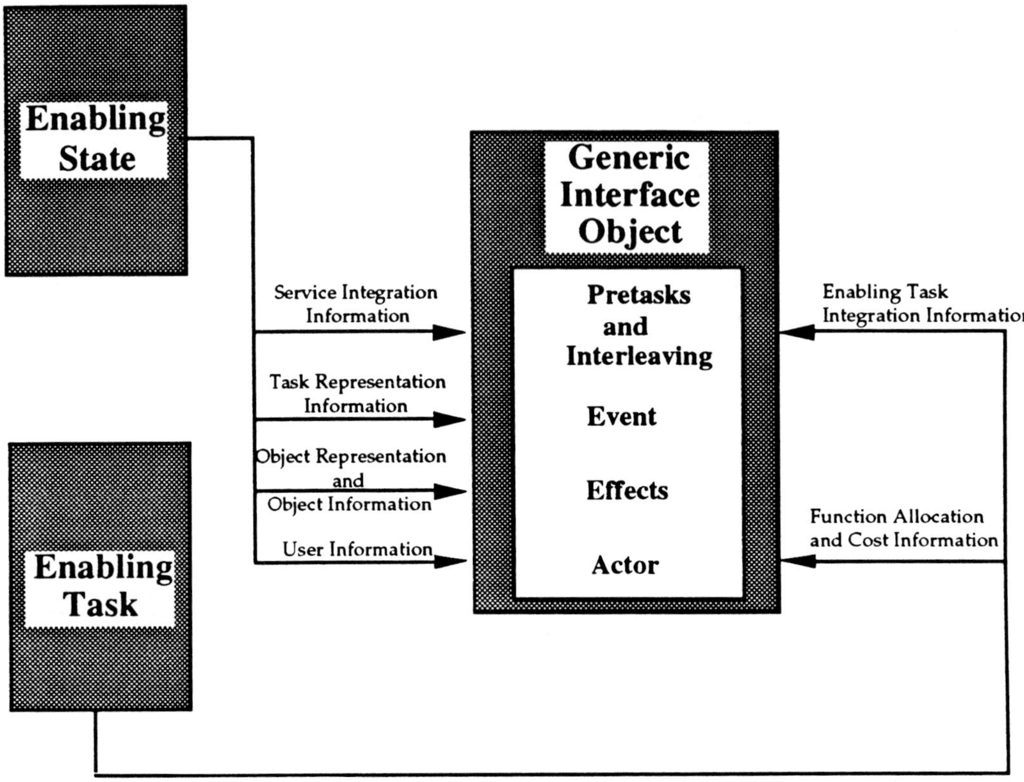

Figure 14: The Flow of Information Derived from the Application of Principles into the Design of Generic Interface Objects

In summary, the use of principles throughout the design process encourages designers to consider issues of service integration and service usability by pointing out pitfalls that a designer might fall into and guides them towards options that have empirical evidence as to their usability.

An example of a full GIO specification table is given in Specification Table 7.

Specification Table 7 An Example of an Enabling Tasks Description for 'Identify Recipient'

Description: If the user has clicked on the edit button and no object has been selected, they get a list of icons to select one of the four available editors (text editor, graphic editor, audio editor or audio/video editor).

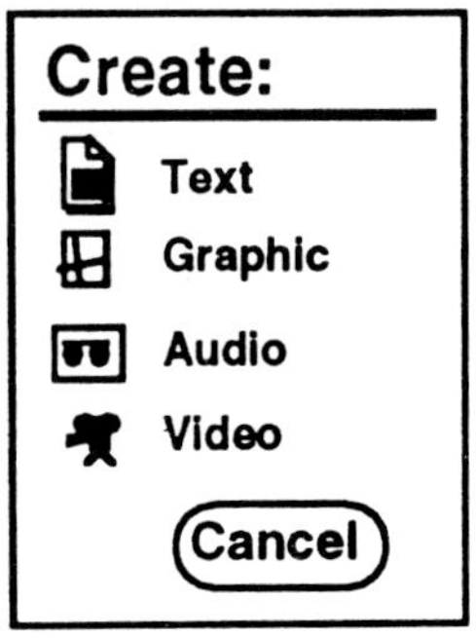

Sources: Enabling States, Enabling Tasks, Goal Description
Class Relations: dialogue window, single choice menu
Design aspects
Interface model: dialogue box
design discrepancies: None
design rationale: The user should be able to quickly find and specify an editor
Properties
name: layout of parts
value: Title on top of the dialogue, below: column of four icons representing text, graphic, audio and audio-video editor, label on the right side of each icon, below: cancel button on the right side
Interaction procedures:
name: editor selection
preconditions: this GIO shown
actor: user
event: clicks on an editor icon
effects: the editor dialogue vanishes, indicated editor invoked
name: cancel button selection
preconditions: this GIO shown
actor: user
event: clicks on cancel button
effects: the editor dialogue vanishes
References
Widgets: editor icons, cancel button

6. Summary: principles result in integrated services

In the specification method the GIOs which are to be used in reaching a particular enabling state are clustered into a *View* for that state (see figure 7), together with the objects which must be present to provide the service relationship defined in that enabling state. All interface objects that are likely to be used together (or in close sequence) and the consequent interaction behaviours are gathered together.

In the GUIDANCE MMMADP prototype (see chapter 3.3), this can be illustrated by reference to the Browser View. This View coordinates access to editing functions for four different classes of media objects (text, graphics, audio and audiovisual), a real time audiovisual communication service and a store-and-forward mailing service, as well as providing overviews of the status and availability of other users of the system and of the workspace (i.e. database retrieval services). All of the different services that these goal tasks require are accessed through this single View (see figure 16).

Figure 16 also illustrates how the different types of principles described in this chapter have contributed to the design of services utilised via the browser. The principles cited are those that have already been described in full in the example specification tables for enabling states and enabling tasks and are given below in abbreviated form providing just the prescription and the rationale.

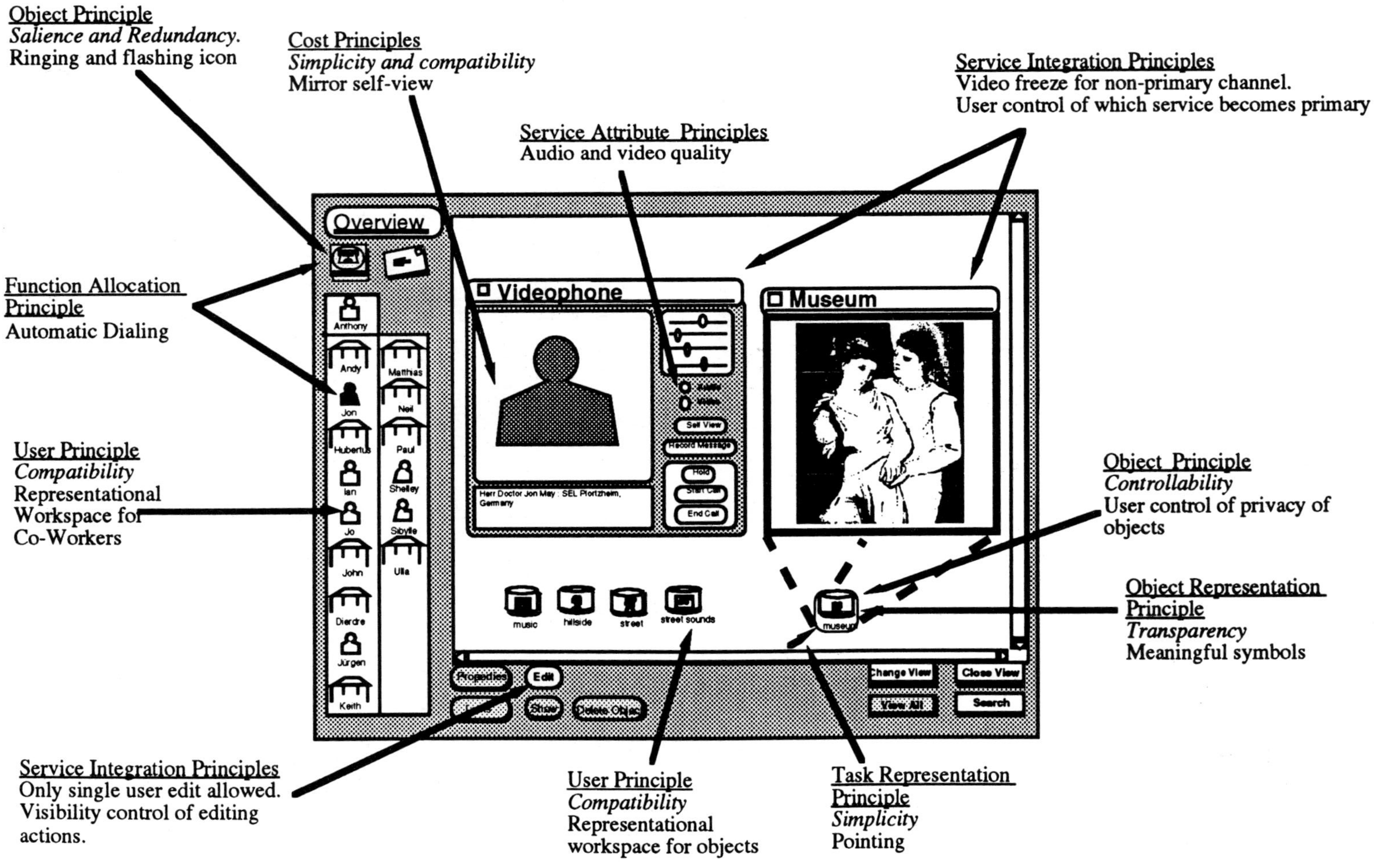

Figure 15 : Principles Result in Integrated Services

Enabling States Analysis

User Principles:

Compatibility : provide the user with a representational workspace which reflects relationships in an open plan office and represents all members of the workgroup and their presence/absence BECAUSE This will allow users to exploit existing knowledge about office work, and the work environment with respect to the availability of persons and the actions which may be performed on these persons.

Compatibility : provide the user with a representational workspace which reflects relationships in an open plan office BECAUSE *(rationale)* This will allow users to exploit existing knowledge about office work, and the work environment with respect to the availability of objects and the actions which may be performed on them.

Service Attribute Principles:

IF audio and moving images are provided simultaneously (videotelephony conversation)
THEN audio transfer rate is <64 KBit/s and standard moving image quality is required.

IF audio only is provided
THEN quality is always voice (3.1—7 kHz)

Object Representation Principles:

Transparency : represent the nature (media type) of each object by a distinct and meaningful symbol BECAUSE *(rationale)* clearly distinguishable object types will reduce user retrieval errors.

User Task Interpretation Principles:

Simplicity : allow the pointer to move in all directions and change the shape of the pointer according to possible actions BECAUSE *(rationale)* This will reduce the number and complexity of actions required, and intended actions will be easy to perform and remember

Object Principles:

Salience and redundancy: use simultaneous audio and visual signals to alert recipient to incoming call BECAUSE *(rationale) combined* audio and visual signals (e.g. flashing and ringing alarms) are most effective in gaining attention.

Controllability : only allow users to change essential attributes (e.g. privacy) of objects which they have created or have been given control of by another author BECAUSE *(rationale)* editing of objects for users engaged in collaborative tasks should be a negotiated or managed process.

Service Integration Principles

Simplicity and controllability : ensure that the number and complexity of actions is kept to a minimum and that users can control which service becomes primary BECAUSE *(rationale)* interference between tasks should be minimised and the locus of control should remain with the user (i.e. the machine should not adopt unpredictable states).

Salience: ensure that the primary service component does not interfere with the secondary service component (e.g. mute a non-primary video service component or mute a non-primary audio component) BECAUSE *(rationale)* machine should assist users in focusing on primary tasks and allow use other input/output modalities for secondary tasks (e.g. when editing a video document users can utilise a free audio channel for conversational tasks).

Controllability: allow only one user to edit the object, and allow the editing user to specify whether his/her actions will be visible on the workstation of the second user BECAUSE *(rationale)* users engaged in collaborative tasks should be able to control what is being presented to other users to ensure privacy

Controllability: allow only one user to edit the object at a time and allow this user to manage the handover of editing capabilities to the other user BECAUSE *(rationale)* users engaged in collaborative tasks should be able to control their own actions and those of their workstation

Enabling Tasks Analysis

Allocation Principles:

Simplicity: providing default settings and make as many adjustments as possible automatically (e.g. focal length) BECAUSE *(rationale)* this will minimise the number of user actions necessary

Simplicity: design the user interface to support the rapid selection of the limited set of user numbers (addresses) without the use of a dialling keypad BECAUSE *(rationale)* the restricted number of communicants makes number dialling unnecessary

Cost Principles:

Simplicity and compatibility: keep the camera static and allows users to adjust their own position, e.g. in a similar manner to using a mirror BECAUSE *(rationale)* reverse control of camera actions demands skill and would be time consuming.

7. Conclusions

The generic principles described in section 2 are intended to aid designers by providing them with IBC service integration support that is applicable across domains. The generic principles were expressed explicitly, but informally, as a limited set derived from the human factors literature. This set is consistent with other proposals in the literature, for example Holcomb and Tharp (1991). This document has shown how these generic principles can be expressed with increasing degrees of formality and parameterisation. The GUIDANCE design team found the principles useful in creating the prototype multiauthor multimedia document production system which is described in chapter 3.3. Their explicit scope and expression with respect to their conditions, prescriptions and rationale reduced ambiguity in design discussions, and contributed greatly to decision-making within the design team.

The specific principles developed on the project were classified and applied within the design method. The method and principles provide an integrated approach to IBC design which encourages designers to think about the design of systems from the users' perspective and provides support to IBC designers in solving the problem of IBC service integration and service usability. The close linking of the principles and the design method offers the designer assistance in identifying the appropriate principles for a given stage in the design process, and provides a schema which can support the development of principles. Thus, principles can be grouped according to their role within the IBC development process with resulting improvements in applicability and availability of design information. By structuring the design sequence so that relevant information is made available at the appropriate point, so that related information is kept together within tables, and so that tables follow a common schema, the task of design is made more manageable, and the likelihood that the designers will be able to make use of the information that is available to them is increased. Communication between individual designers and separate design teams is enhanced by the simple formalism provided by the specification tables, and so groups working on individual services can have more confidence that their designs will remain integrated. Finally, the classification schema encourages designers to use the principles as a checklist or rationale for the design decisions they have made.

Acknowledgements

This work was carried out as part of the R1067 GUIDANCE project funded by the Commission of the European Communities under the RACE programme. The project partners are; Standard Elektrik Lorenz AG, Germany; University College London, UK; British Telecom, UK; Roke Manor Research, UK; and Televerket (Swedish Telecom), Sweden. We would like to thank our colleagues who commented on earlier drafts of the chapter.

Thanks are extended to RACE projects R1065 ISSUE, R1066 IPSNI, R1054 APPSN, R1088 TUDOR and RIC who attended a Usability Consolidation Workshop arranged by GUIDANCE and held in London on 21st and 22nd October 1992.

References

Bornat R. & Thimbleby H. (1989) The Life and Times of ded, Text Display Editor. In J.B. Long and A.D. Whitefield (Eds), *Cognitive Ergonomics and Human Computer Interaction*. Cambridge: Cambridge University Press.

Byerley P., & Bruins, R,. Conceptual Framework for Usage of Telecommunication Services. In: P. Byerley and S. Connell. (Eds). *Integrated Broadband Communications:Views from RACE—Usage Aspects* North-Holland Studies in Telecommunication Volume 18. Elsevier.

Byerley P., Denley I., Melchior E-M., Voigt U-B., Floruß S., and Hermann S., (1993). A unified Method for the design of Telecommunication Services. *Ergonomics. Special Issue on Telecommunication.*

Denley, I., Whitefield, A., Byerley, P., Voigt, U.B., Hermann, S., and May, J. (1992). Design Principles for Improving Service Integration for End-Users in Broadband Communication Systems. In: *Proceedings HCI '92 Conference*. University of York. 15-18th September 1992.

Eason K. (1988) Information Technology And Organisational Change. London: Taylor & Francis.

Esgate, A., Whitefield, A. and Life, A. (1990). Developing Usability Integration Principles for the Design of IBCN Systems. In: *ECCE-5, Proceedings of the Fifth European Conference on Cognitive Ergonomics*, Urbino, Italy, September 3-6th, 1990.

Gardiner M.M. & Christie B. (Eds), (1987) *Applying Cognitive Psychology To User-Interface Design*. Chichester: John Wiley.

Gould J.D. & Lewis C. (1985) Designing for Usability: Key Principles and What Designers Think. *Communications of the ACM, 28*(3), 300-311.

Hammond N., Jorgensen A., MacLean A., Barnard P., & Long J. (1983) Design Practice and Interface Usability: Evidence from Interviews with Designers. In: *CHI 83 Proceedings*, 40-44. Published by ACM.

Hockey, G.R. (1986). Changes in Operator Efficiency as a Function of Environmental Stress, Fatigue, and Circadian Rhythms. In K.R. Boff, L. Kaufman and J.P Thomas (Eds.), *Handbook of Perception and Human Performance (Vol. II) Cognitive Processes and Performance*, New York: Wiley, p. 44.1—44.49.

Holcomb, R and Tharp, A.L. (1991). Users, A Software Usability Model and Product Evaluation. *Interacting with Computers, 2,* p155-166.

Jerke, K.H., Szabo, P., Lesch, A., and Rossler, H.(1990). Combining Hypermedia Browsing with Formal Queries. In Diaper, D., Gilmore, D.,Cockton, G., and Shackel, B. (Eds). *Interact '90. Proceedings of the IFIP TC 13 Third International Conference on Human-Computer Interaction,* Cambridge, U.K. 27th—31st August, 1990.

Kahneman, D. (1973). *Attention and Effort.* Englewood-Cliffs, NJ:Prentice-Hall.

Kinsbourne, M. & Hicks, R. (1978). Function Cerebral Space. In J. Requin (Ed.), *Attention and Performance VII,* New Jersey: Erlbaum.

Laughery, R., Drews, C. & Archer, R. (1986). A Micro SAINT Simulation Analyzing Operator Workload in a Future Helicopter, *Proceedings of NAECON,* New York: IEEE Press.

Lim K.Y., Long J.B., & Silcock N. (1990) Requirements, Research and Strategy for Integrating Human Factors with Structured Analysis and Design Methods: The Case of the Jackson System Development Method. In E.J. Lovesey (Ed), *Contemporary Ergonomics 1990,* p.32-38. London: Taylor and Francis.

Long J.B. & Dowell J. (1989) Conceptions of the Discipline of HCI: Craft, Applied Science and Engineering. In A. Sutcliffe and L. Macauley (Eds), *People And Computers V.* Cambridge: Cambridge University Press.

Melchior, E.M. (1992). Generic User Services Defined. In: P. Byerley and S. Connell. (Eds).*Integrated Broadband Communications:Views from RACE—Usage Aspects* North-Holland Studies in Telecommunication Volume 18. Elsevier.

Mosier J.N. & Smith S.L. (1986) Application of Guidelines for Designing User Interface Software. *Behaviour and Information Technology,* 5(1), 39-46.

Navon, D. (1984) Resources: A Theoretical Soup Stone. *Psychological Review, 91,* 216-234.

Navon, D. & Gopher, D. (1979). On the Economy of the Human Processing System, *Psychological Review, 86,* p. 214-255.

Polson, M. & Freidman, A. (1988). Task Sharing Within and Between Hemispheres: A Multiple Resource Approach, *Human Factors, 30,* p. 633-643.

Rasmussen, J. (1986). *Information Processing and Human-Machine Interaction: An Approach to Cognitive Engineering.* New York:North-Holland.

Rasmussen, J. (1990). Mental Models and the control of action in Complex Environments. Ind D. Ackerman and M.J. Tauber (Eds.), *Mental Models and Human-Computer Interaction I,* Amsterdam: Elsevier.

Sanders, A. (1979). Some Remarks on Mental Load. In N. Moray (Ed.), *Mental Workload: Its Theory and Measurement,* New York: Plenum Press, p. 41-78.

Schneider, W. & Shiffrin, R.M. (1977). Controlled and Automatic Human Information Processing:1. Detection, Search, and Attention. *Psychological Review,86,* 1-66.

Senders, J.W. (1983). On the Nature and Source of Human Error., *Proceedings of the Symposium on Aviation Psychology, 2,* p. 421-428.

Smith S.L. & Mosier J.N. (1986) Guidelines For Designing User Interface Software. MITRE Corporation Report ESD-TR-86-278.

Terrins-Rudge, D., Adamson, S.J., Ionescu, R., Peterson, M. and Hedman, L. (1992). Designing for Designers' Needs. In: P. Byerley and S. Connell. (Eds).*Integrated Broadband Communications:Views from RACE—Usage Aspects* North-Holland Studies in Telecommunication Volume 18. Elsevier.

Walsh P., Lim K.Y., & Long J.B. (1989) JSD and the Design of User Interface Software. *Ergonomics*, 32(11), 1483-1498.

Whitefield A.D., Byerley, P., Denley, I., Esgate, A., and May, J. (1992). Integration of Services for Human End-Users (1): Design principles, Enabling States Analysis and a Design Method. In: P. Byerley and S. Connell. (Eds). *Integrated Broadband Communications:Views from RACE—Usage Aspects* North-Holland Studies in Telecommunication Volume 18. Elsevier.

Wickens, C. (1980). The Structure of Attentional Resources. In R. Nickerson & R. Pew (Eds.), *Attention and Performance VIII*, New Jersey : Erlbaum, p.239-257.

Wickens, C. (1984a). Processing Resources in Attention. In R. Parasuraman & D. Davies (Eds.), *Varieties of Attention*, New York: Academic Press, p. 63-102.

Computers, Communication and Usability:
Design Issues, Research and Methods for Integrated Services
P.F. Byerley, P.J. Barnard and J. May (Editors)
1993 Elsevier Science Publishers B.V.

Chapter 3.3

Prototyping exercises with multimedia multiuser services

Ulla-Britt Voigt*, Sibylle Hermann*, Jürgen Herczeg†, Hubertus Hohl† and Matthias Ressel†

*Alcatel-SEL AG, Lorenzstraße 10, D-7000 Stuttgart 40, Germany

†Research group DRUID, Institut für Informatik, Universität Stuttgart, Breitwiesenstraße 20-22, D-7000 Stuttgart 80, Germany

1. Introduction

Integrated Broadband Communication services (IBC) offer users the integration of previously distributed functionalities, and enable them to work cooperatively with others on different sites, using real time communications and messaging systems. IBC moreover allows the user not only to communicate with text and graphics but also with audio and video between two or more persons, thus allowing multimedia communication. These systems additionally allow the user to jointly navigate in local or distributed multimedia databases and to jointly create or edit multimedia documents. At its most ambitious, a group of users can jointly create Multimedia Documents while having simultaneous multimedia communication.

This increased functionality in comparison to stand-alone computers or connected PCs is not only a sum of its parts. It gains a quality which not only enables the user to do more things in different ways, but it presents the user with new challenges in handling and using such a system. These problems are described in chapter 1.2 and 3.1, and figure 1 illustrates the problem area.

When designing such a system the overall goal is to ensure that the user can handle and use it: to design a usable IBC system. If this is not achieved, or even not regarded as the most important goal, the product will not have any market success. Design is here understood in its broadest meaning, including the whole process from the product idea to eliciting the requirements, preparing the specifications, doing the implementations and in the end incorporating the customers' feedback on the delivered product.

Designing a usable system is a design task which itself is more complex than the design of stand-alone systems. This again is illustrated with figure 1. The crucial thing is to design the interactions properly: the interactions involved in person-to-person communication; the access, retrieval and sending of distributed objects, the creation and editing of distributed, jointly owned objects; and in the combination of these tasks. Additionally, the design of IBC systems can not and should not be seen as the just the design of the interface, but as tackling the design of the work task itself. Deciding which task is to be performed by whom (human or machine) and how has not only consequences for the design of the interface, but also for the work the task is part of (see for example Ulich, 1992).

	Tasks	Objects
Stand-alone	Creation+ Editing of Objects	only locally available, thus they can be accessed, edited and created only by person using specific computer
Integrated	Conferencing Distribution Retrieval and any combinations of these tasks	locally and distributed availabe, distinguishion between private and public objects, public objects with varying degree of being accessible, editable creatable and annotable.

Figure 1: The problem area of integrated service design

But IBC services will not only influence the work. Devices like IBC services are part of or will become part of our culture. People will not only influence 'nature' (in the form of their work) by using devices, but will be influenced by the devices with the result that they change the way that they perceive and conceptualise 'nature' (Olson, 1976). Applying this idea to thinking about the consequences of IBC services in the work environment, it becomes clear that the user, the task and the device are all influencing each other. Using IBC services will thus change the task, but also influence the human who is using with the device to perform the task. These consequences have to be kept in mind, as they represent a requirement for the involvement of psychologists and other human factor specialists in the area of designing devices, especially in the area of designing software. Software, as it is not physically presented but becomes visible or is experienced by using hardware, might even have *more* influences on mental activities than other devices have. The benefit of integrating the results and the knowledge of human factors in design includes being able to anticipate potential consequences of the device being developed, evaluating these consequences and influencing design such that negative consequences are avoided.

Designing a usable IBC service therefore requires the joint effort of human factors expertise and software engineers. For these groups to assist each other to produce a usable design requires the use of a method which enables their close and interconnected work.

This chapter describes the experience of such joint work in the RACE project 1067 'Guidance', which developed the Enabling States Method. Two prototyping exercises are reported, which show how the method was used and refined to better reflect the needs of the human factors personnel and the implementers.

2. Requirements for the design process

Designing a usable IBC service requires human factors knowledge. This includes not only the knowledge of design guidelines or principles, but includes as its most important factor the ability to understand the design of a device as the design of a task and a work context. To anticipate changes of the task and the work context, to evaluate these changes and to assist in the design of these changes, a user oriented task analysis is necessary which allows:

- the description of the task as it is performed at the moment, the user, the devices used and the environmental conditions of the work (ranging from physical environment conditions to social and organisation environment conditions).
- the description of the task as it should be performed with the new device. This includes designing the device and the allocation of functions to user and device.
- the evaluation of the designed device against criteria which were elicited during the task analysis (i.e., what parameters are important in performing a specific task).

Human factors knowledge is operationalised and used for the task analysis and design by a definition of usability which has to be explicit enough to be incorporated in the design method and to be used for design decisions.

Additionally, human factors knowledge should not just come in the end of the design process to evaluate an implemented system, with the opportunity for only minor adjustments, but should be taken into account at the beginning of the design process: in fact, it should *be* the start of the design process. To start with the users' goals and characteristics is the only way to ensure that the design will meet the users' needs and abilities. The following list presents in more detail how a user centred perspective can and should be brought together with the implementers' needs.

Design should therefore:

- start with a description of the user' task, their characteristics, the desired performance and the system environment available (this is a User Analysis).
- use human factors knowledge throughout the whole design process to succeed in breaking down the user analysis for specific system design decisions, and support the designers at crucial design decision points
- enable a system evaluation from the earliest point possible, and integrate the results into the design process

This requires active work throughout the design process, feeding back from design decisions which affect or change previous design decisions, coordinating changes and to keeping to the goal of user oriented design.

Design in general should:

- enable an object oriented approach to design
- enable the communication of the results not only to human factor personnel and implementers, but also to the customer and the users to receive their feedback
- allow an iterative process and rapid prototyping
- provide a protocol of the design decisions and the design procedure

An iterative process allows testing during the design process to define necessary changes as early as possible to include them even if they affect changes to earlier design decisions. The iterative nature of a design method incorporating human factors knowledge, can be described as design cycle, as illustrated in figure 2. Such an iterative design process is supported by object-oriented design, which allows the easy incorporation changes and is also supported by the concept of prototyping, which allows a quick realisation and evaluation of design ideas. Figure 2 also points out that the design process does not end when the product is introduced to the market, but shows that the feedback of customers has to be incorporated in the continuing design cycle.

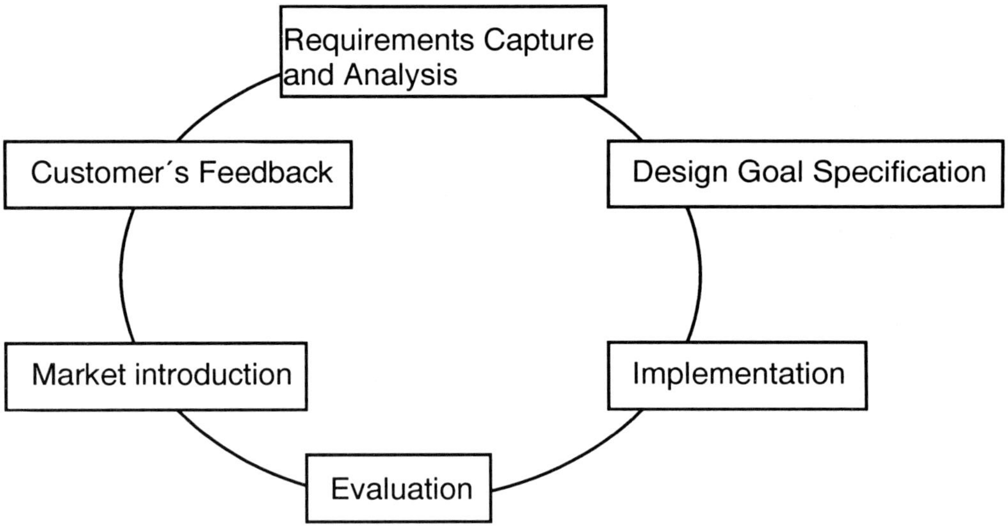

Figure 2: The continuing design cycle

The previous section described how human factors knowledge should be integrated in the design process. This can be understood as a demand on the implementers and their design methods in use. But there is of course also a demand on the human factor personnel to provide their knowledge about the characteristics of a usable system in a form that can be easily integrated into the design process. This means that the analysis of the users' goals and characteristics, their task and work and a definition of usability is not sufficient to determine design decisions if it does not have clear implications for design. It must be expanded and operationalised to give explicit help to designers and allow object-oriented prototyping. It is not sufficient for example to provide a detailed user and usage context analysis by using one of great set of available methods such as VERA (Volpert et al, 1983) and VERA-B (Rödiger, 1988), TBS-GA (Rudolph et al, 1987), TAI (Frieling, 1990), or HAA (Greif, 1991). The results or implications of these analysis have to be made understandable and usable for the implementers by formalising them, so that they map to the methods used later on in the design process. This formalisation also enables easy access to design decisions and thus supports implementation by groups, not only by one person. (Hamborg & Schweppenhäußler, 1992).

Emphasising the user-oriented approach means:

- the provision of a method which enables the use of human-factor knowledge
- the provision of human-factors knowledge in a form that is easily accessible for non-human factors personnel.

3. Prototyping exercises

This chapter describes the use of the Enabling States Method as a method which allows human factors personnel and implementers to work together in designing a usable IBC service. The method is based on the definition of usability given in chapter 1.1. This should determine the design when using the method, but it has also determined the development of the method:

i) ensures that the appropriate enabling states exist for each of the users' goal tasks

ii) reduce to a minimum the costs to the user in reaching the appropriate enabling states for their goal tasks.

The Enabling States Analysis and the Method are described in detail in chapter 3.1. This chapter will show how the final version of the Method was reached by using the experience of designing a multimedia multiauthor document production system. The people involved in this design were a team of psychologists and implementers, but due to the size of the project, time and resources were limited to two psychologists and two implementers for two years.

The characteristics of the method which have not been changed can be described as four steps:

1) the elicitation of user and customer requirements
2) a task analysis to find all relevant Enabling States that the system has to offer the user to enable them to perform their tasks.
3) the analysis of the Enabling States themselves
4) function allocation, performed in the description of the Enabling Tasks.

The prototyping exercise was performed in the area of multimedia tourist information, for which a multimedia multiauthor document production system (MMMADP) was specified, implemented and tested.

This MMMADP consisted of two parts: a 'Reader Interface' and a 'Production Interface'. The Reader Interface allows access to a Multimedia Database about travel information about Baden-Württemberg in southwest Germany, and the Production Interface allows two or more authors to jointly create Multimedia Documents about the region, and to communicate about them with videophone, audiophone and electronic mail.

The hardware environment for the implementation included three SUN UNIX workstations, video cameras, laser disc players, and the emulation of a broadband connection between the workstations. The software used is based on an object-oriented approach using the Common Lisp Object System (Steele, 1990). The implementation of the user interface of the Production Interface is based on XIT (Herczeg et al., 1991), an object-oriented development environment for graphical user interfaces for the X Window System (Scheiffler & Gettys, 1985).

As the development environment influences how the specification is performed, XIT will be described in some detail here.

3.1. Object-Oriented Design

XIT thoroughly follows an object-oriented design. Generality, extensibility, and ease-of-use are the main design goals behind the layered architecture of XIT. In XIT, different system layers provide different kinds of tools for building user interfaces. A low-level toolkit provides a general framework for building many kinds of interface elements, called interaction objects. A high-level toolkit provides an extensible library of common interaction objects, like buttons, switches, menus, sliders, and property sheets, composed of elements from the low-level toolkit. An interactive toolkit provides graphical tools for constructing and manipulating user interfaces interactively, e.g., user interface builders, resource editors, interaction object browsers, and inspectors.

The low-level toolkit provides a powerful framework of classes and methods which describe the properties and the behaviour of basic interface elements. To build complex interaction objects either subclassing and inheritance or structural composition can be used. The framework integrates advanced features for describing automatic screen layout, visual feedback, application logic, and dialogue control, usually not found in other user interface toolkits (Herczeg et al., 1992). All too often, low-level toolkits do not provide appropriate means to modify building blocks or to define new ones. Instead of being general and extensible, they are mostly restricted to a particular look and feel. As a consequence, higher-level interface toolkits built upon them are rather poor as far as flexibility and extensibility is concerned. However, these characteristics are essential for the implementation of applications like the Reader and Production Interfaces. These applications often need specific interface elements like video interaction objects or advanced interaction techniques, as for example drag & drop style interaction.

3.2. User Interface Building Blocks

In XIT, interaction objects are characterised by the following basic categories of properties:

Structure: Each interaction object is organised into a part-whole hierarchy. An interaction object representing a menu, for example, is composed of interaction objects representing the menu entries.

Geometry and Layout: The geometry of an interaction object is determined by its position and size. Besides specifying absolute values, the layout of an interaction object can be described by specifying relations between its geometry and the geometry of its parts. The layout of the parts of a menu, for example, can be described as one below the other.

Presentation: The appearance of an interaction object is mainly determined by presentational attributes, like colour, text, font, image, sound, or speed of video playback, etc. Menu entries, for example, can be presented by individual texts or images.

Reactivity: The behaviour of interaction objects is determined by mapping events onto appropriate actions. Events may be either user input events, e.g. keyboard or mouse events, or internal events. Actions may affect both user interface and application. In the menu example, each menu entry reacts on a button click, triggering an internal event that is sent to the menu object. This event triggers a function of the application and, if required, lets the menu disappear.

Application Logic: Interaction objects may be associated with the objects of the application they represent or on which they perform operations. In the menu example, each menu entry may be associated with an application function, and the menu object may be associated with the application object on which the functions operate.

4. The Reader Interface

The specification of the Reader Interface (RI) was done in respect of a user who wants to obtain multimedia information about holiday locations in Baden-Württemberg, so that they can decide where to go. The user should be able to specify their interests (sports, restaurant, culture, etc) as well as their criteria (beauty, price, etc) thus forming a database query.

4.1. Examples of the specification

The specification resulted in an interface which is presented in figure 3.

The first version of using the Enabling States method consisted of four layers: the goal layer, the enabling states layer, the functional layer and the software (physical) layer. In total, ten steps were included in the specification (figure 4), leading from the description of the user and their goals, through a task breakdown, to a description of the enabling states and enabling tasks. This led to the functional layer, which described the necessary subsystems, tool-boxes and functions, and then to the software layer, which described the views, generic interface objects and widgets. As will be described, not all of these stages remain in the current version of the method, but those that are covered in detail in chapter 3.1.

As pointed out in chapter 3.1, the enabling state layer is the crucial point in the definition of usability and thus in the design of integrated services. In the following case-study we will

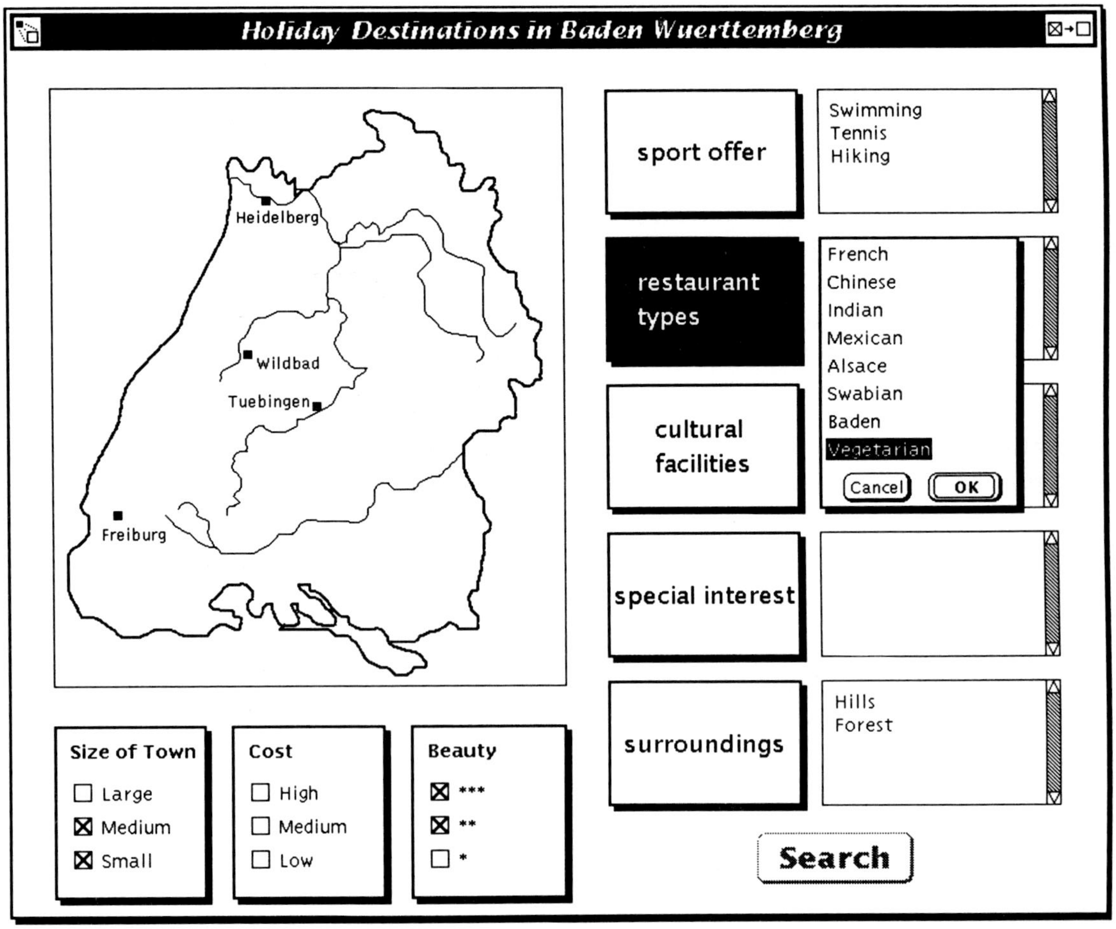

Figure 3: The 'Reader Interface'

therefore concentrate in the presentation of the enabling states specification and its consequences for design decisions.

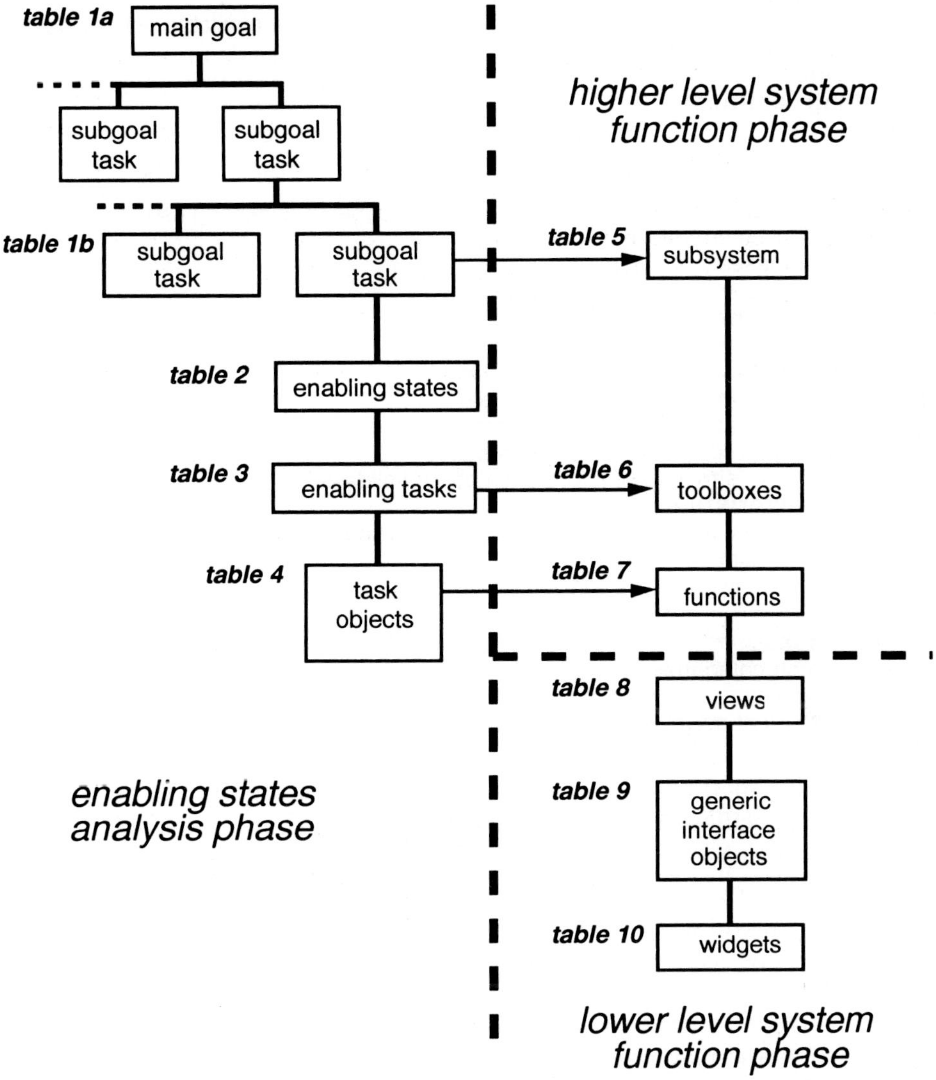

Figure 4: An early version of the enabling states method

Two enabling states (ES) were defined for the reader interface:

1. Overview material available
   (This ES should enable the user to attend to overview information about available holiday destinations).
2. Browsing facilities invoked
   (This ES should allow user to navigate/browse through available set of holiday destinations).

In the first version of the method, the Enabling State Description for the an enabling state consisted of the following information (shown for the first of the states listed above):

ES: Overview material available
Description: Enables user to attend to overview information about available holiday destinations
Source: Subgoaltask: attend to overview
Design Aspects:
Communicating entities:
Users:
Users should be aware that overview information is available and how to obtain it
Workstation
Must have grouping criteria available, must be capable of accepting user input, must know which databases are relevant
Holiday databases
relevant databases must be accessible
Communicating channels
Dynamic characteristics
Function for accessing overview information
Interpreted: Mouse input from the user
database access information
Noninterpreted: contents guide to databases
structure of information about holidays
Static Characteristics
Information types: data, test, graphic still and moving, picture, video, audio
Reliability: medium
Security: read only mode, full access to database
References
Enabling tasks
select search categories
enable overview
construct current set of holidays

Via the specification in the functional layer (subsystems, toolboxes and functions) the physical layer was specified with as a set of views, generic interface objects and widgets. Views are a collection of generic interface objects and widgets, which together form a unique interface element. Generic interface objects are a set of pre-defined interface types, usually built from a number of widgets, for example a window or a menu. Widgets are basic interface elements, such as buttons, icons and scroll-bars.

The view specified was 'overview criteria selector', which presented the criteria information and allowed selection and response. The next tables show the specification of this view, of one of the generic interface objects and of one of the widgets.

View: Selection Form
Description: Form to allow user to indicate search parameters (facilities that holiday must provide and criteria it should fit) and to carry out searches of the database using those criteria
Sources:
Functions:
Group Holidays
Present Form
Complete Form
Select Presentation Style
Present Overview
Design Aspects:

Design Model:
Question and answer dialogues with travel agency
Interface Model: Form -filling
Design Principles:
In dialogue only one decision presented at once, from filling provides no guidance about the relevance or practicality of search parameters
Design rationale:
Present all possible choices to user so that they know what is available and what is not available

User Control:
Preconditions for use:
Region for holiday indicated
Holiday database accessible
MCPR database accessible
User actions: selection of search parameters
Effects of user actions: search parameters updated
Feedback: selected parameters continuously displayed to user
Errors:
Action on errors
User actions: Initiation of Search using current parameters
effects of user actions: Candidate Set reconstructed according to current search parameters
Feedback: Indication of search continuing while it is in progress
Indication that search has been completed and candidate set available by presentation of Overview
Errors:
Action on errors:
User actions: Selection of holiday destination from candidate set
effects of user actions: Hypertext link made to MCPR database at overview page for chosen destination
Feedback: MCPR screen appears
Errors: No link available to MCPR
Action on errors: Display dummy page that has no further hypertext links and is only pseudo multimedia
User actions: Quit view by clicking close box
effects of user actions: end of session with query system
Feedback: window closes (perhaps after confirmation dialogue: Do you want to quit the query system?')
Errors: User did not intend to quit system
Action on errors: none possible

Device Action:
Conditions: Following completion of search, no holidays fit search parameters and so candidate set is empty
Effects: Display error messages overlaying GIO 'map' with text 'No holiday destinations fit your requirements. Please alter them and try again' together with OK button
Parameters: Candidate Set
Conditions: Following Completion of search, some holidays fit search parameters and so candidate set is not empty
Effects: display members of candidate set on GIO 'map' (ie provide overview)
Parameters: candidate set

Properties:
Name: size
Values: approx. two thirds of screen, but must be able to contain appropriate GIOs and widgets
Controlled by: system
Name: position
Values: initially central on screen if this will not obscure secondary views, otherwise offset as necessary. Subsequently under user control
Controlled by: system/user

References:
Generic Interaction Objects
map
criteria list
facility list
pointer
Widgets:
selection form border
selection form title
selection form close button
search button

Generic Interaction Object: map
Description: Map of region that the user wants to search for holiday destinations in, showing destinations that fit current search parameters
Sources:
View: Selection form
Design Aspects:
Design principles:
Design discrepancies:
Design Rationale:
Device Actions:
Conditions: View 'Selection Form' opened
Effects: Map displayed
Parameters: Current Region
Conditions: Changes to Candidate Set
Effects: New members of set indicated on map, members no longer in set taken off map
Parameters: candidate set
Properties:
name: contents
values: map of current region
controlled by: database
name: position
values: upper left of parent view 'selection form'
controlled by: system
name: size
values: minimum of 7 cm horizontal 5 cm vertical. Sufficiently large scale so that no two destinations will overlap if simultaneously selected.
controlled by: system
Interaction procedures
Activity: user clicks on holiday destination
action: hypertext link to MCPR database at overview page for that destination
References:
Widgets:
map picture
city button
city name

Widget: Facility OK button
Description: Closes the facility list currently being viewed, recording selected facility values as search values
Sources:
Generic Interface Object: Facility list
Usage Aspects:
Frequency of use: medium
User: expectations: one click will make facility list vanish, revealing selection form, and the values of the facilities selected will appear in facility list.
Design Aspects:

Device Actions:
Pre-Conditions: Activation by Widget 'Facility button'
Effects: Makes this widget appear
Properties:
Name: size
values: to fit contents
controlled by: system
name: contents
values: 'OK'
controlled by: system
Name: position
Values: bottom right of widget facility box, to right of widget facility cancel button
Controlled by: system
Name: border
values: highlighted, emboldened or otherwise more obvious than widget Facility cancel button
controlled by: system
Interaction Procedures
Activity: clicked on
Action: close display of selectable facilities, pass selected values to search system

4.2. Implementers' Feedback on the Specification of the Reader Interface

This specification was the first use of the method and important information about its usability was gained by using the specification for implementation. The main issues will be reported here.

4.2.1. The Table Hierarchy

Of all tables only tables 8-10 were understood as primary input for the implementation. Ideally they should contain all the information necessary for the implementers. However, some information was more clear when read in connection with some of the higher level tables. Especially the relation and the interaction between several interface parts gets lost within the numerous pages of the tables 8-10. A look especially at tables 7 (functions) could sometimes clarify the desired interaction. Further assistance has been provided with the storyboards.

4.2.2. Pictures, Storyboards

In the original form of the tables pictures were included only very sporadically. This sometimes made it hard to understand the desired properties of an user interface object, especially the visual appearance.

4.2.3. *View, Generic Interaction Object, Widget*

The implementers did not fully agree with the definitions of View, Generic Interface Object and Widget. They suggested that a View is an author defined user interface object, mixing a number of instances of generic interaction objects and widgets to form a unique interface element. Generic interaction objects (GIOs, formerly called generic *interface* objects) are predefined interface types, usually built from widgets or less complex GIOs. The definition of a GIO includes properties common to all its instances. Actual initialisations of these interface types should state: number, type, properties, and layout of their parts. Examples are scrollable windows, menus, a set of buttons. Widgets are basic interface elements, that are not considered as compounds (although the actual implementation may break them down into smaller parts). Examples are a button, an icon, a scroll bar. The discrimination of GIOs and widgets is not well defined and therefore arbitrary. It should be noted, however that a view is something completely different, namely an instance, while GIOs and widgets are classes.

4.2.4. *Object-Oriented User Interface Design*

The tables did not fully support an object-oriented user interface design. The following section explains why the description of a user interface is best done following an object-oriented approach and how this can be applied in designing the tables. It tries to present the minimum knowledge designers should have, if they want to write a table in a manner most useful for implementers. This knowledge has also been incorporated in the representation of interaction objects with the implementation tool XIT (cf. section 3.3).

User interfaces can be described as a hierarchical network of user interface objects connected by a part-of relation.

The Reader Interface, for example, consists of a map, a number of criteria menus, a number of facility buttons and facility lists, and a search button. The map contains small buttons as subparts, representing cities. The criteria menus are built up from a label and several menu entries, which, in turn, consist of a check box and a text.

Typically there always exist several similar interface objects. In an object-oriented description you define the properties shared by similar objects in a so-called object class. Objects belonging to a certain class are called instances of this class. For every property (often called slot, examples are size, position, colour) a class may or may not specify a default values; if an instance of that class is created (instantiated) and no value for this property is explicitly given, this default value will be used.

The class of 'facility button' could define its instances to have a drop shadow, with a default value of 5 (pixels) for the 'depth' of the shadow. So every facility button instantiated will have a drop shadow of depth 5 without having explicitly define the shadow for each button.

Another characteristic of object-oriented design is inheritance, by which the properties of one or more classes may be inherited by a subclass. The subclass not only inherits the slots from its superclass, but also the defaults, although these defaults can be redefined by the subclass stating its own defaults. Thus, a new field 'inherits from', describing the hierarchy of inheritance, should be introduced in the tables for widgets and GIOs. An additional backpointer 'inherited' by may be helpful.

A simple application is the definition of buttons. The class button defines properties for the overall look of a button, such as form, thickness of border, layout of text, and will additionally give suitable default values for these properties. Furthermore, it will define a property 'action', which will describe the action that shall be taken if the button is clicked. There is no default value, because every actual button will have a different action associated with it, and so the value of this slot has to be given when defining an instance of button. The classes 'OK-button' and 'abort-button' can now be defined as subclasses of class 'button', with the defaults for the text set to 'OK', and 'abort', respectively. Then, e.g., every instance of 'OK-button' will get the defaults defined by 'button' and carry the label 'OK', no matter what default for the label text is given in the definition of 'button'. Nevertheless, several 'OK-buttons' with different actions may be instantiated for different purposes.

4.2.5. Defining behaviour

In this first version of the tables, operations on the user interface object could only be stated as actions of an interaction procedure. Unfortunately, every interaction procedure already associated an action with an activity. While operations were defined for a class of integration objects, it is not normally known what activity (or, more general, what event) will trigger these operations. It was recommended that operations were specified in a new field (e.g. called 'operations') in the tables defining object classes (i.e., tables for Widgets and GIOs). This closely corresponds to the object-oriented paradigm, where methods describe operational properties of objects.

For example, certain classes of user interface objects canusually be moved with the mouse over the screen. In this way you can specifyexactly how the movement should 'look and feel' (moving a wire-frame or the actual object), but you may leave the specification of how the move is triggered (e.g. by grabbing the objects or by selecting the operation move from a menu) to the definition of the instance.

Activities (events) are then associated with actions (operations) in the field 'interaction procedures', as before. Note that a default for this association can be given in the class definitions, as for properties. The default given for any activity, however, may be overwritten by the specifications of an instance. The field class tables should be called 'default interactions procedures', in order to differentiate it from the field 'interaction procedures' in View tables.

The action field of an interaction procedure for a given object may also contain an invocation of an operation on a different object. Any interaction procedure involving more than one object should be stated in the table of the most specific object receiving the event (eg mouse click), given in the field 'activity'.

It was also thought that it might be helpful for the authors of the user interface specification, and improve clarity, to include an optional 'conditions' field in the description of an interaction procedure. Thus it could be specified under what circumstances an activity should trigger an action. This is not a critical extension, because conditions have already been incorporated in the specification of either the activity or the action.

4.3. Modification of the Method

These points of criticism were discussed by the specification team. The first point (that tables 8-10 did not contain all necessary information, and therefore it was necessary to refer to tables 1-7) typified a common lack of communication between specifiers and implementers: since it is impossible to provide a fully specified description of a system on the first pass, tables 1-7 ought to be read by the implementers, so that they can gain an understanding of the context of the design task, which is essential in user-oriented design. Specification will not, nor ever can be, complete, so the implementers have to take design decisions which must take into account the rationale of previous design decisions.

The second criticism, the lack of pictures and storyboards, was modified in the second version of the method. It turned out that the authors of the specification almost always made sketches or storyboards during the specification process, and it was agreed that, for the benefit of the implementers, the sketches should be incorporated into the description of user interface objects. This improves the comprehensibility of the specification by the implementers, and as a side effect, they also found it easier to locate an individual table in the specification, since they could recognise it by its 'picture'.

The modification of the method also incorporated the proposals in respect of the representation of Views, GIOs and widgets, and the properties proposed were included.

To better serve the idea of object-oriented design, the tables were enlarged to include information about 'classes' of Widgets and GIOS, and names were given to interaction procedures to integrate the idea of object orientation for this issue.

5. The Production Interface

The specification of the Reader Interface was followed by the proposed changes of the method and specification of the Production Interface (PI). The specification of the PI was made for users who want to cooperatively create a multimedia document. To support the joint creation, communication facilities (real time and asynchronous) have to be provided. This application was chosen as it represents one of the main aspects of IBC: computer supported cooperative work (CSCW). This is also the greatest challenge and one of the greatest tasks for designers who want to design a usable system.

The Production Interface is a collaborative work environment that supports multiple authors in planning and producing hyper-media documents. A document consists of a set of interlinked information nodes, each node containing exactly one type of information: a video clip, an audio clip, a picture or text. The links connecting information nodes are represented as buttons on the user interface. The origin of links is restricted to picture and text nodes. Links can be followed by selecting the associated button, e.g. with the mouse. The destination nodes will then be presented visually (text, picture), audio-visually (video) or acoustically (audio).

Functionality A workgroup of authors, usually about a dozen people, is supported in simultaneously producing a shared hyper-media document. Producing a document as a group involves three main activities: communicating, editing, and browsing. The user interface components of the system correspond to these activities. Collaborative group activities are facilitated by supplying both synchronous and asynchronous communication services: videophone and electronic mail, respectively. Editing and browsing activities include creating new information nodes and links, modifying the contents and the properties of existing nodes and links, deleting nodes and links, and searching for or focusing on specific nodes. Typically, authors work on the information nodes they have created and they are responsible for. However, care is taken to ensure that authors working on the same information can not edit it simultaneously.

Figure 5 shows the implemented user interface of the system. The workgroup overview on the left lists all members of the workgroup and their current work status as distinct icons. The communication icons above are used to invoke realtime calls to other users (by videophone) or to send messages to them (by electronic mail). Each workgroup member has a personal view of the shared document structure. This overview which contains the icons representing information nodes of a specific document covers most of the screen. Workgroup members may arrange the information nodes on their personal document overview independently from other workgroup members. Below the document overview, buttons for invoking editing and searching activities are provided. Communication services and editors for inspecting and editing the different information types (text, picture, audio, and video) pop up only when invoked, overlapping the overview.

Interaction and Visual Feedback User-system interaction in the Production Interface consistently follows a noun-verb syntax. First, objects are selected, then valid operations are invoked on them by clicking on the appropriate buttons. For example, the recipient of a videophone call is chosen by selecting the icon of the person. Complementary visual feedback techniques are provided to inform the user about the availability of operations and the status of objects. For example, the visualisation of a button always reflects the availability of the underlying operation on the selected object: highlighted buttons represent operations in progress, greyed out buttons represent operations not available at the moment.

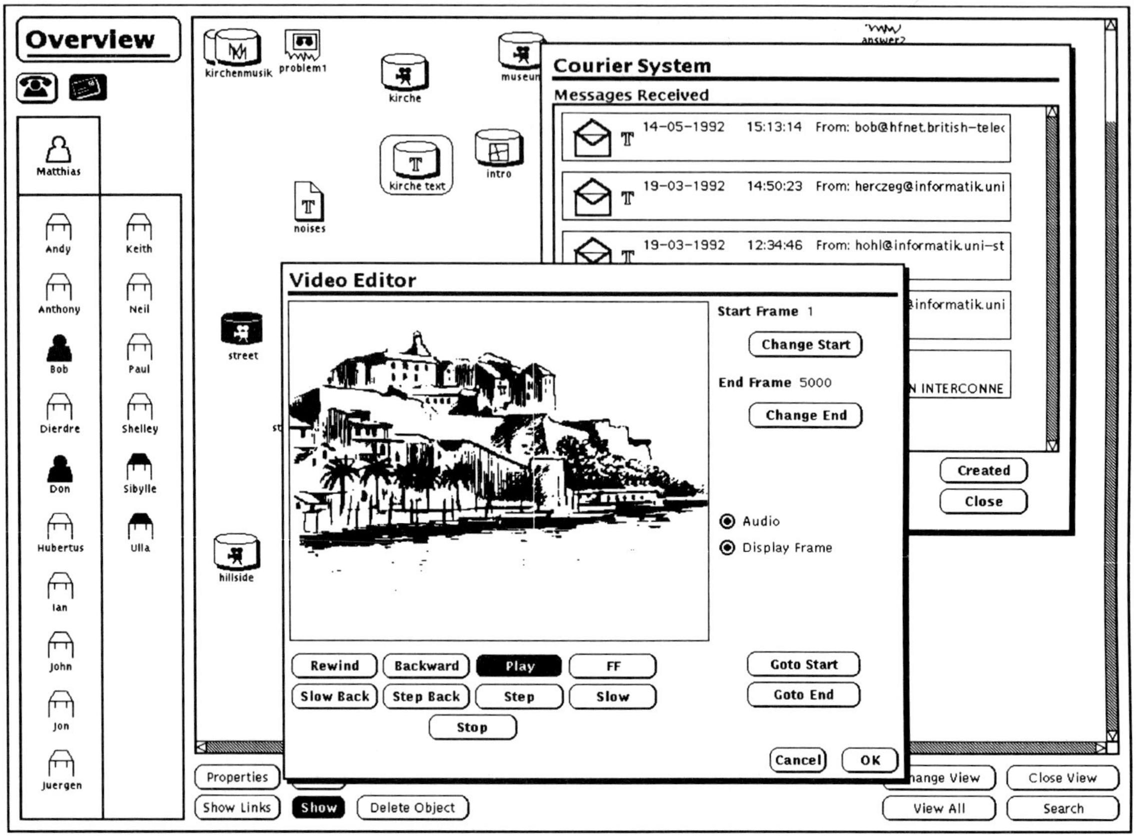

figure 5: the Production Interface

The Production Interface also supports drag & drop style of direct manipulation. Sending a document object as a mail message, for example, can be achieved by dragging the object's icon over the icon of the person it is to be delivered to.

Most of the available activities like creating, viewing, editing, linking or destroying document nodes can be performed simultaneously by different users. As soon as more users are involved, each user is dynamically made aware of the other users and their activities. The workgroup overview always displays the current status of the workgroup members. The presence or absence of authors is indicated by the form of the icon. An icon in form of a human silhouette represents a user who is currently logged in, whereas an empty desk represents an absent user. Users are immediately notified of incoming mail: the icon for the mail system starts flashing and the visualisation of the user's mailbox is updated automatically. Analogously, users are notified when there is an incoming videophone call. As opposed to mail notification, users are forced to accept or to reject the call as soon as they invoke the videophone system. Synchronous editing of different parts of the document database is supported. Document nodes that are currently edited by other users are made unavailable for editing operations, and marked as busy by drawing a frame around them.

5.1. An Event-Based Client-Server Architecture

The collaborative multiauthor environment of the Production Interface is based on a distributed client-server software architecture. It provides an environment for communicating between client applications and for accessing remote or shared information services. Each client represents a member of the workgroup who is currently running a Production Interface at their workstation.

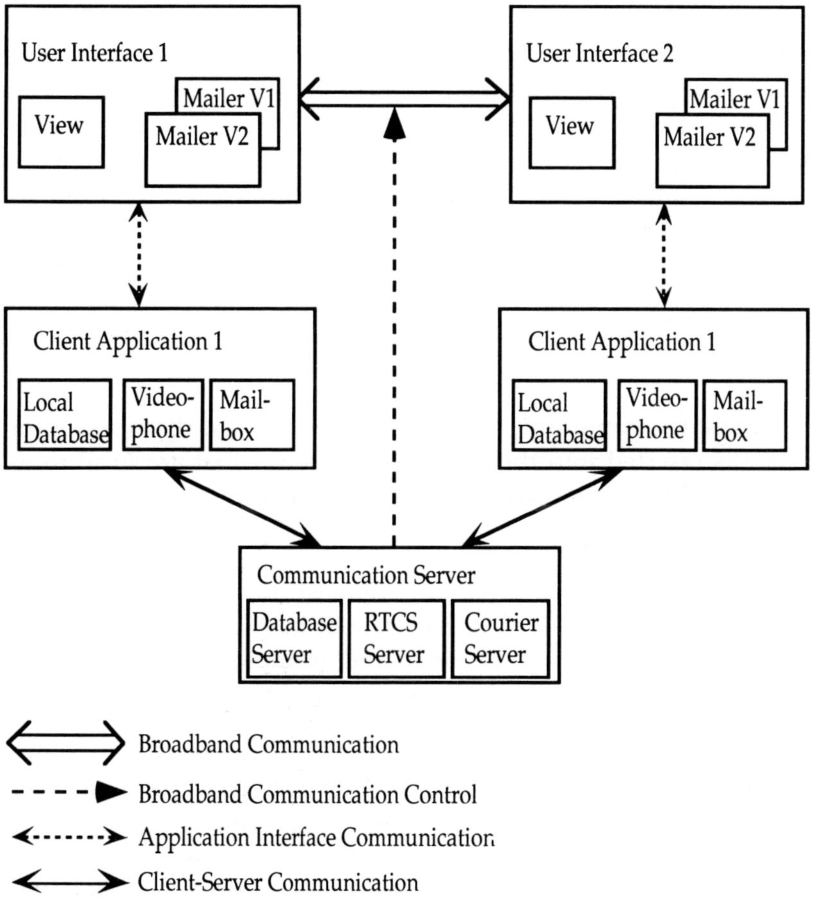

Figure 6: The layered architecture of the client-server environment

Figure 6 shows the layered architecture of the client-server environment. The workgroup members communicate with their client application programs via the user interfaces, which contain the views for all the components of the Production Interface; namely workgroup overview, document overview, communication tools, and editors. Two workgroup members can communicate directly using the broadband communication channel, i.e., by videophone. The communication between the server, the application clients, and the user interface is asynchronous, i.e. the processes basically run in parallel. They communicate to one another by sending messages and requests. Messages originating in the user interface are usually called events. These can be user events like mouse clicks or key presses or events automatically created by the window system, e.g. to request a window to update its contents when it becomes visible.

The *communication server* provides several services for the Production Interface:

- Database Service for managing hypermedia and workgroup information
- Courier Service for sending mail
- Real Time Communication Service (RTCS) for controlling videophone connections

The *database server* maintains a shared, persistent object database, which stores the hypermedia documents and workgroup information. Since hyper-media documents must be shared between the members of the workgroup, the database server is responsible for managing remote database queries requested by one of the clients. A locking protocol ensures consistency of shared data when multiple clients simultaneously try to access the same information for writing. By propagating object changes to all clients simultaneously, the server always ensures that database objects are presented consistently in all client user interfaces. For efficiency reasons, each client application caches crucial parts of the shared data in a local database, which also includes information about the user-specific layout of information nodes in the document overview.

For the *Courier Service* the communication server handles the distribution of mail. It takes care that such mail is inserted into the personal mailboxes that are local to each application process. It also ensures that the receiver of mail is notified immediately if they are present in the system.

For the *RTCS* the communication server handles requests to establish a videophone connection between two workgroup members. It must take care of busy lines and rejected calls. The actual broadband connection is provided by the video and audio service emulation implemented by the Guidance project.

The implementation of the client-server environment is based on 'sockets', which are the basic building blocks of Inter-Process Communication (IPC) in UNIX. On top of the low-level socket interface a generic, protocol-based event manager has been implemented that supports event-driven communication using application-specific protocols. It supports both synchronous and asynchronous client-server data transfer. Specific protocols have been defined to support the interclient communication services for the Production Interface.

5.2. Examples of PI specification tables

As in the discussion of the reader interface specification, an enabling state description will be presented, followed by the specification of a view, a generic interface object and a widget. It is intended to use similar specification objects to allow a comparison with the earlier specification method. The various differences are discussed out after each table. The Enabling State described deals with the provision of services to the user, which enables them to communicate with others about comments they have on a specific multimedia object.

Enabling State : Requirement comments communicable

Description: Comments made on the requirements under consideration can be communicated to another human user, via mail system, with or without the presence of real-time face-to-face communication

Sources: Subgoal 4g Communicate feedback

Design Aspects

Entities

Users :

Task: knowledge of personal role
knowledge of comments available for communication
knowledge of roles of other users
knowledge of roles to communicate requirement to
knowledge of availability of other users
knowledge of mail to be examined
knowledge of subject / context of mail
knowledge of presence of communication
knowledge of subject / context of communication

Processing: able to select comments to be communicated
able to specify recipients
able to communicate
able to attend to task

Workstations

Task: knowledge of available recipients
knowledge about database
knowledge about mail system
knowledge of presence of mail for user
knowledge of presence of communication
knowledge of identity of user
knowledge of role of user

Processing: able to list comments that can be communicated
able to access database
able to access other workstations
able to access mail system
able to alert user to presence of mail
able to alert user to presence of communication
able to receive communication
able to receive mail
able to present communication to user

Database:

Task: knowledge of entities with access permission
storing previously created comments

Processing: able to transmit data to other entities quickly
able to access database contents quickly
able to store communication contents

Mail-system:

Task: identities of recipients
identity of data in database to be mailed

Processing: able to read data database
able to integrate database records into mail message

Channel 1a: User—Workstation

Static Characteristics:

Communication Type: Human-Machine Bidirectional
Information Types:
Text (keyboard)

Data (mouse)
Audio (speech)
Video (moving)
Reliability: High
Security:
Dynamic Characteristics
Channel Control:
selection of media
authorisation of workstation as recipient
acceptance of incoming communication
designation of outgoing communication
acceptance of mail
Entity Control:
identification of recipient
identification of database record to mail
other mail message contents
face-to-face communication
navigation through mail messages

Channel 1b: Workstation—User
Static Characteristics:
Communication Type: Machine-Human Bidirectional
Information Types: Text (screen)
Audio (speech/sounds)
Video (moving)
Graphics (still)
Reliability: High
Security: Workstation should only present some of the communication and mail contents to authorised users
Dynamic Characteristics
Channel Control:
material summary
selection of appropriate media
incoming mail / communication alerts
Entity Control:
face-to-face communication
contents of mail message
feedback of mail message entry
details of possible recipients

Channel 2: Workstation—Database
Static Characteristics:
Communication Type: Machine-Machine Unidirectional (RETR)
Information Types: Data
Reliability: High
Security: only users participating in communication allowed later access or knowledge of stored communication
Dynamic Characteristics
Channel Control:
start/stop recording
delete records
Entity Control:
requirement summary info
face-to-face communication
mail messages

Channel 3: Workstation—Mail System
Static Characteristics:
Communication Type: Machine-Machine Unidirectional (DIST)
Information Types: Data
Reliability: Low
Security: Only addressees able to access message

Dynamic Characteristics
Channel Control:
bandwidth required
error checking
Entity Control:
identification of recipients
identification of data to send
other mail message contents
Channel 4: Mail System—Database
Static Characteristics:
Communication Type: Machine-Machine Unidirectional (RETR)
Information Types: Data
Reliability: Low
Security:
Dynamic Characteristics
Channel Control: error checking
Entity Control:
identification of data
contents of data
Channels 5a: Mail System—Workstations
Static Characteristics:
Communication Type: Machine-Machine Bidirectional
Information Types: Data
Reliability: Medium
Security: Workstation must have been authorised by a recipient
Dynamic Characteristics
Channel Control: feedback about progress of mail message
Entity Control: contents of mail message
Channels 5b:Recipient Workstations— Mail System
Static Characteristics:
Communication Type: Machine-Machine Bidirectional
Information Types: Data
Reliability: Medium
Security:
Dynamic Characteristics
Channel Control: authorisation by addressee
Entity Control:
Channels 6: Workstation—Workstation
Static Characteristics:
Communication Type: Human-Human Bidirectional
Information Types:
Audio (speech)
Video (head & shoulders / document)
Reliability: High
Security: Users should be able to check each others' identities.
Dynamic Characteristics
Channel Control::
Entity Control: face-to-face communication
References:
Enabling Tasks
(a list of Enabling Tasks is given, which is not presented here for space reasons)

The specification of the Enabling State has changed so that the description of the task and the processing requirements for the user, the workstations, and the databases have been enlarged and differentiated. This gave a better basis for defining the Enabling Tasks, which should prepare this enabling state. So, for example, if the user has to have knowledge about their personal role, there has to be an Enabling Task providing that information. This

Enabling task then has to be allocated: does the user have to find out what their personal role is or is the information provided by the machine?

Additionally, the different channels are presented in more detail with information about their static and dynamic characteristics, so that it is easier to define a specific channel and determine the necessary enabling states and thus the necessary enabling tasks. This version of the method led to a detailed description of Enabling States and a long list of Enabling Tasks, which were clustered under 5 Super-Enabling Tasks: give user knowledge of task domain, prepare and maintain communication channels, prepare workstation, prepare and maintain databases, and prepare courier service.

View:—Overview

Description: The overview displays all people in the workgroup and all objects (plans, documents, comments). The user can invoke realtime calls to other people or begin courier messages to them. The user can examine and edit the properties of objects, and examine the properties of people (ie roles and privacy property). The user can invoke the editing of objects, looking at objects (using editors without write permission), and can delete objects (if permitted). The links to and from an object can be shown and the link editor invoked. Busy objects are identified, and they can only be looked at, if they are text or graphic. The user can limit the overview to specified classes of objects, and can search for objects by class and name.
(see appendix for representations of this View)

XIT ref.: (a code number for an XIT element)

Sources:

Function:

F 201 Open workgroup overview
F 202 Close overview
F 203 Select style of overview
F 204 Change privacy level
F 205 Identify object properties
F 206 Present object overview
F 207 Retrieve object

Design aspects

interface model: Open plan office; the user can look around and see who is in, what people are doing, and what work objects are available.
design principles: Compatibility, Simplicity, Salience
design discrepancies: Representation of 'office' is not visually realistic; people are not shown next to the objects they are working on.
design rationale: Information that the user needs in order to plan actions and construct goals should be available continuously, although it should not become distracting. The legitimate actions possible from any state of the overview should be readily apparent.

Properties

name: size
default value: quarter to two thirds of screen
or just large enough to display Widget Overview button and Communication buttons
controlled by: device
name: position
value: anywhere on screen
default value: starting from the right upper corner
controlled by: user
name: layout of parts
value: object overview on the right side, workgroup overview on the left side, Overview and communication buttons in top left above the workgroup overview, editing and view buttons under the object overview
default value:
controlled by: device
name: contents

value: object overview, workgroup overview, button arrays
controlled by: device
name: border-with
value: :medium
controlled by: user
name: foreground colour
value:
controlled by: device
name: background colour
value: white
controlled by: device
Interaction procedures:
name: opening of browser
preconditions: browser is closed
actor: device
event: browser becomes open
name closing of browser
preconditions: browser is open
actor: device
event: change in contents of workspace (ie new objects created, objects deleted)
effects: contents of visible workspace updated to reflect changes
References
GIOs:
object overview
workgroup overview
edit buttons
communication buttons
view buttons
search dialogue
personal properties dialogue
show dialogue
busy dialogue
delete dialogue
Widgets:
Overview window
Overview button

As suggested and described above, the proposed changes were incorporated: a reference to the development environment (XIT) was given if applicable, the Interaction procedures were named and described, and properties such as colour were described.

GIO: Object overview
Description: Representation of all of the plans, documents and comments in the workspace
XIT ref.: (a code number for an XIT element)
Sources
View: Overview
Design aspects
interface model: desktop
design principles: salience
design discrepancies
design rationale**:** properties and relationships between objects in worksystem apparent from their user defined groupings within workspace
Properties
name: layout of parts
value: Vertical scrollbar at the right side, horizontal scrollbar at the bottom
controlled by: device
name: position

value: inside View overview, right of GIO workgroup overview and above GIOs communication, editing and view buttons
controlled by: device
name: size
value: full width of View overview, leaving room at left for GIO workgroup overview and below for GIOs communication, editing and view buttons.
controlled by: device
name: border-width
value: medium
controlled by: device
name: background-colour
value: white
device
Interaction procedures:
name: clicking on background
preconditions: object selected
actor: user
event: click in background
effects: selected object becomes unselected;
References:
GIOs
workspace horizontal scrollbar
workspace vertical scrollbar
Widgets: object icon

As in the description of the views, a reference was made to the XIT development environment. The interface model description in the 'design aspects' enables the implementers to develop a broad idea and refers to models already known. Again, the interaction procedures were described and named which allowed a class to be developed for them.

Widget: OK button
Description: a button that can be clicked as part of a dialogue
XIT ref.: (a code number for an XIT element)
Sources
GIOs:
Busy Dialogue (A & V)
Busy Dialogue (T & G)
Busy Dialogue (links)
Delete Dialogue
Search Failure Dialogue
Show Dialogue
Personal properties dialogue
Personal properties violation dialogue
Class Relations:
superclass dialogue button
Design aspects
inherited
Properties
name: contents
value: 'OK'
controlled by: device
Interaction procedures: as specified in GIOs

This specification of the widget shows the consequences of applying object orientation in the specification: the 'OK button' inherits as much as possible from its superclass 'button', and only its specific characteristics, (the word 'OK') now needs to be specified.

5.3. Implementers' Feedback

Once again the specification underwent some criticism by the implementers, reported in this section.

The first specification tables, e.g. for the Reader Interface, had the problem of redundancy. Information was specified repeatedly. Each OK-button to leave a dialogue would have been specified individually for each possible dialogue. Further, information spread among different specification tables made it hard to find inconsistencies.

As a consequence, the specification tables for the Production Interface make extensive use of abstractions like the 'OK-button'. More general abstraction are generic buttons, text entry fields or icons, which encapsulate the properties common to all members of these classes. Tables for OK-buttons or cancel-buttons may then reference these more abstract tables adding more specific information.

The abstractions used in the tables map straightforwardly onto the interaction objects of the object-oriented toolkits of XIT. The basic widgets match almost exactly the object classes for buttons, editable text fields, and icons that are defined in XIT. Specialised interface elements could be easily implemented by means of the mechanisms offered by the low-level toolkit of XIT.

The hierarchy of abstractions for widgets and GIOs matches the hierarchy defined by the interaction objects in XIT. The reduced redundancy in the tables is reflected by the opportunity of reusing code by means of inheritance along the class hierarchy. The table entries specified for an interface element closely correspond to the categories of properties of XIT interaction objects, especially structure and presentation, geometry and layout, and reactivity. These (mostly descriptive) properties map easily. The more functional properties needed to specify the links to and the effects on the application were more difficult to represent. One reason for this was that the internal application functionality which had to be implemented had not been specified as detailed as the user interface.

To summarise, the implementers believed that implementing the user interface for the PI from the specification tables substantially benefited from the correspondence of the descriptive properties between the tables and the representation of interaction objects in XIT and also from the fact that the object-oriented approach used for the implementation matched the abstraction hierarchy used in the tables.

The specifications of descriptive properties like structure, geometry, visual presentation, the interaction procedures, and the hierarchy of interface elements map easily onto the corresponding concepts used in an object-oriented implementation. The pictures and storyboards further complete the specification of the visual aspects and interdependencies of the interface objects in a natural way. They also may help to detect inconsistencies in the user interface. Unfortunately, there is almost no help in implementing the internal functionality of the applications involved. To apply the specification tables to the user interface of an application, the internal functionality of the application should either be available already or should have been specified separately.

6. Experiences from the prototype

In the course of the evaluation of the Production Interface it became obvious that the method developed did not deal sufficiently enough with the issues of integration of various functionalities and the support of cooperative work in creating multimedia documents. For example, the specification of the interface for joint editing of video object during a videophone call did not mention the fact that the simultaneous display of two video sources was not useful, as the user can only watch one video source at one time. It therefore had to be respecified, to indicate which sources are displayed during a videophone call if objects are displayed which include audio and video sources.

Additionally, it turned out that the design of Goal Tasks (what the user wants to do) and Enabling Tasks (what the user of the machine has to do) did not differentiate enough between these two tasks. Analysing the method thus concentrated on:

- how to represent better the integration issues of the design tasks during the design
- how to differentiate better between the design of a Goal Task and an Enabling Task.

The review of the method concluded that the bridge between the Enabling State layer and the functional layer did not provide the basis for a complete specification. It became obvious that the structure employed was 'too structured', such that it helped the specification team to break down their tasks in small units, but it had the disadvantage that the integration aspects vanished. There were no means provided to include the integration aspects later on. Also, the clustering of the Enabling Tasks into Super-Enabling Tasks, which was done to reduce redundancy, had the consequences that detailed but important information was lost and that the design for the enabling tasks became too generic and not adjusted adequately for the specific task in question.

Therefore the method was changed to its current form which tries to promote the integration issues at the Enabling States layer, and incorporates the understanding of IBC services as including user, service relationship and object. This is the form of the method that is described in chapter 3.1.

Starting with the task layer, the enabling states descriptions follows directly from the task breakdown. The enabling states description is still the crucial point where design decisions are made. It includes the specification of an Enabling State for a specific task, of Integrated Enabling States for tasks which must be performed simultaneously, followed by the specification and allocation of the resulting Enabling Tasks.

These specifications are intended to include all necessary user-oriented specification and operational requirements. Following this stage, design can use existing design tools that allow the creation of some parts of the interface by drawing. Many tools (e.g., XIT, UIMX, NeXT interface builder etc) allow designers to drag predefined widgets (such as menus, buttons and text boxes) into place to create dialogue boxes and windows. Other systems allow a limited form of actual drawing to create widgets (Peridot) and some graphical objects (Lapidary; Marchionini & Sibert).

The use of these tools will expand in future, and so a method has to provide a bridge to the use of these tools as well as to existing libraries or other development environments, such as for XIT in the PI examples.

The specification tables given below present an example of the current method for real time face-to-face communication. The integration issues are dealt with by integrating Enabling States within the specification, which include the Enabling States of specific tasks, which should be able to be performed simultaneously. Additionally, the principles for design are integrated in the Enabling States Specification such that the design decisions made here are based on human factors knowledge.

ES 1: face to face communication with one person is performable.

Source: communicate with one recipient from face to face

Description: The user can communicate with one recipient via videophone.

Description of the user

Enabling State Description of the User

The user has to know how the recipient will see them, they have to look in the camera and sit in the right place so that they can be filmed.

The user has to be in a state where they can start and end face to face communication.

The user has to be in a state where they can control or change the quality of the communication provided.

The user has to be informed about incoming calls (subject and caller).

The user has to be in a state where they can handle incoming calls while they are communicating: hold, resume, forward them.

User Principles applied:

IF the task is to communicate with other persons in the workgroup

AND users already have knowledge of working with the system to generate multimedia documents

AND the service environment supports same interactions and presentation styles in different tasks

AND the performance should minimise users ´s costs: mental workload to reduce errors

Then: COHERENCE

provide the user with a communication service which is as similar as possible with other services provided in the system

Because: this will allow them to use knowledge from other applications and reduce the memory load

IF the task is to communicate with other persons in the workgroup

AND users already have knowledge of communicating with others using telephone

AND the service environment supports similar representation of the task as on telephone

AND the performance should minimise users ´s costs: mental workload to reduce errors

Then: COMPATIBILITY

provide the user with a representation of the communication service which is as similar as possible to telephone

Because: this will allow them to use existing knowledge and reduce the memory load

References:

Enabling Tasks to be performed for the User

give user self view

make user look in camera
make user sit in the right place
give user possibility to start and end communication
give user possibility to control quality of communication
inform user about incoming calls (subject and caller)
give user possibility to hold, resume and forward calls.

Description of the Service Relationship

Description: Service relationship type: 1 (uninterpreted task actions, nonrepresented object)
Involved Entities: user, workstation, recipient
Communication Type: Human-to-Human Conversation
Channel characteristics:

Information content: the head and shoulders view of the participants and their voices are transmitted.
Service components: Audio and video service components are required.
The audio service component must be **mandatory**, the video service component must be **optional**.
Synchronisation requirements: lip synchronisation is required for the combination of audio and video service components.
Realtime information transmission is required.
Format preservation of audio and video service components is not required; therefore the transmission of compressed information is acceptable.
Quality of audio service component: Since the participants can see each other and use visual cues, a low audio quality, comparable with standard telephone quality, is sufficient.
Quality of video service component: A standard video quality (4:3 aspect ratio, 625 lines, 120 000 pixel, 20-40 MBit/s peak bit rate) comparable with NTSC, PAL or SECAM Standard Resolution TV quality, is required.
Quality of the service: Adequate quality of loudness, brightness, contrast. Graceful degradation of the service is required in case of system breakdown.
no distractions

Service Attribute values recommended:
IF audio and moving images are provided simultaneously (videotelephony conversation)
THEN audio transfer rate is < 64Kbit/s and standard moving image quality is required.

References:

Enabling Tasks to be performed for the Service Relationships:
establish standard video quality
establish audio quality with transfer rate < 64Kbit/s
establish synchronous transmission
reduce distractions on secondary workstations-user channels, especially if they are audio or video
determine quality of service (loudness, brightness, contrast..)

Description of the object

Description: The recipient is the object of the communication.

Enabling State Description of the Objects

The recipient has to be identified. The recipient has to be attentive and informed about medium of call, caller and subject of call. The recipient has to be able to decide how to handle the call (accept call). They must be able to decide about visibility for the user.

Object Principles applied:

IF the task is to receive calls from other persons in the workgroup while working on multimedia documents
AND recipients have no special needs
AND the service environment supports audio and/or video output
AND the performance should minimise recipient's costs: mental workload to reduce errors
Then: SALIENCE and **REDUNDANCY**
provide the recipient with a simultaneous audio and video signal to alert their attention
Because: combined signals are most effective in gaining attention.

IF the task is to receive calls from other persons in the workgroup while working on multimedia documents

AND recipients wish to decide about accepting the call
AND the service environment supports messages which inform about medium of call, caller and subject of call
AND the performance should minimise recipient's costs: affective costs (not to be disturbed while working on important paper)
Then: TRANSPARENCY and **CONTROLLABILITY**
provide the recipient information about medium of call, caller and subject of call and enable them to accept or ignore call
Because: this will enable the recipient to work undisturbed and receive only important calls.
IF the task is to receive calls from other persons in the workgroup while working on multimedia documents
AND recipients wish to decide about being viewed by others (keep privacy)
AND the service environment supports to show or show not the video/picture of the recipient
AND the performance should minimise recipient's costs: affective costs (not to be presented)
Then: CONTROLLABILITY
enable the recipient to show or show not video of themselves
Because: this will enable the recipient to communicate without frustration

Enabling Tasks to be performed of the object
identify recipient
inform recipient about call
inform recipient about medium of call
inform recipient about caller
inform recipient about subject of call
attract recipient's attention.
give recipient possibility to show or show not their image

Integrated Enabling State 3:
Real time communication can be performed at the same time as a document is being edited. The editing is viewable to both the user and the recipient.

Description of the user

Enabling State Description of the User
The user has to know how the recipient will see them, they have to look in the camera and sit in the right place so that they can be filmed in case of video phone.
The user has to be in a state where they can start and end RT communication.
The user has to be in a state where they can control or change the quality of the communication provided
The user has to be informed about incoming calls
The user has to be in a state where they can handle incoming calls while they are communicating.
The user has to be in a state where they can start and end editing video documents
The user has to know what the recipient/other users can see while they are editing
The user has to be able to hand-over editing rights

User Principles applied:
IF the task is to communicate with other persons in the workgroup in real time and edit video documents cooperatively simultaneously
AND users already have knowledge of working with the system and working in groups
AND the service environment supports same interactions and presentation styles in different tasks
AND the performance should minimise users ´s costs: mental workload to reduce errors
Then: COHERENCE
provide the user with services which use the similar functions for similar tasks to be carried out: use buttons to invoke functions, use the same way of button layout for indicating availability etc.
Because: this will allow them to use knowledge from other applications and reduce the memory load, it will especially reduce memory load and number of errors and time to perform task if the user wants to include more than one goal task at one time
IF the task is to communicate with other persons in the workgroup in real time and edit video documents cooperatively simultaneously

AND users already have knowledge of communicating with others about documents in an open plan office
AND the service environment supports similar representation of the task as on telephone, fax or E-mail
AND the performance should minimise users ´s costs: mental workload to reduce errors
Then: COMPATIBILITY
provide the user with a representation of the communication service and the editing service which are as similar as possible as editing in an open plan office
Because: this will allow them to use existing knowledge and reduce the memory load

References:
Enabling Tasks to be performed for the User
give user self view
make user look in camera
make user sit in the right place
give user possibility to start and end RT communication
give user possibility to control quality of RT communication
inform user about incoming RT calls
give user possibility to hold, resume and forward calls.
give user possibility to start and end video editing
give user information about availability of RT and Video editing services

Description of the Service Relationship
Description:
Service relationship type:
1 (uninterpreted task actions, nonrepresented object) and
2 (interpreted task actions, not represented object)
Involved Entities: user, workstation, recipient
Communication Type: Human-to-Human Conversation and Human-to-Machine Send
Channel characteristics:
Information content:
A+V: the head and shoulders view of the participants and their voices are transmitted
Audio: only the voices are transmitted
(for video document editing: moving image is completely represented)
Service components:
AV: Audio and video service components are required.
A: audio service components are required
The audio service component must be **mandatory**, the video service component must be **optional**.
(for editing: moving image service component)
Synchronisation requirements: lip synchronisation is required for the combination of audio and video service components.
(Synchronisation is required for the display of the video document)
Realtime information transmission is required for videophone and audiophone.
Format preservation of audio and video service components is not required; therefore the transmission of compressed information is acceptable.
Quality of audio service component:
AV: Since the participants can see each other and use visual cues, a low audio quality, comparable with standard telephone quality, is sufficient.
Audio: IF the participants can not see each other, audio quality, comparable with standard telephone quality is required.
Quality of video service component: A standard video quality (4:3 aspect ratio, 625 lines, 120 000 pixel, 20-40 MBit/s peak bit rate) comparable with NTSC, PAL or SECAM Standard Resolution TV quality, is required.
Quality of the service: Adequate quality of loudness, brightness, contrast. Graceful degradation of the service is required in case of system breakdown.
no distractions:
Quality of data service component: A standard data transfer rate (64 KBit/s) is required, errors are not acceptable.

IF videotelephony is provided simultaneously with video document editing, the video communication's moving image is frozen and only the moving image of the video document is presented. The audio information from the video document is reduced in volume to allow the user and the recipient to talk to each other. It must be possible for the user, to unfreeze the video document image and to freeze the video document image and to reduce and increase volume of the audio sources.

Service Attribute values recommended:

IF audio and moving images are provided simultaneously (videotelephony conversation)
THEN audio transfer rate is < 64Kbit/s and standard moving image quality is required.

IF only audio is provided (audiophone conversation)
THEN quality is always voice (3.1-7kHz)

TASK

Task Interpretation: Indication of document to work on is done by selection, this is done in using the mouse (movements and actions represented by arrow on screen), possible action indicated by different display of arrow (hand/arrow). Editing is done by using the keyboard or other input devices.

Principles applied:

IF the task is to edit a video document and communicate in real time from face-to-face simultaneously
AND there are at least two users working cooperatively
AND the service environment supports audio and video communication and includes a multimedia editor
AND the performance should be quick and error free
Then: SIMPLICITY, CONTROLLABILITY
ensure that the number and complexity of actions is kept to a minimum and that users can control which service is primary
Because: interference between tasks should minimised and the locus of control should remain with the user

IF the task is to edit a video document and communicate in real time simultaneously
AND there are at least two users working cooperatively
AND the service environment supports audio and video communication and includes a multimedia editor
AND the performance should be quick and error-free
Then: CONTROLLABILITY
allow only one user to edit the object, and allow the editing user to specify whether their actions will be visible on the workstation of the second user
Because: users engaged in collaborative tasks should be able to control what is being presented to other users to ensure privacy.

References:

Enabling Tasks to be performed for the Service Relationships:

present document
place document on the screen
integrate editing tool
establish standard video quality
establish audio quality with transfer rate < 64Kbit/s for videophone and audio quality of 3.1.-7 kHz for audiophone
establish standard moving image quality
establish synchronous transmission for audio and video
reduce distractions on secondary workstations-user channels in video editing task: freeze video communication and reduce audio volume of video document
change default for reducing distractions
determine quality of service (loudness, brightness, contrast..)

Description of the object

Description: For this integrated enabling states description, there are two objects: the recipient of the audio or videophone conversation, and the document being edited.

Enabling State Description of the Objects

1. recipient

The recipient has to be identified.

RTCS: The recipient has to be attentive and informed about medium of call, caller and subject of call. The recipient has to be able to decide how to handle the call (accept call). They have to be able to decide about visibility for the user.

2. document

The document has to be indicated. The document has to be retrieved. The attributes of the document should be changeable (Attributes: see table 3, 'document').The object has to be stored. Visibility of the document for the recipient has to be indicated.

Object Principles applied:

References:

Enabling Tasks to be performed of the object

identify recipient
inform recipient about medium of call
inform recipient about call
inform recipient about caller
inform recipient about subject of call
attract recipient's attention.
indicate document
retrieve document
change attributes
indicate object's visibility
differentiate pointing devices for users

7. Conclusions

The specification work reported here has been carried out on paper. This was an advantage on the one hand, as no machine based constraints could hinder the work, but was a disadvantage on the other hand, as a lot of repetitive work had to be performed by hand. A computer-based version of the method is under development (see chapter 3.1). Avoidance of repetitive work should not be the only reason for the development of a computer-based version, for it could also support the building of classes and hierarchies of tables at various stages of the method. This is extremely important to make the complex design task more manageable, especially in supporting the description of Enabling Tasks.

The cooperation between human factors personnel and implementers is such that with the current version of the method, less freedom is given to the implementers to interpret the specification. This has to be improved to make use of their experience and creativity, and can be done by supporting the management of changes by relating them to higher level, earlier, design decisions.

It would also be beneficial to have a prototyping tool which can easily use the enabling state description to produce something that can be tested with users, providing continuous feedback to the specification. More emphasis must also be paid to helping the implementers understand the method, so that they can use the specifications better. Although the steps in the method are stable, the precise content and structure of the specification tables should not be regarded as inflexible. These can be modified to suit the particular requirements of the design organisation and the implementation methods that are available. The flow of design information through the sequence, and the timely and interpretable availability of principled human factors support are is of primary importance.

References

Frieling, S. (1990). *Analyse weist den Weg*. Bundesarbeitsblatt 10/90

Greif, S. (1991). Organisational Issues and Tasks Analysis. In: Shackel,B. & Richardson, S. (Eds). *Human Factors for Informatics Usability*. Cambridge:University Press.

Hamborg, K.-C. & Schweppenhäußler, A. (1992). *Expertise: Argiets und Softwaregestaltung. Manuskripte 64*. Hans-Böckler-Stiftung, Düsseldorf.

Herczeg, J., Hohl, H., & Schwab T. (1991): XIT - A Multi-Layered Tool for User Interface Design. In: Bullinger, H.-J. (Ed.), *Human Aspects in Computing: Design and Use of Interactive Systems and Work with Terminals.*

Herczeg, J., Hohl, H., & Ressel M. (1992). Progress in Building User Interface Toolkits: The World According to XIT. In: *Proceedings of the ACM Symposium on User Interface Software and Technology.*

Olson, D. (1976). Culture, Technology and Intellect. In: Resnick, L. *The nature of Intelligence.* New York: Wiley & Sons.

Rödiger, K.H. (1988). Das Arbeitsanalyseverfahren VERA/B in der Softwareentwicklung. In: Nullmeier, E. & Rödiger, K.H. (Eds) *Dialogsysteme in der Arbeitswel.* Mannheim:Wissenschaftsverlag, pp. 185-205

Rudolph, E., Schönfelder, E. & Hacker, W. (1987) *Tätigkeitsbewertungssystem-Geistige Arbeit* (TBS-GA). Göttingen:Hogrefe.

Scheiffler, R. W., & Gettys, J. (1986) The X Window System. *ACM Transactions on Graphics, 5(2)*

Steele, G. L. Jr. (1990). *Common LISP: The Language*. Digital Press, 2nd edition.

Ulich, E. (1992). *Arbeitspsychologie*. Stuttgart: Schäffer-Poeschl Verlag.

Volpert, W., Oesterreich, R., Gablenz-Kolakovic, S., Krogoll,T.& Resch, M. (1983). *Verfahren zur Ermittlung von Regulationserfordernissen in der Arbeitstätigkeit (VERA)*. Köln:Verlag TÜV Rheinland.XS

Computers, Communication and Usability:
Design Issues, Research and Methods for Integrated Services
P.F. Byerley, P.J. Barnard and J. May (Editors)

Chapter 3.4

Evaluating System Design Realisations

Frank Wilson* and Anne Clarke†

*London HCI Centre, University College London, 26 Bedford Way, London WC1H 0AP, UK.

†HUSAT, Loughborough University of Technology, Loughborough, LE11 1RG, UK.

1. Introduction

Evaluation is a natural part of the 'ideation, specification, realisation, evaluation, iteration' cycle evidenced in the majority of design processes. In the main, Human Factors (HF) practice has focused on evaluation of design realisations, i.e. evaluation of 'concrete' representations of the intended system. Design realisations can range from simple graphics illustrating design concepts for interaction screens, through functional prototypes (whole or part), to near complete or early versions of systems. The initial focus of this chapter pays special attention to the developmental status of HF practice in system evaluation by examining selected evaluation approaches in order to highlight general issues for the practitioner. It then goes on to consider current methodology in HF evaluation of design realisations although, for the sake of completeness and contrast, includes some coverage of evaluation activities which do not rely on design realisations. Some case examples are then presented in order to illustrate selected procedural level considerations which reflect on issues identified. The conclusion summarises our observations on the practice of evaluating design realisations and provides comment on the optimisation of evaluation practice.

2. Evaluation as a system development activity

Consideration of the HF literature suggests that system evaluation practice is greatly varied, sometimes emphasising preferred procedures (e.g. Clegg et al 1988), sometimes general principles (e.g. Whiteside et al 1988). No attempt is made here to prescribe a unifying framework, but consideration is given to the characteristics of evaluation practice, along with some identified issues for practice, as orientation prior to consideration of current methods and case examples of their application.

System evaluation can be defined as 'the assessment of the conformity between a systems actual performance and its desired performance' (Whitefield et al, 1991). Contemporary HF views of 'system' generally include consideration of the user as a key system component, since the concern of HF is with interactive work systems (e.g. Dowell and Long, 1989), wherein the expression of performance concerns the effectiveness with which the system accomplishes tasks in terms of the quality of the products and the incurred resource costs for both the user and the device (eg computer, videophone, etc). Given such contemporary concerns, it is necessary both to be able to identify *desired* performance and to be able to estimate *actual* performance with some level of guarantee. It is also necessary to be able to compare them in some reasoned manner, and therefore to have available the necessary information to support design decisions as part of an integrated design activity. Evaluation methodology, then, should be able to support such goals by providing the practitioner with suitable concepts, procedures, and operational guidance such that questions about conformance, and costs of conformance, can be addressed.

Until recently, evaluation has mainly relied on the experience of the practitioner since there has been little in the way of formal support, or tools to aid evaluation practice. The experiences of the authors, along with accounts in the literature, suggest that the variation in practice is partly due to the absence of any accepted general framework for evaluation offering the necessary rigour and hence guarantees. Nonetheless, there are established traditional approaches which, whilst not yet meeting the needs of the practitioner fully, offer some level of support and allow access to issues worthy of attention in method development. Some of these are considered in the following subsections before considering their implications for methods.

2.1 Evaluation practice

Probably the most widely identified general orientation towards system evaluation is to be found within *design for usability* or *usability engineering*. This is a general approach to integrating human factors in system design, and has various contributors from academia and industry. The approach concerns system development in general, and includes various propositions concerning evaluation. Whilst not being a unified, widely adopted approach, it has a good level of consensus yet maintains interesting contrasts in emphasis (for example, see Shackel 1986, Whiteside et al 1988).

The design for usability approach has been influential on practice, and researchers from major developer organisations (e.g. DEC, IBM) have contributed to its development, as have user organisations (see Gardner and McKenzie 1988). An interesting result of the implementation of this approach by developers is their investment in usability laboratories where the primary focus is usability evaluation in support of system development.

The fundamental features of design for usability are :

- User Centred Design—focus from the start on users and tasks;
- Participative Design—with users as members of the design team;
- Experimental Design—with formal tests of usability in pilot trials, simulations, and full prototype evaluations;
- Iterative Design—design, test and measure, and redesign as a regular cycle until results satisfy the usability specification;
- User Supportive Design—help systems, manuals, selection and training should be treated as part of the design activity.

Evaluation in this approach is characterised by an emphasis on criterial assessment, experimental testing with users, and inclusion of these in an iterative design process. The most important feature of the process is that usability goals are set and re-expressed as performance criteria for iterative testing and redesign. The following subsections consider features of design for usability, along with identified critical issues for evaluation practice.

2.2 Evaluation focus in design for usability

Design for Usability has a strong emphasis on criterial assessment as a provider of focus in evaluation. This requires the establishment of usability goals, as a specification of desired performance, and the subsequent derivation of acceptability criteria relative to the goals. This general scheme is operationalised via a 'factorial' approach whereby usability is decomposed to its primary factors of effectiveness, learnability, flexibility and attitude. These factors are then further decomposed to provide reference and indices for criterial measures. For example, learnability is expressed as 'within some *specified timeframe*, based on a *specified amount of training and support*, within some *specified relearning time* for intermittent users'. The terms in italics indicate indices or subfactors against which criterial levels of acceptability must be set, in numerical terms, for a particular system or set of users (and also for different levels—best, planned, worst case). Goals for the other usability factors are set in similar ways.

The reliance on criteria requires the ability to pre-specify and then to measure the various factors. For easily quantifiable measures such as learning time this may be easy, but for those such as 'acceptable levels of human cost in tiredness, frustration etc' (Attitude) it is less clear, and may indicate the need for a qualitative assessment. Usage of the approach is further complicated by the lack of support to evaluators in specifying criteria and measuring against them. Furthermore, the emphasis on quantitative performance ignores attention to the behaviours underlying performance. These issues highlight some of the difficulties which may be experienced in applying the approach. Designers may need to know about the features of, for example, computer and user behaviour which actually produce the observed performance—the exact nature of the problem, which parts of the task were easy or difficult, and what provided the constraints on performance. It appears, therefore, that there is a need for a *diagnostic* approach to evaluation to complement the criterial/measurement approach advocated by design for usability. This issue seems to be supported by the observation of Whiteside et al (1988), who observe that 'not all, or even most, aspects of usability experience can be given operationally defined criteria and measured'. Indeed these workers do not limit the use of usability factors or their criteria, and propose designing the scope of the evaluation to suit the purpose at hand. For example, if we consider the work reported in chapter 1.1, we can see two clear opportunities for extending the criterial view by examining whether all appropriate enabling states are reachable, whether they are easy to reach, and whether they actually enable the user's goal tasks to be performed.

Thus, in the definition of focus for evaluation, there are issues concerning scope of the factors, measurement vs diagnosis, and quantitative vs qualitative assessment. Whilst there is no formal support for dealing with these, the evaluator must reason about which factors or issues are to be addressed as part of the evaluation focus, how to obtain data which allow understanding of the behaviour underlying performance, and how to devise measures which support comparative assessment. Since there is the identified problem of not being able to pre-specify and operationalise all possible factors / criteria, there is the danger of resorting to a 'broad brush' approach (measure everything). This usually results in a mass of data which record a snapshot of the world without delivering much meaning—the evaluator must still identify 'what are the relevant questions'. General factors such as 'learnability' can be grouped together to enable a 'general' evaluation approach, however in dealing with emerging and identifiable issues in design, there is often a need to develop a 'specific' evaluation focus. This is dealt with further under methods and in the examples provided later.

2.3 Including the users in evaluation

In line with the general spirit of human factors approaches, design for usability proposes 'experimental testing with users'. From the literature covering this topic, there appear to be two key issues. This first concerns the 'inclusion' of the end user in the design and evaluation process—since, in design for usability, there is an emphasis on the literal inclusion of users as part of the design team. Whilst such a strategy may work for some organisations, and for some system types, it may not work for all and ignores the more general requirement to *represent* the end user in the design process (Wilson, 1988). Inclusion assumes that users are generally able to articulate or usefully disclose what they know; that they have a clear and conscious appreciation of what they know; and that they are able to recruit what they know to support development in appropriate ways. However, users may not be able to represent themselves in ways useful to or appropriate for design. This is not to belittle the potential contribution of users in the design process, but does point out that there are representations of users, eg conceptual models, which may be recruited as, or instead of, the user's physical presence in the design activity (see chapter 2.1 for a fuller discussion of the use of User Models).

A further issue for design for usability, and the inclusion of users, concerns the propositions concerning testing. It is generally unclear what is meant by 'experimental' in that approach, and indeed it might seem that 'comparative testing' and 'formal and empirical studies' indicates full control and manipulation of relevant variables. However, as will be illustrated later, there is a whole range of test methods, such as informal observation, questionnaires and interviews, in addition to controlled experimentation, which can be of great benefit to, and are commonly used to support, evaluation. As proposed by Whitefield (1991), 'to equate evaluation with fully controlled experiments is to confuse the goal with the methods of reaching that goal'.

2.4 Iterative process

A traditional weakness of evaluation activities is that they were often carried out too late in the design process to render any real benefit to development. The 'too little too late' problem is one of presenting redesign opportunities when commitment and investment in development are such that it is unlikely that opportunities for improvement of system usability will be taken up. Design for usability, therefore, needs to ensure a timely input to the design process. The strong emphasis on iteration, starting with early design representations, and continuing through to full prototypes or early versions of the system, attempts to ensure a continuity of 'test redesign and test again' until the usability specification is satisfied. Our identification of issues concerning 'usability specification' notwithstanding, it is also important to understand the limitations of iteration in design for usability also.

Iteration is generally recommended in most system development contexts. It can pay dividends, especially in the absence of good human factors design methodology, however it is not without problems. For example, iteration may be difficult for very large systems especially where formal methods are employed (e.g. in safety critical applications). In addition, it does not alter the type of input from human factors activities, and simply repeats the same activities as late evaluation in a shorter timescale. Thus, some of the weaknesses of late evaluation are found with iterative evaluation, and in addition, the resources of late evaluation can be substantial. A solution, then, would be to seek different kinds of input at different stages of development, depending on the focus of the design and the identified issues (see 2.2, general vs specific evaluation focus).

In general, then, it is proposed that evaluation can happen any time that a suitable representation of the system is available. Evaluation should begin as early as possible, and should be repeated to test redesigned features and/or increments and changes to design as development progresses. There are related issues such as timing, resourcing and planning for greatest benefit, and these are considered along with methods, and in the examples.

2.5 Issues for practice

Given our consideration of design for usability as a representative of general evaluation practice of the traditional variety, it can be seen that whilst there are many attractive features which offer support to development by ensuring focus on user issues, there is also room for improvement and progress in evaluation practice. Many of the issues arise because the relevant concepts (for example performance) and the relations between concepts (for example, what criterial evaluation can address about performance) are insufficiently defined. Overcoming such weaknesses will require a more rigorous framework for evaluation, in terms of the completeness and coherence of its conceptual underpinnings. This is one area where there is some evidence of progress in support of practice. For example, Whitefield et al 1991 propose a new framework for evaluation practice which, whilst being incomplete and not yet generally accepted, sets out to identify the requirements for such a framework and propose part of its structure. Other areas include the development of analytic methods and the development of tools to support evaluation planning.

One drawback for contemporary approaches to evaluation is the reliance on the availability of a concrete representation of the intended system, whether it be prototype or early version of the system. Since it is proposed that evaluation should occur as early as possible so as to influence design before commitment and expense obviate suitable change, then the development of methods allowing evaluation prior to the realisation phase must be attractive for human factors practitioners. Techniques are needed which can operate on models, specifications or other descriptions of intended systems, and which can manipulate them so as to describe or predict a system's behaviour. This is a relatively new area where methods are not yet well established and, indeed, it is proposed that the work so far has had little impact on practice (Bellotti 1988)—see also the sections in this chapter on methods and examples.

One particularly interesting area which promises progress is the development of tools to support evaluation practice. The development of tools aims to support the practitioner by formalising part of the evaluation activity, or by supporting evaluators with little experience on which to draw. Examples include the development of standard checklists and questionnaires for wide-ranging evaluation of a company's current or proposed technology (Clegg et al, 1988. Ravden and Johnson, 1989), along with operational schemes for selection and configuration of methods (Wilson, 1988) and computer-based tools support for Evaluation Planning (Denley 1990). Little of this work is yet well established or well accepted, but there is a clear demand since so much human factors evaluation work currently relies on the actual presence of highly trained human factors professionals with experience in this area.

2.6 Operational issues

We previously identified and discussed selected issues in evaluation practice, and identified a need for progress in certain areas. In addition, there are a host of operational level issues which have to be taken into account when planning evaluation of design realisations. We have already addressed the issue of determining focus and scope for the evaluation. Whether a factorial approach is being used or not, there is a need to ensure that the evaluation activity is properly focused. This will be covered to an extent in the examples given later, but for the moment we emphasise the need to establish the information requirements of the evaluation. What is it setting out to do, what issues are to be addressed, what usability factors are in focus, what explicit questions are being asked, and so forth. These will provide the direction for defining the measurement needs in the evaluation activity, whether it be pro-actively criterial or not, and hence the definition of the data set which is to be delivered as output.

Having a clear definition of the information requirements for the evaluation activity has further benefits. For example, a good specification of the data set will enable a reasoned selection and configuration of evaluation methods. Similarly, having explicit evaluation goals expressed as a detailed information requirement allows definition of the necessary system state for achieving such an output—it may be necessary to enable or simulate added functionality in order to make the required observations. Defining the information requirement will also involve establishing the focal tasks, that is to say those tasks which are necessarily a part of the evaluation focus for the attainment of evaluation goals. Having well defined focal tasks, it is usually a straightforward matter to determine the necessary user and equipment requirements for the enactment of tasks in the evaluation. At this stage, and in order to provide an appropriate context and setting for evaluation, one which will 'ring true' for the evaluator and the user, it is common practice to design 'evaluation task scenarios' whereby the execution of tasks is carried out in a realistic and reasonable setting in terms of the surroundings and the rationale for 'why we are doing what we are doing'. People are generally good actors and can provide a realistic estimate of real-world system performance 'in role' if the evaluator supports them with appropriate physical and conceptual props—a realistic task setting, and a good rationale for why they are doing what they are doing. This is especially important, since the list of interactions which must be viewed to fulfil the information requirement may be large, and may render a meaningless series of events encouraging unrealistic responses from users, unless they are 'stitched together' into realistic task scenarios for the evaluation. More is said on these issues in the examples in the case studies which follow, but first some consideration is given to the general nature and characteristics of selected evaluation methods.

3. Evaluation methods and approaches

Several schemes for grouping evaluation methods have been proposed. However, either their basis is not made explicit (e.g. Long and Whitefield 1986), the range of methods they include is limited (e.g. Meister, 1986, ignores Analytic Methods; Lea, 1988, ignores Specialist Reports), or their terms are unclear (e.g. Lea uses 'formal' to refer to non-empirical methods). It is suggested (Wilson 1988, Wilson and Whitefield 1989, Whitefield et al 1991) that methods can be differentiated with respect to certain characteristics to understand better the methods themselves, how they operate, their scope for application, and how their outcomes differ.

The principal attributes which can be used to discriminate methods, to identify logical groups, and to assess their utility and suitability in a given application context, are their operational characteristics. Operational characteristics include any attributes or attribute states which determine the range of operational capability, and examples include: System State (real vs representational); Information Output (e.g. measurement vs diagnosis); Data Type (e.g. errors, time, memory probes); Data Analysis (formal vs informal); Information Reliability; Process Time (i.e. time required to utilise the method); Data Interpretation (subjective vs objective); Design Stage (e.g. analysis, specification, prototype, implementation); and Resource Requirements.

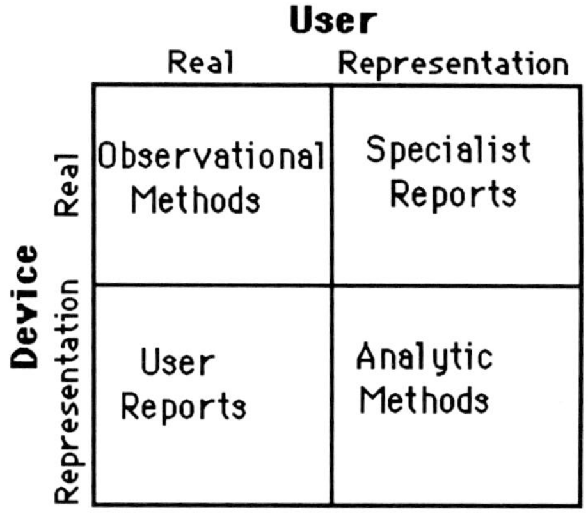

Figure 1: Categories of methods by system state

System State, where the system comprises a user and a device such as a computer or a videophone, can be parameterised for each system element as real or representational. Real refers here to implemented programs, devices, simulations, prototypes, and mockups of designs, as well as to actual or approximate user populations. Representational refers to symbolic descriptions; these include formal notational models (e.g. notations used in program development), models of users (either explicit, or implicit as held by, for example, human factors engineers), as well as users' mental representations of computers and other interactive devices. System state is used as a basic classification scheme for methods, and the resultant groups are described as Analytic Methods, Specialist Reports, User Reports and Observational Methods (Wilson and Whitefield, 1989). Each of these classes is now considered briefly so as to identify some of the operational characteristics which may have consequences for evaluations of design realisations.

3.1 Observational methods

Observational methods involve the observation of both real users and real computers. They range from the informal observation of a few users, such as User Observational Studies (UObS), to experimental observational studies with appropriate subject populations and rigorous control of variables (ExObS). These methods are highly empirical, but often require theoretical support for the hypotheses under examination (especially ExObS). In terms of information output, they support both performance measurement and diagnosis, and they provide data amenable to both formal and informal analyses. Observational studies typically involve the accessing of 'live' systems in use, or the setting up of realistic simulations of usage.

Procedures generally require the arrangement of an 'observation session' where predefined tasks and task structures are initiated to allow observation of their execution. Data capture often relies on the use of video and audio recording of sessions, coupled with the use of concurrent or post hoc verbal protocol elicitation. This latter element gives the evaluator access to the thoughts and feelings of the users, and may be supported by a structured interview, questionnaire, or ad hoc investigation of recorded events after the task session is completed. Further information supporting fine grain analysis of interactions can be obtained by data-logging where user inputs to the system are logged, as are system responses and any internal system events of interest.

The scope of observational studies is wide and it is necessary to carefully plan the execution and data capture if the opportunities are to be used to best effect in fulfilling the information requirement of the investigation.

3.2 Specialist reports

Specialist reports involve human factors or systems engineering professionals, who are not necessarily target-users of the system in question, using representations of the user (either written or mental) in assessing a version of the real computer. The evaluation can be structured or unstructured. Example specialist report methods are Task Walkthroughs and the use of Guidelines (both explicit and implicit).

Walkthroughs have been shown to compare well with User Observational Studies (Hammond et al, 1985) in terms of the problems predicted by the specialist, and they can be quick to carry out. However, these methods only weakly support performance measurement, and are better at diagnosis, i.e. they do not easily support the comparison of observed performance with criteria since real users are not involved in the performance. However they do facilitate the examination and prediction of system behaviour and likely interaction problems. In addition to these comments, the issues are generally the same as for observational studies.

3.3 User reports.

User reports involve real users and a representation of the device, which will probably be in the users' memories, but may be external. The class of methods includes rating scales, interviews and questionnaires, and probably offers the least reliable source of data in terms of scope for validation. Methods such as questionnaires can, however, benefit from formal analyses aimed at increasing rigour, and the methods are generally easy to apply and quick to use. The value of user reports is that they give access to data which is not accessible by any other means, since they are concerned with the elicitation of attitudes concerning the system in question. Since user attitudes are important in determining both user comfort with the system, and the likelihood of voluntary usage of systems, the resulting data is of great value. User reports can also be used to sample a population for reports of difficulties and errors experienced by users, and in this manner can be used to reveal estimates and population statistics concerning usage problems. However, since such methods rely heavily on user interpretation of events, and user memory of events, the reliability of the output is highly constrained and further consideration is usually given to the contents of user claims. These can be further investigated in evaluation activities, by statistical analysis, or by submission to a rationalisation or validation process where the content of user claims is assessed against independent observations of system attributes.

3.4 Analytic methods

For the sake of completeness some consideration is given here to Analytic Methods, although their usage is not appropriate to the evaluation of design realisations. Analytic Methods rely on recruiting representations of both the computer and the user, and involve manipulating these representations to predict performance. Current examples of interest are covered in chapter 2.1, and other historical examples include the Keystroke Level Model (Card, Moran and Newell, 1983) and Whitefield's (1989) Blackboard Design Model. Such methods are in the early stages of development (see Whitefield, 1989), and their suggested advantages are that they require few resources, they can be used early in design, and are quick to complete. Their disadvantages are that the modelling techniques are not suitably mature and the reliability and accuracy are not well validated. They also require the representation of all system elements in a formal manner, and so do not suit development paths where formal design methods are not implemented.

4. Evaluation methods in practice—some general issues and examples

Given the considerations in this chapter so far, it should be clear that there is a strong reliance on practitioner expertise in current evaluation practice. Various workers, such as those cited, contribute different views on evaluation relative to their specific interests. However, there is as yet no complete unifying scheme for system evaluation which makes explicit and formalises all of the considerations without reliance on the practice and experience of the evaluator. The following subsections consider, firstly, some general issues in operationalising methods for evaluation, and then go on to consider two examples derived from current research and development work in Europe which characterise practice embedded within particular system development contexts. The reader can thereby examine how different practitioners deal with operational issues so as to enable effective evaluation activities in particular contexts.

A general issue in evaluation practice is the establishment of the evaluation focus—what are the questions ?, what is to be measured ?, how can measurements be made ?, and so forth. The design issues may be specific to a developer or user company, or may be generally addressed within some overall scheme such as 'design for usability'; i.e. the evaluation may be general or specific. In a similar manner, the approach used may be general, such as the 'usability factors' view, or may be particular, requiring the identification of specific usage factors of interest for a given system. Alongside these issues, there is that of the general approach; should it be blindly iterative, or should it be sensitised to particular stages where increments to system functionality, or access to users and usage scenarios, provide a natural opportunity ? How ever the evaluation approach is ordered, it can be seen as akin to any problem management process, or quality tracking process, and requires an explicit view of performance, a reasoned approach to conformance testing, and some means of testing against explicit goals whilst leaving room for a reasoned enquiry into the evident performance characteristics of the system and the behaviours underlying observed performance.

The timing of the evaluation activity may follow the system development path, and be opportunistic, or may choose to drive design by specifying a usability test schedule which may then become a critical design track. This may require the delivery of design realisations at times which suit evaluation, and not only the engineering activities. Thus there may be a consequent need to develop prototypes or other design realisations to allow testing independently of the build schedule. Whilst placing a demand on resources, this can be extremely valuable since it provides the opportunity to feed design in a more proactive manner, and allows the interface and interaction design issues to be de-coupled from a build driven implementation strategy.

On the issue of what should be observed/measured, criterial assessment allows a strictly measurable comparison between desired and observed performance levels, but demands pre-specification of all performance factors of interest, their criterial levels, and suitable metrics. In order to feed design, however, there must be a good understanding of what drives performance—the behaviours of devices and people acting in unison—and so the information capture in the evaluation must be rich enough to allow reasoning about what is actually going on in their interaction during task execution. So, while a factorial/criterial approach supports explicit specification of desired performance relative to issues of intertest to design, it needs to be coupled with a broader view, allowing coverage of the interests of other stakeholders (e.g. users, managers, customers etc), and including scope for a more contextually determined investigation of emergent issues. Thus, a general information requirement can be specified which includes all stakeholder interests and main areas of enquiry.

Specifying the information requirement allows a definition of the data set which is to be delivered from the data-capture activities into the core evaluation process—the measurement and judgement of conformance to requirements. Having this explicit data definition allows the evaluator to identify accessible system usage contexts, or to design suitable scenarios, and to incorporate coverage of focal system tasks which will allow delivery of the necessary data. Thus, having achieved goal definition for the evaluation, along with information requirements, usage scenarios and data / task definitions, the evaluator is in a position to select, combine and configure evaluation methods which will deliver the necessary information in a suitable format. Operationalising evaluation, then, involves a planning activity where opportunity and the requirements of the design activity provide a set of variables which have to be balanced to deliver suitable answers. As stated earlier, and depending on requirements, answers may vary from simple statements of performance levels (e.g. error rates) to very detailed exposure and explanation of the drivers of performance and their parameters.

The following examples show how two teams of system evaluators dealt with two very different system evaluation problems.

5. Case Study 1—Evaluation of a multimedia information system

This example of a case study is taken from RACE project 1065—ISSUE, and is provided to show how evaluation plans might target specific opportunities.

5.1 Background

A national chain of travel agents is planning to introduce a new multimedia information system into their high street offices. The system will supplement the existing printed brochures and offer the customer more information about the company's holidays, plus an easier means of searching for holidays which match their requirements. The system will provide access to the central computer database which includes full motion video descriptions of the holiday resorts along with details of the hotels. Access will be provided by a high speed broad band connection and the information regarding holiday availability will be updated continuously. If the introduction is successful then it may be developed further to allow customers to make their own reservations and payments interactively, and extended to provide remote access by customers from their own homes. The company hopes this system will be attractive to customers and increase their satisfaction with the process of purchasing holidays.

5.2 Definition of system objectives

One of the major goals of the system was to reduce the waiting time that customers experience. When the office is busy they can use the system rather than wait for one of the staff to become available. This benefit will be totally lost if they have to ask the staff to help them interact with the system, so it will have to be easy to use. In addition, it is believed that the computer system will offer the company an advantage over their competitors by confirming their market image as one of technical innovators. Thus the objectives identified for the evaluation are that the system should be:

- attractive;
- easy to use for the majority of their customers;
- capable of providing the information that the customers would like.

5.3 Evaluation context

The design project was proceeding to plan, and the hardware infrastructure was soon to be installed. The systems database had been constructed and the design team had a specified amount of labour and resources to produce and test an attractive user interface within a 3 month period. As the interface software had not been coded yet there was considerable scope for the design team. However, on-line access to the central database was not yet available so a subset of the database has been specially copied to a laser disk for the evaluation. The disks limited capacity allowed only complete holiday information for one country to be included, but this was considered adequate for the purposes of the evaluation.

Although there was no specific human factors experience within the design team one of the software engineers had experience with using a prototyping tool which could produce graphical interfaces quickly.

Given the resources available the design team decided to develop a prototype interface to simulate on-line access and then take the prototype system to offices in a number of towns for evaluation by potential customers in a realistic setting.

5.4 Aims of the evaluation

Given the aims of the system the aim of the evaluation were to:

- gauge initial reaction to the system—would it be perceived as attractive to use?;
- establish how easy complete novices found the system to use on the first occasion;
- confirm that the system could provide the information wanted by customers.

5.5 User definition

The company had a clear understanding of the demographic characteristics of their customers based on data collected by their marketing division. Although information regarding experience with computers was not available, the company had information regarding their customers' socio-economic status, education and age. On the basis of this information the company concluded that the system users were likely to be young adults of above average educational attainment, and it was felt that they may well be attracted by the prospect of using the terminal. However, in terms of relevant skills, it was decided that the general public baseline should be assumed. Although a significant minority of the company's customers were aged 60+ it was felt that they were unlikely to wish to use the system and would prefer personal attention from one of the office staff.

5.6 Task requirements

The design team interviewed 10 office staff drawn from a number of different branches to understand the nature of the information exchanged in a typical interview with a prospective holiday purchaser. A semi-structured format was used which allowed the interviewee ample opportunity to suggest those aspects which they considered important. It was concluded that the most important considerations were readily identifiable (e.g. country, geographical location, price bracket, accommodation type and local facilities) and there was agreement about the relative importance of these factors. The design team also noted that the sales staff frequently claim that an attractive brochure photograph can have a significant influence.

This information was confirmed by way of brief interviews with 20 customers arriving to select brochures or book holidays at one of the offices. In addition to asking them what factors were important in determining their decisions, the customers were also asked about the way in which they used the brochures, what they found useful, what they found difficult and what they would like in the way of additional information.

5.7 Prototype development

The initial prototyping simply involved producing a series of high quality drawings of key screens in the interface. These images were produced using the information collected during the initial interviews and represented different ways of presenting the information and structuring the interaction. The drawings were presented to a selection of customers who were asked to rate them in terms of ease of understanding and usefulness (i.e. the degree to which the system would enable the customers to carry out likely tasks). This exercise was completed quite quickly and gave the design team useful evidence for proposing (and rejecting) possible interfaces.

On the basis of this information a working simulation prototype interface was developed. Apart from the restricted holiday database and video clips, the information available was comparable with the on-line system.

The need to display video sequences, the assumption of minimal computing experience and the requirement for an attractive and easy to use interface indicated the use of a graphical user interface with a colour screen.

Two styles of user interaction were considered: touch screen and mouse. Although the likely level of wear and tear did not indicate that a touch screen was the only viable input device, it was decided that the hand-eye co-ordination necessary for mouse usage should not be assumed.

In order to simplify the users interaction with the system, it was decided that there would be no need for the user to input any text, thus obviating the need for a keyboard. Interaction was enabled by selection of screen options represented by icons and command buttons The buttons were rendered to simulate an impression of depth, and their operation was accompanied by an audible click to offer strong feedback to the user.

Video segments relating to the destination countries and resorts were shown in a video window which had three buttons: start, pause and zoom (a facility to expand the video window to the full size of the screen and then shrink it when finished). These buttons used the pictograms commonly found on domestic video recorders.

Interview data suggested that the key search criteria would be location, hotel cost, hotel facilities and activities available at the resort. To keep the interaction simple, the user was required to set these criteria one-at-a-time, rather than allow multiple criteria to be chosen at once. Choosing selection criteria was done through a top level master menu of options. For attractiveness, use was made of images and graphics rather than simple text menu options. The selection options were budget, medium cost and luxury, and the user was presented with three still images of hotels and could make a selection by simply touching one of the images, and then pressing an OK button. This dialogue style was termed 'picture menu' by the designers. Having chosen the selection criteria, the user could press a button to view the hotels. Picture menus showing hotels matching the user's selection criteria are then presented, and they simply touch a hotel's image to see its details. These are held in pages so as to promote the impression of an electronic brochure.

At any time the user could (i) return to a previous stage in the interaction by pressing a step back button, (ii) cancel a previously selected option or (iii) get some help information by pressing help.

5.8 Planning the evaluation

Since one of the primary goals of the evaluation was to gauge the reaction of customers to the system, a naturalistic approach was adopted in which potential customers were asked to first browse the database, and then try to select holidays using the terminal. The restricted nature of the test database resulted in a degree of structure being imposed on the evaluation. Thus, each participant was asked to try to find holidays which met a number of requirements e.g. self catering holidays within a particular price bracket, located away from popular tourist areas and which offer certain leisure activities. This restriction also allowed comparisons to be made more easily between different participants attempts to use the terminal.

The design team had already collected information concerning the ways that customers used the company's printed brochures and wished to determine whether the terminal could provide a similar level of access. For this reason, the participants were first asked to browse through the database for five minutes and make any comments that they wished concerning both the interface and the holiday information presented. They were then asked to attempt a simple search task and, on completion, given a brief questionnaire with rating scales. A maximum time of 15 minutes was allocated for each participant since this was considered to be the most that many customers would be prepared to commit at short notice.

A significant amount of time was given to writing instructions for the evaluators to give to the participants. These instructions detail the purpose of the evaluation, the roles of the participants and the evaluators, the importance of the information and exactly what the participant would be required to do. The instructions and tasks were piloted with a number of individuals not directly involved in the project. A few unanticipated problems were identified and appropriate revisions to the interface, task and instructions were made.

A number of measures of the participants performance were devised. A screen capture routine was incorporated into the computer's operating system software to allow unobtrusive recording of every user input and every change in screen display. This facility would allow the design team to establish:

- the commands given by the participants,
- the sequence of screens displayed,
- relevant screens missed completely,
- any screens repeatedly displayed,
- the search targets identified,
- the time taken to complete the search task if successful.

This information was to be supported by notes made by the evaluator present during the test sessions.

5.9 Selecting the participants

Three different offices were selected as evaluation test sites. They were typical of the company's offices in terms of location and business turnover. Eight participants were chosen from the customers visiting each of the branches on a given week day, giving the total of 24 participants.

Care was taken to ensure that those asked to participate were selected without obvious bias. Thus half the sessions were conducted during the lunch hour to allow those normally at work during the day to be included, and brief biographical details were requested (in confidence) so that the representativeness of the sample could be confirmed.

5.10 Procedure

The terminal was set up in an appropriate location in the offices selected for the evaluation. This location was in the public area adjacent to the racks of printed brochures, but the terminal was positioned so that the participants could not be watched by other customers whilst completing the evaluation tasks. The evaluators had comparatively little difficulty in persuading customers to take part in the study but found it necessary to stress that the session would be brief, and that the participant's ability was not what was being assessed. They noted that adults with young children were more likely to decline their offer, and so included consideration of this bias in the analysis of results.

5.11 Data analysis

The data from the three different sites were initially considered separately, but when no consistent differences were discovered the data were combined.

The first stage of the analysis involved reviewing each participant's initial comments along with their exploratory browsing activities. A brief description was written which included the subject's first impressions, assumptions about the system's functioning and contents, and misunderstandings regarding the interface.

The second stage comprised a detailed analysis of each participant's search strategy when attempting to solve the problem. The appropriateness of the actions, the use of facilities, the errors made, the time taken, the information apparently considered and the comments made were all noted.

Finally the interview and rating scale data were analysed. The participants replies to the open ended questions were examined and categorised. Areas of common agreement as well as singular views were noted. Possible explanations from the other data were sought where apparently contradictory views were expressed by more than one participant.

The rating scale data were summarised (mean values and standard deviations) and then plotted graphically. Aspects of the system and its interface given particularly low ratings were considered in the light of other related data.

5.12 Recommendations

The overall response to the system was favourable with the majority of participants expressing a willingness to use a similar system in the future. The ability to see video clips describing the various holiday resorts was rated very favourably and seen as a distinct improvement over the limited information presented in the standard brochure. A minority of the users claimed that they would even be prepared to use such a terminal to make the final holiday reservation and deposit payment by credit card.

Based on all of the data a list of recommendations was drawn up; some of these are relatively simple to implement and include revisions to:

- the choice of command names,
- the logic of the dialogue structure in the search facility,
- the size of the characters used for the textual descriptions,
- the size and positioning of the command buttons,

Other recommendations were more demanding but could still be considered for implementation within the time frame of the implementation exercise. These include:

- the creation of an overview which gives the user a better idea about what they have seen and what remains to be seen whilst browsing ;
- introducing a notebook or marker facility to allow easy review of, or return to, specific database records.

Some recommendations are substantial and would require resources beyond the scope of the current review. For example some customers said that they would like extensive video material concerning the facilities available for each hotel rather than the selection of colour photographs provided. These comments were noted and kept for consideration when the success of the implemented system is reviewed.

6. Case Study 2—Evaluating videotelephony-based support services

6.1 Background

RACE project 1054—APPSN has launched application pilots as field trials in six countries using nine pilot sites as far afield as the south of Portugal and the north of Finland. These application pilots set out to design and test the feasibility and practical viability of providing various support services in a distributed manner using videotelephony as the primary communication medium. Support services included simple enhancements to current service provision, such as enhanced telephony for deaf people (including visual communication and translator/relay services) and video-enhanced alarm services, as well as novel service offerings, such as daily contact between elderly people and remote service centres, online help and counselling, and remote supervision of medical and physiotherapeutic regimes (LHC 1992).

Each pilot system was loosely built around existing service infrastructures, and made use of staff and expertise in residential homes for elderly, alarm centres, organisations for deaf people, and so forth. The technology implemented was essentially prototypical in nature, being adapted from commercial applications such as videoconferencing, or specially designed for situations requiring a high degree of customisation; for example home terminals for elderly people built into standard domestic television receivers, and special videotelephony terminals for deaf users with visual indications of status and functions for delivery of text to complement visual communications.

The services were delivered by service organisations such as local welfare authorities, deaf organisations, and residential homes for elderly. Given the novelty of the technical implementation, and the experimental nature of the undertaking, the service organisations were supported by network provider companies and terminal developers throughout the pilot period.

6.2 System users and their design requirements

At the start of the project, the prospective system users were candidates from various categories identified as having 'special' needs, i.e. those not regularly taken into account in system design activities, and included elderly people keen to retain an independent lifestyle, but requiring novel forms of support in order to do so, hearing impaired people wishing to gain equality of access and opportunity relative to the benefits offered by telecommunications access to information and services, and groups of mobility and visually impaired users whose support requirements could not otherwise be met either due to their being geographically remote, or due to there being too few support resources on offer in their local area. In these cases, videotelephony (VT) based services were seen as offering opportunity to make better use of professional resources without wasting them on travelling, and without transferring the travel problem to the user. In all cases the service design issues were paramount, i.e. how to design services for transfer to the new medium, and in addition, it was envisaged that the terminal design issues would require special attention both for the end user, given their special needs, and for the service provider staff, given the potential for impact on working practice.

6.3 Evaluation contexts

Given the 'pilot' nature of the design exercise, there was little opportunity to perform iterative evaluation in the traditional manner—nor would this have been desirable. The aim was to design, as well as possible, a set of services for testing in the field. This required a field test period without design changes if possible, or if avoidable, and thus required the delivery of a set of services, and associated hardware, networks, and software, in a fairly robust and usable condition at the start of the field test phase. For this reason, a two phase evaluation activity was implemented, the first being based around evaluation of prototypes in a laboratory setting, and the second being the longitudinal evaluation in the field. Iteration would be possible to the extent that the first phase would specify design modifications to support the second phase, and any really serious design problems emerging in the second phase could be judiciously dealt with where necessary.

6.4 Performance setting for design evaluation

To support a rigorous approach to design and evaluation, the pilots engaged in a formal definition of user requirements at the outset. This covered user requirements for support, given their special needs, requirements for the expression of support as organised services, given their particular needs, and requirements relating to the hardware, networking and task issues. The requirements definition document included a complete list of tasks which the user should be able to accomplish in the transaction of specific services, the functionality required to support such tasks, and the specific design requirements relating the special needs issues to the usage of such functions. Thus the user requirements definition provided the initial focus for performance specification.

The performance requirements identified in the user requirements specification were then recast as performance criteria, whereby the various pilots identified the requirements relevant to their particular group and, in negotiation with the various stakeholders, established acceptability criteria relative to these. So, for example, it was determined that 'the system should be usable by at least 80% of the target population', 'the client should be able to use the terminal without using an instruction manual', and 'the service centre should be able to handle up to 100 clients'. Thus the criteria were a mix of achievement levels relative to some variable (e.g. % population) along with desirable states (e.g. no manuals) and absolute measures of effectiveness (e.g. up to 100 users). In addition, the design specifications for each pilot system made explicit reference to tasks and requirements from the requirements definition document, and explicitly stated the mapping between the system features to be implemented and the tasks/functional requirements responded to. Thus the evaluators were able to conduct a form of early evaluation at the specification stage (we are reluctant to consider this completely 'Analytic' since it is not wholly formal, but point it out as an instance of early evaluation where the completeness of the design specification is tested with respect to the requirements definition).

From this point in the case example we will follow through *one pilot* example, except where general issues need explaining, since to cover nine pilots simultaneously would be unduly laborious. The site is chosen as being typical of the rest in its activities, although some of the detail did vary.

6.5 Evaluation phase 1—prototype evaluation

At an early stage in the design process a 'minitrial' was conducted as part of an integrated test activity whereby the hardware, software, networks, services, service tasks, and user performance and attitudes could be examined prior to design intent being accepted for implementation.

6.5.1. Evaluation objectives

The objectives of the evaluation were to test the design assumptions at the earliest stage possible (nine months into a three year build and test programme, before the hardware had been commissioned). It was required that the results provide strong predictive indications of the likely success of the pilot systems in the field in terms of all performance features.

6.5.2. The prototype design

The prototype designs conformed completely to the technical specification on the 'outside'. That is to say, they were presented as 'real' systems, even though many of the components had to be simulated using equipment which would not be included in the final system. They were embedded in a prototype 'usage environment' whereby a 'service centre' and the 'front room' of a typical elderly person living on their own were simulated in a laboratory setting.

Verisimilitude was seen as being important to help the candidate users and service providers 'role play' system usage in a very realistic manner.

6.5.3. Evaluation Planning

The tasks and requirements lists from the requirements definition were studied, and a simple form of hierarchical task analysis used to identify 'base tasks', or task components which would have to be executable to guarantee execution of the whole range of tasks. These were then recast as a set of realistic 'task scenarios' whereby the candidate users could transact services for the delivery of a range of support types covering the focus of the intended pilot. Scenarios included examples such as 'Enquire of your local service provider how your elderly neighbour can get financial support for new spectacles', 'Contact your local service centre to get advice on exercises (and practice) to relieve stiffness in the joints', 'Enquire of the service centre for times of trains to allow you to visit your children in another town', and 'Get your local service provider to explain what you must do with the new form for local taxes which has just arrived'. Thus, the task list covered all types of support service, and all of the system functions available such as call set-up, document display, and so forth. A questionnaire was then designed addressing general factors such as learning, ease of use, user satisfaction and so forth, as well as particular issues for the designers such as feelings of 'social presence', eye to eye contact, turntaking strategies and other parameters of conversation control. A strategy was devised for applying the questionnaire as a 'semi structured interview' so that the evaluator could obtain the required data within a more natural discussion throughout the evaluation activity.

In order to reduce the workload on the principal evaluator, and to allow comparison of judgements where task difficulties or problems were not very clear, provision was made for a third party observer to capture and document problems as they occurred. This also allowed the evaluator a more natural interaction with the client (demonstrator role). In addition, video and audio recording of the sessions was planned so that a complete record of events was available for analysis in case any problems, difficulties, or user issues could not be fully documented and understood during the live session, or in case conflicts of interpretation occurred in the analysis. A sample of ten prospective elderly users was then selected to balance gender, age and background, thus permitting a degree of representativeness with the target population.

6.5.4. Evaluation Conduct

For each prospective client (evaluation subject), the routine of the session was standard. They were brought to the simulated 'living room' which had been furnished in typical fashion for elderly people living in that area, thanked for participation in the trial, and made to feel at home. Coffee and cakes were provided, and the evaluator used this time to introduce the planned service systems, the terminal (domestic TV set with special remote control), and some description of how the services were intended to aid independent living. When the client was settled, they were informed that the evaluator wished them to try out the system, and to 'role play' a set of tasks. Prior to the execution of the first task they were given a short demonstration and introduced to the service provider (via the videophone) who was a local social worker positioned in the simulated service-centre in another part of the same building. (A similar evaluation activity was going on there, but that is not covered in detail here).

Each evaluation task was executed in turn, after a brief description of the task scenario during which the client could ask any questions of clarification, and the evaluators then observed task performance and noted any procedural errors and difficulties as they occurred. After each task was completed the principal evaluator talked with the client about the task, covering all of the items in the questionnaire, and making notes of their ratings against the various scales, along with other relevant user comments about the system. On completion of the task set, the clients were asked about their general impressions and any comments made concerning usability or acceptability (including their rationale) were noted. This part of the session also allowed the evaluators an opportunity to understand the candidate user's predictions concerning likely uptake of services, acceptance issues, and other topics relating to service opportunity.

6.5.5. Results obtained

The results from this study provided an initial view of the clients' responses to the service system. Along with their general feelings concerning the underlying concepts (e.g. we would have to be sure of security since someone might be watching us discussing our business), much information was obtained concerning general service issues (e.g. I would have to get to know the servicer well before I could talk about my personal affairs), specific design issues (e.g. I don't like this modern TV set, it would have to be like the ones which I and other old people choose), and the fulfilment of the user requirements. This latter issue depended on observations such as whether people could read forms displayed by document camera, whether they could execute the call set-up and call closure procedures easily, and whether they could position themselves in a satisfactory manner using the self-view function. Data derived from these observations also drove design changes. For example, the spatial relationship of the function keys on the handset caused some of the users to select the wrong key to end a call—they selected the one adjacent to the call set-up function, assuming some semantic relationship implied by position and colour.

In general this evaluation activity was very successful. It provided a very realistic prototype long before the actual system functionality would allow such detailed testing, and allowed users to execute very realistic tasks such that they could envisage what the real system would be like when available. The methods used to supply the information requirement were a mixture of Observational Methods (Tasks plus direct and AV Recording) and User Reports

(Questionnaires, Semi Structured Interviews, and Discussions of User Views after system usage). By investigating the user views immediately after the session tasks, it was felt that a clear access path to accurate reports of difficulties and attitudes was enabled in contrast with traditional approaches to asking users off-line, where much forgetting might have taken place. In addition, expert walkthroughs (Expert Reports) were used when difficulties could not be well understood from the session data. Here, human factors experts conducted walkthroughs of the tasks to identify the likely sources/causes of difficulty which had been seen.

The results allowed design changes to be implemented prior to the service systems being launched in the field, and many of these changes, especially where the problem definition did not point to an obvious solution, were retested with a small sample of candidate users to verify the new design (selective iteration relative to specific design problems).

6.6 Evaluation phase 2—field trials

Following on from the results of the mini-trials, the engineering change requests were specified, the system designs re-engineered in line with these, and the implementation versions of the system built for usage in the field trial. During this period a field evaluation was planned and designed, as will be described in the following subsections.

6.6.1. Evaluation objectives

The field evaluation set out to determine the degree to which the pilot systems conformed to the design objectives of the project. These included delivery of useful and attractive support services to people with special needs living in their homes, providing service delivery strategies which were easy for users to negotiate and attractive to use, providing hardware and software which supported the service transaction tasks in an easy and satisfactory manner, and implementing designs in accordance with the explicit requirements and performance criteria set for the various user groups. The specific objectives of the evaluation activity, then, were to observe and measure all attributes of system performance which would contribute to the satisfaction of the design objectives.

6.6.2. Evaluation Planning

Information Requirement—The evaluation planning activity was initiated by identifying the various stakeholders in the service systems and their respective concerns. Since the service system pilots were being funded by the EC RACE initiative as part of their R&D effort in advancing the development of European IBC and its usage, RACE were identified as one of the primary stakeholders in the evaluation. In support of this objective, analyses were carried out concerning the information needs of the various part I activities in RACE whose task it was to deliver a more general view of usage requirements across all application and technology sectors. This provided a highly detailed information requirement for RACE covering issues such as usage profile, costs, market sector definition, and so forth.

The end users and their service provider organisations were also identified as primary stakeholders, and similar analyses were conducted. Interviews and meetings with provider organisations exposed the organisational issues of interest, the questions which had to be answered for these organisations, and hence their specific information requirement. In the case of the end users, the user requirements definition provided a primary focus, and meetings with experts in the various care sectors provided further user-level issues which presented as being worthy of inclusion in the information requirement for the evaluation planning.

A further stakeholder group were identified as the technology developers, including terminal and network developers. Representatives of these groups were present in the project consortium, and were used as a source of expertise concerning the general design issues of interest to technology developers.

By the above means, the objectives of the evaluation were specified for all stakeholder groups and this rendered a very broad range of issues and questions to be answered by the evaluation activity. These were then recast as a structured document whose structure was agreed by the evaluators in the different pilot groups as providing a consensus view of the information requirement (there was a degree of overlap between the requirements of different stakeholder groups). This gross structure was then decomposed to reduce the questions to a data level which allowed specification of the necessary data capture in terms of the metrics to be used to answer criterial evaluation issues, and the measures necessary to allow diagnostic and more open ended enquiries as necessary.

Specifying Data Acquisition—having developed the information requirement document to the level of data structures to be delivered by the evaluation activity, the evaluators then analysed each pilot in turn to understand the contextual factors which might constrain information acquisition. By this means they were able to identify opportunities for data acquisition in each pilot, and to agree a general approach and structure. A three stage evaluation activity was agreed. Firstly, early evaluation would allow examination of learning, service uptake and initial user reactions to service system issues. Secondly, a mid stage evaluation would allow comparison of results with those of the early stage after users had been actively using the system for around six months. Finally, a late stage evaluation would allow examination of longer term usage and usability issues. In addition, a detailed longitudinal view was included, whereby the system activity profile would be logged to make possible a detailed examination of system activity on a day-by-day basis. In order to obviate data loss due to memory and other features, strategies were developed for capturing user problems and issues arising in the periods between these intensive evaluation activities.

These included various 'visiting' regimes where data on usage problems were gathered informally as part of a general client contact activity.

Developing the Evaluation Plan—by relating the general information requirement, via an expression of the data set necessary to allow an accurate response to the evaluation goals, to the opportunities and constraints for evaluation, it was then possible to define a selection of methods for each topic/issue/goal from which a specific method could be selected to suit local circumstances. The *information requirement* document was then progressed to a complete *evaluation planning* document wherein the specification of information item, data structure, acquisition stage and methods were related within a structured document which provided the general evaluation plan across all pilots. A small sample from this information requirement definition is shown in Table 1.

Focus and Information Required	Data Type	Stage Acquired	Methods Applicable
•*User perceptions and attitudes*			
ratings by users	user statements	all, esp'L	UR
issues limiting acceptance and usage	user statements direct observations	all, esp'L project reports	UR /UObs
....etc. etc.			
•*Usage levels*			
frequency	frequency count	continuous	O(data log)
duration	time distribution	continuous	O(data log)
usage pattern	detailed log	continuous	O(data log)
etc. etc.			
•*Implications for IBC*			
IBC enhancement of existing services	user statements / descriptions/ direct observations	all	UR, UObs, SR, project reports
changes to current practice to accommodate IBC	user statements/ descriptions	all	UR, O, SR
benefits from using IBC	user statements/ descriptions	M,L	project reports UR,O,SR
....etc. etc.			

(Methods covered in example shown here—User Report (UR), User Observations(UObs), Analysis of project reports (project reports), Specialist Reports(SR), Indirect observations (O, e.g. activity logs) nb user = end user and service provider)

Table 1 : Example extracts from Evaluation-Planning Document showing Information Requirement Definition

The value in having a common evaluation plan was mainly in ensuring common coverage, and enabling a common reporting structure which would facilitate direct comparison of results. For each pilot, the evaluation plan was then recast as a low level procedure taking account of local circumstances.

6.6.3. Evaluation conduct

In the example pilot a continuous logging facility was designed which would capture all network events as a continuous record showing date, time of connection, caller, recipient, closure time, and call duration (computer based). In addition, service centre workers logged each call by type of service required, and noted any problems or difficulties evident for themselves or their clients. For each of the three main evaluation stages, questionnaires and semi-structured interview formats were developed for user and service provider interviews, and a frequent 'visitor' system devised to allow sampling of user activities and attitudes throughout the twelve month study period. Direct observations of calls were made frequently, and any observed problems or difficulties discussed with the parties involved to allow retrospective protocol analysis. Particularly difficult problems were further supported by having human factors experts provide analyses, walkthroughs or in-depth interviews as appropriate.

For each stage the data set acquired, including the portion of the logs covering that period, were integrated for all pilot sites. An initial analysis of all of this data was conducted and shared with all pilot site evaluators to sensitise each to new problems or issues arising elsewhere.

6.6.4. Samples of results obtained

Terminal Design and Usability Issues —Terminal Controls—From the observations of usage, and the interviews and self reports from end users and service staff, a range of problems and design issues were identified. For example, difficulties in fitting and operating 'personal switches' worn to allow alarm calling from remote locations were reported early in the pilots. Adjustments to these designs allowed iterations around the design problems until an appropriate solution could be found. Another 'controls' example concerns a general problem for service provider staff who, having several tasks to execute as part of service provision (e.g. getting data from the PC, selecting forms for display, etc), found difficulty in using a handset. In many cases, requests were made for a 'hands-free' mode of operation which, due to some echo and feedback control problems, was not easily rectifiable.

Display and Display Quality—More than 50% of client statements, during parts of structured interviews which covered the display element, were positive declarations concerning the added value of being able to see the service provider 'as if present'. Clients were generally pleased at being able to develop a personal understanding with people they could see and get to know 'as a person'. No real problems were evidenced which diminished this satisfaction in any meaningful way since, in the pilot site example here, they were provided with a high quality image (analogue broadband) using service centre equipment which gave 'eye-to-eye' contact.

Call Initiation and Closure—In this pilot site there were early problems due to the acoustic signals used for call status being inconsistent with those used for normal telephony. Clients were generally unable to understand whether a line was ringing, busy, unobtainable or whatever. By adjusting these signals to conform to normal telephony signals the problems disappeared.

Terminal Appearance and Form—Many clients in this pilot site felt that the terminal was too big and not very attractive. It was a very modern and high quality display set, but it did not suit their rather 'traditional' taste in domestic appliances. In addition, they felt that the additional 'topset' (camera housing) made the set look less pleasing than if it were integrated so as not to be so obvious.

Service System Usage Profile and User Attitudes— The service system at this pilot site was generally well used and the clients were very enthusiastic about the system. The following diagram (figure 3) shows the usage levels for 11 users over the trial period

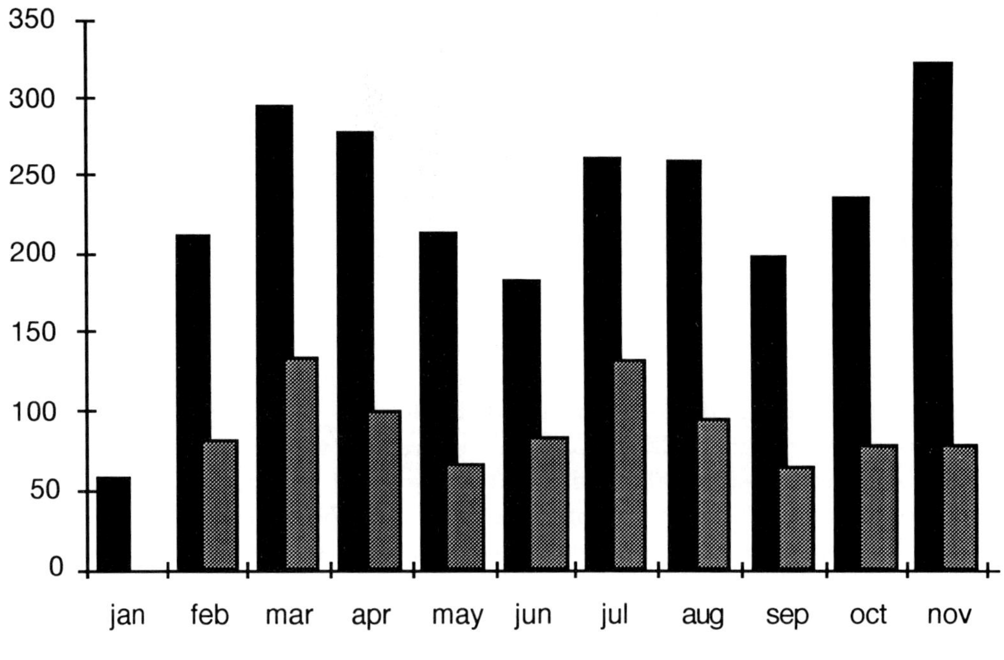

Figure 3. Call frequency by month.

The activity logs showed that there was an initial sharp increase in service usage from the outset, and that this stabilised at near saturation level except for two significant 'dips'; one caused by a heatwave affecting network equipment, the other due to client and staff holiday periods and some terminal problems. Connected call activity is shown by black bars, with grey bars representing instances of callers receiving an 'engaged' response (probabilistic).

Investigation of client attitudes and experiences (UR) reveal that, whilst they had some initial reservations, they quickly got to 'know' the service providers, and were able to rely on them for the very personal forms of support normally obtained in the person-to-person style of service delivery in the home.

Further investigation of the data logs revealed a significant change in the profile of call duration over the trial period. In the first month the average call time was around 2 minutes, whilst by the end of the trial this had increased to nearly 20 minutes. Clients, who at first had been cautious about dealing with personal issues, were now in a position of having formed a degree of trust in the service provider and the system, and could now deal with personal issues with ease. Indeed, some said that the control they had over the session made it much easier to deal with interactions of this kind.

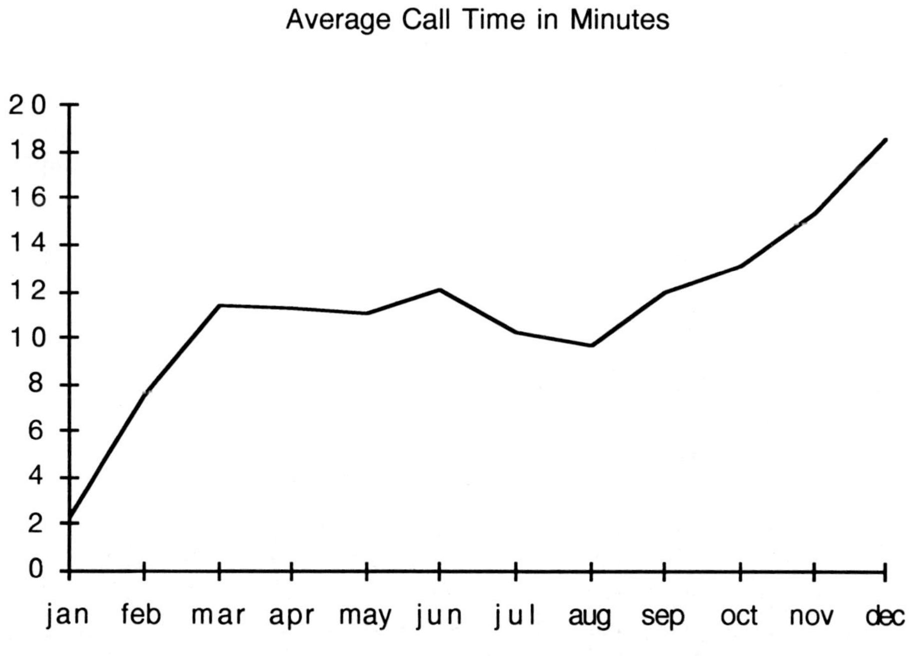

Figure 4. Average Call Duration by Month

The depressions in the call length profile for this sample site (figure 4) indicate equipment and network problem periods where the service providers had to 'ration' time since the full service period could not be provided.

6.7 General observations on evaluation at this site

Only a small sample of the findings from this pilot have been considered, and only some examples of the requirements, evaluation criteria, and information requirement to which the evaluation activity was attempting to respond. This evaluation activity undoubtedly gained benefit from an explicit evaluation planning activity which made clear the relationships between the goals of the evaluation (questions to be answered and explicit information requirements), the criteria defined and their attendant metrics and data structures, the needs for understanding of behaviours of the networks, equipment and people which determined any performance observed, and the range of applicable methods which could operate on these in the context of the pilot and its operational circumstances. A whole range of methods were used in combination, and each had to be configured to suit local circumstances to some degree. For example, since the clients were very elderly (aged 72—84) the User Reports strategy used a 'semi structured interview' approach, based around a questionnaire format to guarantee coverage and attention to detail, but allowing the interviewer to obtain that coverage within a more natural, and less pressured, freestyle interaction. System logs provided a gross view of 'activity' which could then be refined by understanding the behaviours underlying the performance profile. Additionally, frequent observations by the evaluators, especially at the service provider end, allowed them to observe the nature and detail of problems and their cause. And finally, by having a stable reference to design goals in the form of an explicit requirements definition, including performance criteria, the evaluators were able to track conformance to criteria, in terms of performance levels, and conformance to requirements not expressed as criteria, in terms of presence/absence of desired functions and the implementation style.

The delivery of a very rich data set allowed the evaluators to identify not only the nature of the design issues, but also the interactions of various factors contributing to the usage and uptake profiles of interest to the various stakeholders in the evaluation activity.

7. Conclusions on evaluation of design realisations

In our initial considerations of evaluation issues we observed that in order to enable effective 'conformance testing' it is necessary to be able to specify both desired performance and actual performance in common terms for comparison and judgement of degree of conformity. The case examples suggest that whilst much of performance can be pre-specified using current methods, it is not possible to predict all possible system behaviours and their desired levels of performance. For example, in the APPSN case it was noted that one unexpected effect of system implementation style was the forming of trusting relationships with care workers at a distance. The predictability of such behaviours notwithstanding, it may have been very difficult to specify criterial levels for acceptable performance in this regard. This example, along with others in the text, argues for an evaluation approach which allows a broader view than that which might be engendered by the constraints of a purely criterial approach.

Further to the above point, our consideration of contemporary approaches, as mooted in the literature, addressed the issue of relating the criterial approach to the scoping provided by a factorial view. It was argued that whilst a so-called 'general' scheme with limited factors mitigated against particular focuses in the evaluation, a truly general approach would encourage the ad-hoc measurement of everything possible, resulting in a major analysis and interpretation task for the evaluator which might waste resources on inconsequential matters. The cases considered here provide examples of the development of 'special' views, expressed as both 'usability factors' and otherwise, where a very clear and specific brief is provided to the evaluation planning activity detailing the interests and issues identified by the stakeholders in the design and evaluation activity. In addition, they show how the experienced evaluator is able to recognise opportunities for investigation of emergent issues and to reconfigure the methodology dynamically to address the questions raised by such phenomena. In consideration of the expression of criteria relative to factors, we earlier argued that 'fixed' expressions of criteria (e.g. learning expressed as time) may be inappropriate, and indeed our examples show that acceptable learning was expressed, for example, as 'usage without manuals'. The business of 'definition of criteria' has much to do with the general design goals of the developers and users, but needs to be parameterised with respect to features which are particularly meaningful to the stakeholders.

Our earlier consideration of evaluation issues also introduced the problem of over-emphasis of performance relative to its measurable parameters, whilst ignoring the underlying behaviours which determine performance levels. The case examples show that evaluation in practice relies on the ability of the evaluator to not only capture a 'rich picture' of system behaviour (e.g. audio visual recording plus system logging), but also the analytic ability to relate performance features to underlying behaviours (e.g. the increment of call duration over time) either through ad hoc investigations of the available database, or through complementary investigations of features which present as interesting (e.g. additional visits to clients to investigate changes in call pattern). Related to this theme, there is also that of 'inclusion' versus 'representation' of users in system design. In the examples we have seen that methods such as 'expert walkthroughs' are used to good effect since, whilst they cannot predict performance levels, they can undoubtedly support prediction of likely user problems or, used post-hoc, allow predictive estimation of likely causes of observed user problems which are not yet clearly expressed.

As stated in our consideration of general features of contemporary approaches, there is a tendency to assert 'iteration' as a necessary feature of design for usability. Nonetheless, the examples show that not all design paths can be driven from a user interface perspective (i.e. not planned around prototype iteration), but that opportunities for iteration may be built in if there is a sound rationale for such a demand (e.g. minitrial vs pilot testing, and the observed need for design changes to personal alarm switches in the APPSN case). That is to say, even when iteration is possible, it may be as much opportunistic as planned.

Two further points seem worthy of inclusion here. First, it was argued that the focus for the evaluation would determine the what, when, and how of evaluation—what do we measure, when can we measure it, and how should it be measured. It was argued that the provision of a specific focus would enable detailed evaluation planning, thus increasing the level of guarantee concerning the suitability of the outcome. Our examples have shown that the establishment of evaluation focus can be extremely varied, and should be determined relative to the needs of the design activity—the specification of focus can vary from the definition of a simple set of goals and measures (Multimedia Terminal example) to an extremely detailed information requirements encompassing a whole range of specific factors and questions for the evaluation (APPSN example).

Finally, the traditional literature on evaluation, even when confident of its advised processes, says little to the evaluator about situating the evaluation activity in the context of the real world of system development. Since the evaluation activity is essentially about estimation of world states which cannot be measured exactly (the sampling problem), it seems necessary when estimating 'actual performance' to consider the circumstances in which such performance can be accurately approximated. The examples show a variety of sampling activities where the estimation of actual performance relied on either an approximation of the world in the laboratory (APPSN minitrials), an expression of likely usage of implementation (the Multimedia Terminal tests in the Travel Agency), or an arrangement of actual usage over a protracted period (the APPSN 1 year field trials). In all cases the evaluators had to deal with the problem of responding to a particular information requirement, and it appears that the opportunities for data acquisition (circumstances or context), along with a view of an appropriate task set for observation in the evaluation, provided the scoping for a definition of appropriate 'task scenarios' and hence the methods which will suit information acquisition in that context. This latter point is also relevant since, from the candidate user's perspective, the task scenarios in which they contribute behaviours (situated actions) to the estimation of the real world is a primary determinant of the evaluation results.

In conclusion, it appears that there a number of issues not yet well addressed by the human factors literature on evaluation. Some scope for improvement in the general conception, approach, and methods for evaluation have been considered relative to case examples of evaluation, as have a number of operational considerations. It appears that despite the absence of a well formed theoretical approach, some evaluators, in some instances, can surmount problems such as those discussed. We might doubt the generalisability of such success to all cases, and point out the reliance by such evaluators on experience in the form of tacit or implicit knowledge. We propose, therefore, that the advancement of evaluation practice, through the development of better conceptions, approaches and methods, might do well to examine and study evaluation practice in more detail in order to make explicit and better formalise the elements of successful practice in evaluating design realisations.

References.

Bellotti V., (1988) Implications of Current Design Practice for the Use of HCI Techniques. In D.M. Jones and R. Winder (eds), *People and Computers IV*. Proceedings of HCI'88. Cambridge: Cambridge University Press.

Card S. K., Moran T., & Newell A., (1983) *The Psychology of Human Computer Interaction,* Hillsdale, New Jersey: Lawrence Erlbaum.

Clegg C., Warr P., Green T., Monk A., Kemp N., Allison G., Lansdale M., Potts C., Sell R., & Cole I., (1989) *People And Computers: How To Evaluate Your Company's New Technology*. Chichester: Ellis Horwood.

Denley, I., (1990) *The Evaluation Planning Aid—A computer-based tool to aid evaluation practice*. London: London HCI Centre, University College London.

Dowell, J. & Long, J.B., (1989) Towards a conception for an engineering discipline of human factors. *Ergonomics*, 32(11), 1513-1535.

Gardner A., & McKenzie J., (1988) *Human Factors Guidelines for the Design of Computer Based Systems*. Ministry of Defence and Department of Trade and Industry.

Hammond N., Hinton G., Barnard P., MacLean A., Long J., & Whitefield A., (1985) Evaluating the interface of a document processor: a comparison of expert judgement and user observation. In: B. Shackel (ed): *Human-Computer Interaction—INTERACT'84*. Amsterdam: North-Holland.

Lea M., (1988) Evaluating user interface designs, in T. Rubin, *User Interface Design For Computer Systems*. Chichester: Ellis-Horwood.

LHC, (1992) *Interim Evaluation Report on the Service Pilots*. RACE Project 1054 deliverable 54/ucl/lhc/C/527/b1. London: London HCI Centre, University College London.

Meister D., (1986) *Human Factors Testing and Evaluation*, New York: Elsevier.

HUSAT, (1992) *Usability Evaluation: Guidelines for IBC Service Designers*. RACE Deliverable 65/HUS/-/DS/A/014/b1. HUSAT: England.

Ravden S., & Johnson G., (1989) *Evaluating usability of human-computer interfaces: a practical method*. Chichester: Ellis Horwood

Shackel B., (1986) Ergonomics in design for usability. In: M.D. Harrison & A.F. Monk (eds), *People and Computers: Designing For Usability*. Proceedings of HCI'86. Cambridge: CUP.

Whitefield A.D., (1989) Constructing appropriate models of computer users: the case of engineering designers. In: J.B. Long & A.D. Whitefield (eds): *Cognitive Ergonomics and Human-Computer Interaction*. Cambridge: Cambridge University Press.

Whitefield A.D., Wilson F., & Dowell J., (1991) A Framework for Human Factors Evaluation. *Behaviour and Information Technology* 10, 65-79

Whiteside J., Bennett J., & Holtzbatt K., (1988) Usability engineering: our experience and evolution. In: M. Helander (ed), *Handbook Of Human-Computer Interaction*. Amsterdam: Elsevier Science Publishers.

Wilson F., (1988) *Human factors evaluations in the development and maintenance of interactive computer systems*. Report LHC/EXT/ALV/EV/1.1F. London : London HCI Centre

Wilson F., & Whitefield A.D., (1989) Interactive Systems Evaluation: Mapping Methods to Contexts, in E. Megaw (ed.), *Contemporary Ergonomics 1989*, London, Taylor and Francis.

Computers, Communication and Usability:
Design Issues, Research and Methods for Integrated Services
P.F. Byerley, P.J. Barnard and J. May (Editors)
1993 Elsevier Science Publishers B.V.

Chapter 3.5

Usability and Standards

Keith R Dickerson* and Leif R Hedman†

*RACE Consensus Management Office, 165 boulevard du Souverain, B-1160 Brussels, Belgium.

†Telia Research, Box 85, S-201 20, Malmö, Sweden.

1 Introduction

This chapter identifies the standards relevant to the usability of telecommunications services and information systems, including international, regional and national standards, and also those proprietary standards that have a wide impact on the usability of systems and services. The way these standards are used by designers of products and services is examined, together with the impact that the standards have had and will have on designers in the future. The benefits arising from the standardisation of usability, and the commercial considerations which may work against this are discussed. A glossary of abbreviations and terms used in the standards world is appended for reference purposes.

'Usability' is the ability of a system or service to be used effectively by its target users to carry out a specified set of tasks in specified environments. It extends and formalises the principle of 'user friendliness' that is popularly applied to products and services that are generally thought to be 'easy to use' compared with alternatives. Usability has long been important to *users* of telecommunications services. It is also becoming increasingly important to *suppliers* of telecommunications services and terminals because:

a) users are demanding more usable systems,
b) service providers see improved usability as a way of differentiating their products from the competition,
c) bodies like the CEC through European Directives such as that on Visual Display Terminals, are requiring a specified level of usability to be built in to future products and services.

Through the European Directives, and similar procurement directives in the U.S., standards that are referenced by these directives are having an important impact on telecommunications services and information systems. All usability standards that are either referenced or have had an impact on these directives are examined in this chapter.

The process by which international standards are defined is described in section 2. There are many things that contribute to making a system or service usable and the standards bodies most relevant to these are introduced in section 3. Although usability is still a relatively new area of standardisation, a consensus had been reached on the essential meaning of the term, and this definition from ISO 9241 is described in section 4. Several other standards bodies are also active in the definition of usability standards and these are covered in sections 4-7, including proprietary user interface standards in section 8.

But how useful are usability standards to designers of products and services? There is evidence to show that many designers are not aware of many relevant usability standards, even though they would provide a useful contribution to improving the design of a product or service. There is also evidence that many designers see standards as a hindrance that *constrains* design rather than as an *aid* to improving it. It is also relevant to examine how standards are actually used by designers in practise, the benefits obtained, and the problems that are encountered. These issues are covered in section 9.

The results of the standardisation of usability and the benefits to users and suppliers of telecommunications equipment and services is covered in section 10. The standards required to ensure the usability of future integrated services are discussed in section 11. An extensive glossary of terms used in standardisation and references to all standards mentioned is included at the end of the chapter.

2 Workings of the standards bodies

The types of organisation producing standards, recommendations or 'pre-standards' are:

1. International standards bodies, such as ISO and CCITT. These produce standards and recommendations applicable throughout the world.
2. Regional standards bodies, such as ETSI. These produce standards applicable to large parts of the world.
3. National standards bodies, such as BSI, SIS, DIN and Afnor. These produce standards applicable to a single country.
4. Pre-standardisation bodies, such as ECMA, RACE and SPAG. These produce 'pre-standards' or 'functional specifications'. These are not standards in the legal sense but are often used as the basis of standards that are later adopted by a regional or international standards organisation.
5. Proprietary 'standards' which have a wider application than in the companies that produce them. Examples of these in the usability area are standards defining the 'look and feel' of a user interface.

The flow of information from research in human factors and related areas (such as software engineering) to the standards bodies relevant to telecommunications services is shown in figure 1 (see the glossary at the end of this chapter for the full name and a short description of each of the standards bodies shown).

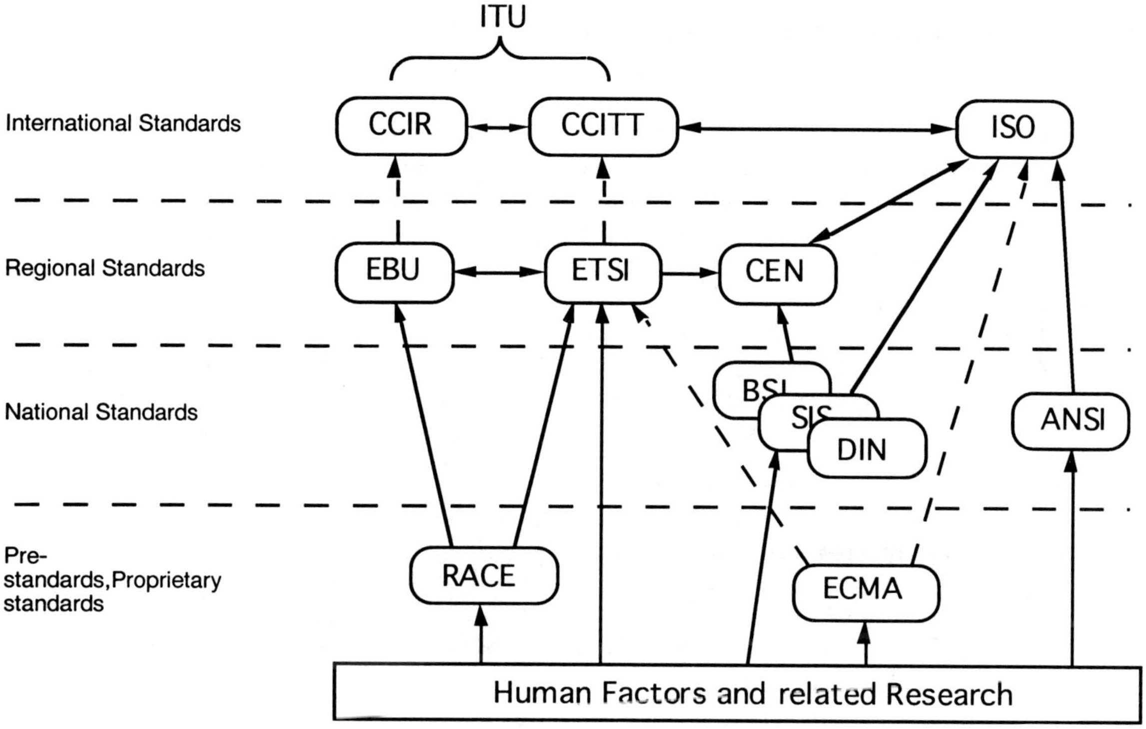

Figure 1. The relationship between standards bodies.

Contributions are made to meetings of the working groups of each standards body. Generally contributions flow upwards in figure 1, from research organisations and companies to national standards bodies, to regional standards bodies, to international standards bodies. However, contributions can also be made directly to an international standards body by a research organisation or company if it is member of that standards body, without going through the national and regional standards bodies.

In fact contributions to a working group can generally *only* be made by a member of that standards body. It may therefore be necessary for a contribution based on the consensus achieved in a national or regional standards body to be submitted by a member of the standards body rather than from the standards body itself. In this way, an important consensus can sometimes be achieved in a regional standards body prior to submitting it to an international standards body. For example, in Europe a consensus is often achieved in ETSI before one of the member countries submits this to CCITT. This has been formalised, so that a consensus agreed in ETSI cannot be opposed by another member of ETSI when a contribution based on this consensus is made to CCITT. A similar position applies to the pre-standardisation organisations. For example, in RACE a consensus can be achieved on a particular technical issue prior to submission to ETSI or CCITT. This consensus is expressed in the form of a RACE Common Functional Specification (CFS), and contributions based on this are made to the standards bodies. A list of the RACE CFSs on usability is given in section 7.

3 Standards contributing to usability

The organisations active in defining usability standards for telecommunications services are ETSI in Europe, and CCITT internationally. ISO is also active in defining usability standards applicable to computer systems. As computers and telecommunications merge, the standards bodies have to work together to an increasing extent, and joint technical committees have been set up between, for example: CCITT and ISO. The 'core' usability standards are defined by the bodies shown in figure 2.

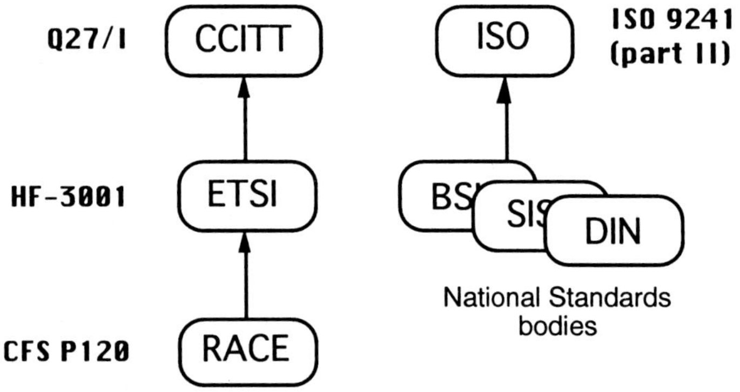

Figure 2. Standards bodies defining 'core' usability standards.

In addition to the 'core' usability standards, there are many other standards for both telecommunications and computer systems that will have a significant effect on the usability of the services offered to users. The functional specifications of a telecommunications network or of the components of that network will affect the services that can be offered over that network. For example, the maximum information transfer rate specified at the user-network interface will affect the quality of the information that can be presented to the user (eg: the image quality). This is shown in more detail in figure 3. This shows:

a) the issues that have an impact on the usability of a telecommunications service,

b) the place of usability in the issues that contribute to the overall usage of a telecommunications service.

All the issues shown in figure 3 will have some effect on the usage of a telecommunications service. The example is shown by the thicker dashed line in figure 3. This shows the effect of cell transfer delay (in an ATM network) on service usage. It can be seen that cell transfer delay affects overall delay, which in turn affects response time, service quality, usability, acceptability, demand and overall service usage. Some of the generic service integration principles given in chapter 3.2 can also be seen in figure 3. For example, feedback is included as contributing directly to usability. Coherence and compatibility are shown as contributing to memorability and therefore to learnability and usability. Links to cost have not been shown because all of the above issues will have some impact on the cost of providing the telecommunications service. As recognised in the ETSI concept of utility (here generalised as 'acceptability'), a highly usable product may be more expensive than a less usable one and so be less acceptable and receive a lower share of the market.

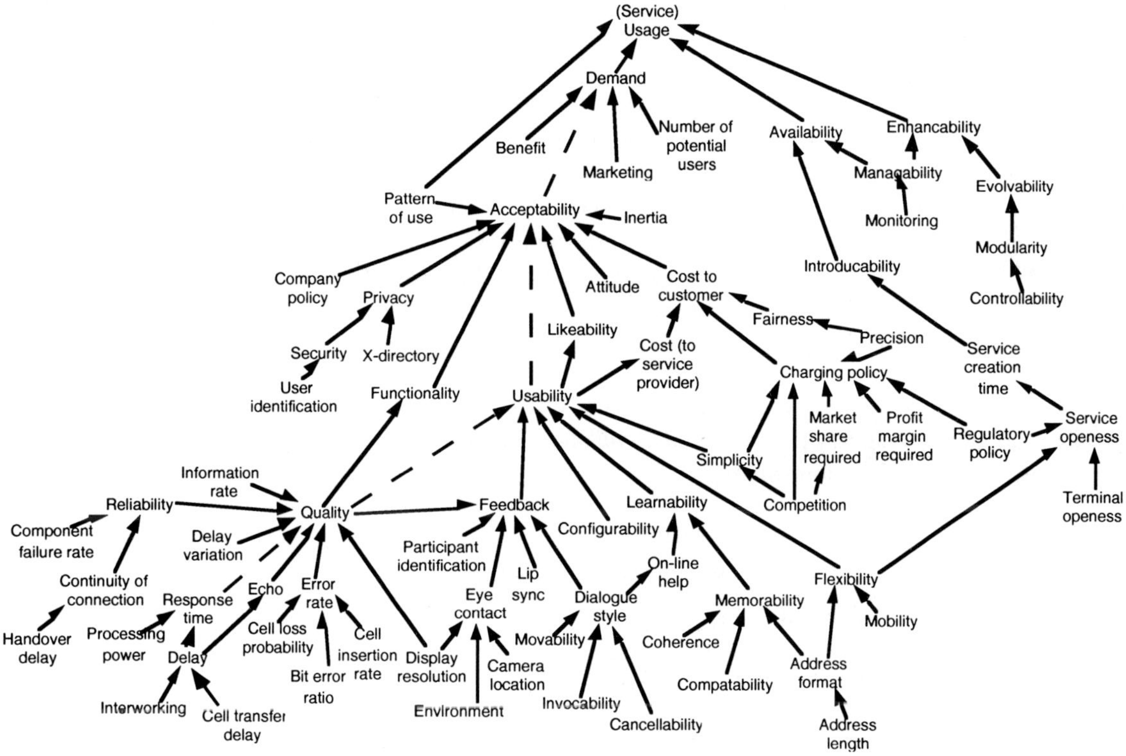

Figure 3. Issues contributing to *usability* and *usage* of a telecommunications service.

For these reasons, many telecommunications standards will affect the usability of the telecommunications services. These could be called 'enabling standards' because they *enable* usable telecommunications services to be offered to customers. The route for 'enabling standards' in telecommunications is that shown on the left of figure 2. This route is generally from work done by research organisations and programmes such as RACE, through to ETSI for European standards (T1 for USA), and then to CCITT for recommendations applicable worldwide. This chapter concentrates on those standards necessary to provide multimedia services as these will have the greatest impact on the usability of telecommunications services compared with the current situation. The standards organisations relevant to usability and their existing or emerging standards are covered in more detail in the following sections.

4 International standards

The international-level standards bodies that have produced recommendations on usability are ISO for computer systems and CCITT for telecommunications systems.

4.1 International Organisation for Standardisation (ISO)

ISO was founded in 1946 as a voluntary, non-governmental body. ISO standards cover all fields except for electrical and electronic engineering (covered by IEC) and telecommunications (covered by ITU). The technical work of ISO is done in technical committees (TCs) and their subcommittees (SCs) and working groups (WGs). The main work on usability is done in TC 159 *Ergonomics,* with the full working group reference ISO TC159/SC4/WG5 *Software Ergonomics and Man-Machine Dialogue..* This is drafting ISO 9241 described in detail below. Relevant work is also done in ISO TC159/SC1/WG2 *Ergonomic Principles related to Mental Workload.* A committee draft on design principles is available.

ISO also has a joint technical committee with IEC with the reference ISO/IEC JTC1 SC18/WG9 *User System Interfaces and Symbols.* This specifies a range of user interface items including the following standards:

- cursor control for text editing,
- symbols used to represent keyboard functions,
- objects and actions,
- graphical symbols used on screens.

ISO 9241 Ergonomic Requirements for Office Work with Visual Display Terminals (VDTs)

This is a 17-part standard which is not yet complete. The sections already approved are listed below, with target dates for completion of the other sections:

Part 1	General Introduction	Approved
Part 2	Task requirements	Approved
Part 3	Visual display requirements	Approved
Part 4	Keyboard requirements	Draft available
Part 5	Workstation layout and postural requirements	Draft 1/93
Part 6	Environmental requirements	Draft 5/93
Part 7	Display requirements with reflections	Draft 5/93

Part 8	Colour of screen images	Draft 3/93
Part 9	Non-keyboard input devices	Draft 5/93
Part 10	Dialogue principles	Draft 2/93
Part 11	Usability	Draft 2/93
Part 12	Presentation of information	Draft 10/93
Part 13	User guidance	Draft 10/93
Part 14	Menu dialogues	Draft available
Part 15	Command dialogues	Draft 6/94
Part 16	Direct manipulation dialogues	Draft 6/94
Part 17	Form filling dialogues	

After the drafts have been produced and agreed in committee they must be approved by ISO, and so it is unlikely that the full text of ISO 9241 will be approved until 1995. However, Part 11 *Usability* is likely to become stable during 1993. This proposes the following definition of usability and its components:

Usability: 'The effectiveness, efficiency and satisfaction with which specified users can achieve specified goals in particular environments'

Effectiveness: 'The accuracy and completeness with which specified users can achieve specified goals in particular environments'

Efficiency: 'The resources expended in relation to the accuracy and completeness of goals achieved', and

Satisfaction: 'The comfort and acceptability of the work system to its users and other people affected by its use'.

It also states that Effectiveness and Efficiency are often referred to as 'performance measures'. There is no reference to the enabling state definitions of usability given in chapter 3.1. However, the two sets of definitions are consistent and the enabling states definitions can be considered as extensions of the ISO definition. For example, it is unlikely that users will be able to achieve their specified goals effectively or efficiently unless the enabling states exist and can be reached at reasonable cost to the user.

At present this standard only applies to VDTs used for office tasks, but it has been agreed to extend ISO 9241 to terminals used to access telecommunications services. National legislation enacted in member countries to comply with the EC directive on VDTs will reference ISO 9241. This will include the following requirements:

- Employers must perform an ergonomic analysis of workstations; plan the work so that users can have periodic interruptions; and pay for any spectacles the user needs in order to work with the display.
- Users must be fully briefed on all health and safety issues; be provided with an eye test if they request one, both before starting work and at regular intervals; and be trained in the use of the workstation.

- The display must be easy to read, not appear to flicker or jitter, have easily adjustable brightness, contrast, tilt and swivel controls, and be free of glare.
- The keyboard must be able to tilt and be separate from the screen, have a matt surface to avoid reflective glare, be well laid out and well labelled, and have a space in front of it to provide support for the hands and arms.
- The desk must be large enough for flexible arrangement of the computer equipment and paper documents, have a low reflectance to avoid glare, and include a document holder. Users should have enough space around them to be able to change position.
- The chair must be stable, allow the user to move freely, be adjustable in height, and have a back rest that is itself adjustable in tilt and height. Users are entitled to a footrest if they want one.
- Lighting must be satisfactory, taking into account both the type of work and the user's visual ability, and glare and reflections must be prevented.
- Noise from equipment must not distract attention or disturb speech.
- All radiation must be reduced to negligible levels, and equipment must not produce excess heat that might cause discomfort to users.
- Software must be suitable for the task; easy to user; and feedback should be provided.

When it is complete, ISO 9241 will provide quantitative support for the above requirements.

4.2 International Telegraph and Telephone Consultative Committee (CCITT)

In CCITT, work is carried out in Study Groups (SGs). These are numbered by roman numerals from SGI to SGXVIII. Recommendations are produced which are not mandatory on members. Recommendations on Usability are drafted by SGI and are described in this section. However, in addition to these, many important recommendations are drafted in other study groups that will have an affect on the usability of telecommunications services offered to customers as explained in section 3. Therefore some recommendations drafted by SGX and SGXVIII are covered in this section.

All work in CCITT is done in response to a series of questions defined at the beginning of each study period and numbered as Q1..Qn. Study periods last four years, and at the end of each the recommendations produced are published as a coloured book. The latest is the Blue Book published at the end of the 1984-88 study period. The next book will be published in 1993 covering recommendations produced during the 1989-92 study period.

To allow CCITT to react more quickly to produce urgently required recommendations, accelerated procedures have been introduced. These can be approved before the end of the study period.

CCITT SGI Services

Working Group 2 covered usability during the 1989-1992 study period. The following questions relevant to usability were considered:

- Q27 *Usability*
  On providing customer satisfaction and efficiency when using worldwide telecommunications.
- Q28 *Symbols, pictograms and keypad layouts*
  Various national, regional and international standards bodies are responsible for the design, evaluation and standardisation of symbols, pictograms and icons. In order to coordinate the work of these bodies it is desirable to follow a well defined methodology. Draft Recommendation F.910 describes procedures for designing, evaluating and selecting symbols, pictograms and icons.
- Q29 *Customer Control Procedures in the PSTN and ISDN*
  Telecommunications Operators have been asked to submit contributions directed at the choice of digits to be used for such services, the method to select a carrier and the method to access a supplementary service. Work is in progress concerning the effect of long dialling sequences on usability.
- Q30 *Tones and Indications for a Human User*
  This specifies the feedback provided to the user on the status of a telecommunications service.
- Q31 *Human Factors Aspects of Voice and Non-Voice Services using Public Terminals*
  Recommendation E.134 on *Generic Operating Procedures* will be submitted for approval in March 1993.
  Recommendation E.135 on *Human Factors Aspects of Public Telecommunications Terminals for People with Special Needs* is currently being drafted. This will include the requirements of the aging population for telecommunications services.
- Q32 *Human Factors issues of new and updated voice and non-voice telecommunications services.*

CCITT SGX Network Management

User interfaces for network management functions are important to the usability experienced by the end-user. This is because the efficiency by which the network can be managed will affect the service provided to the user. Therefore usability is important in user interfaces for network management as well as for end users. CCITT SGX specifies the user interface and particularly the man-machine language to be used for network management. This is called MML No 1.

CCITT SGXVIII Digital Networks including ISDN

CCITT SGXVIII is responsible for making ISDN recommendations worldwide, including those covering B-ISDN. Work on multimedia services began in 1989 and the network capabilities required to support these have been defined in CCITT Framework Recommendation I.374 *Network Capabilities to support Multimedia Services*. Much further work is required on this during the next study period (1993-1996) to ensure that the network capabilities provided by B-ISDN will be adequate to support multimedia services.

This recommendation will be important to the usability of multimedia services in the future because appropriate network capabilities must be present in the network to enable multimedia services to be offered effectively to customers. As well as sufficient information bit-rate to provide adequate media qualities and response times, there must be adequate capabilities present to enable synchronisation of media streams at the receiving terminal and so avoid problems such as inadequate lip synchronisation. See the discussion on ETSI NA-12409 later in this chapter for more information.

5 European regional standards

In Europe, usability standards are produced by ETSI and CEN. ETSI produces Technical Standards (ETSs) and Technical Reports (ETRs). Neither of these are mandatory in their own right, but they can be referenced by the CEC in European Directives, which then become mandatory on all European equipment suppliers. CEN produces standards covering both EC and EFTA countries.

5.1 European Telecommunications Standards Institute (ETSI)

Standards in ETSI are drafted by 12 Technical Committees (TCs), each divided into a number of Sub-Technical Committees (STCs). Participation in standards committees is from volunteers amongst the member organisations, and it is often difficult to find sufficient expertise to draft ETSI technical standards. In order to progress work on important standards more quickly, ETSI also has the ability to set up project teams. Participants are paid to draft technical standards or reports, although these must subsequently be agreed in and are subject to modification by the nominated technical committee.

The main ETSI technical committee working on usability is TCHF, the Technical Committee on Human Factors. This was set up in 1989 to ensure that telecommunications services have a high level of usability by optimising both performance and user acceptability. TCHF has 3 sub-technical committees:

HF1 *Human Factors of Telecommunications Services.*
This deals with the terminal to user interface, including user procedures for a range of telecommunications services including videophone and UPT. A large part of the work of HF1 recently has been concerned with generic user procedures and their application to videotelephony. Several reports specify the user interface to telecommunications services. It is accepted that only 'minimum' users interfaces will be standardised. This will allow a level of consistency of operation of services to make them easier to use, for example by specifying the use of the * and # keys on a telephone keypad, but will allow manufacturers to provide added features to differentiate their products.

HF2 *People with Special Needs.*
This covers the requirements of People with Special Needs for telecommunications services, including the design of telecommunications terminals. The main recommendation on usability that has been approved by HF2 is ETR029 *Access to Telecommunications for People with Special Needs.* HF2 have also recently completed ETSI Technical Report HF-2006 which gives an overview of European standards on telecommunications facilities for people with special needs.

HF3 *Usability Evaluation.*
This deals with all aspects of the usability of telecommunications services including how to measure usability and how to determine the required usability of a service. HF3 is particularly concerned with the definition of usability and has extended the ISO definition as described below.

In addition to TCHF, several other ETSI technical committees are producing recommendations that will have an effect on the usability of the telecommunications services offered to users due to the relationships shown in figure 3. In particular, the recommendations covering services and terminals will affect the usability of the services offered to customers, and so some of these are covered in this section.

The recommendations important to usability are listed below. Until recommendations have gone through the public enquiry process and been fully approved, they are specified by a 2-letter and 5-digit number. The two letters indicate the TC responsible for the recommendation. The digits indicate the work item number in that TC. Hence HF-01002 is a work item in STC HF1. NA-12409 is a work item in STC NA1. These numbers are replaced by a permanent document number once the recommendation has been approved.

ETR029 Access to Telecommunications for People with Special Needs

This is a permanent document as it has been fully approved in ETSI. As its subtitle ('Recommendations for improving and adapting telecommunication terminals and services for people with impairments') suggests, it describes the various types of impairment suffered by people with special needs, including impairments affecting vision, hearing, speech, reading, language comprehension, mobility, growth and intellect. It also explains the difficulties these impairments cause in the use of telecommunications services and terminals. Finally it lists a set of proposed recommendations that have been or will be drafted based on the difficulties identified in this report.

These recommendations will have an impact on a number of issues in service and terminal design, including the design of telephone booths, specification of time-outs, colour visual displays and operating instructions.

HF-00002 Human Factors Guidelines for ISDN Terminal Equipment Design

This is a comprehensive 200-page handbook providing guidelines that, if followed, will ensure a minimum level of usability from a terminal connected to the ISDN. The first section describes human factors as part of the design process. The second section describes the range of users of ISDN terminal equipment and the types of tasks they have to carry out. The third section covers design issues for ISDN terminals including telephones, facsimile machines, videophones and multimedia terminals.

However, the bulk of the handbook is an A-Z listing of design guidelines relating to various user interface issues for ISDN terminals. These guidelines range in application from acoustic signals, to camera location for videophones, and to pointing devices and screen layouts for multimedia terminals. All-in-all, this is a comprehensive handbook which is well worth obtaining.

HF-01002 Human Factors Guidelines for the Design of Minimum Phone-based User Interfaces to Computer Services.

This recommendation provides guidelines to ensure that a minimum level of usability can be obtained from a phone-based user interface to a computer service. Phone based interfaces use the basic telephone keypad digits 0 to 9 and * and #. They can be used to access applications such as voice mail, home banking, information retrieval and teleshopping.

General usability guidelines are given such as 'User must be aware of what they can do' and 'Users must know that they are dealing with a machine'. More targetted guidelines follow on how to construct the dialogue and the specification of time-outs and response times. Functions that are recommended to be provided from the keypad include a help function that can be accessed at any time by the user, and a cancel function to correct errors in data entry.

HF-01003 Recommendation on Tones generated locally in the Terminal.

Tones are an important form of feedback to the user on the status of a telecommunications service. However, throughout the world there are over 100 different tones providing this feedback, which is causing problems for international travellers as they find it difficult to recognise the appropriate tones. This recommendation aims to harmonise these tones, which should result in services that are both easier to use and more consistent between countries. The first stage is to recommend the tones that should be generated locally in a terminal to indicate different status conditions. Following this, a recommendation will be made on the tones generated in the network. A third recommendation will extend the scope to cover indications other than tones, including visual feedback using displays.

A problem at present is that this recommendation is only attempting to 'average' the tones used by various administrations and does not attempt to address the fundamental question of how best to inform the user of call progress. The rationale for this is that there will then be less confusion in the short term between the tones generated locally by the ISDN terminal and the tones generated by the network.. Specifying 'average' tone values will probably also make it easier to get a standard agreed internationally, as fewer changes will be required to telecommunications equipment. However, an opportunity may have been lost to provide a set of tones that are easier to recognise and distinguish than the current ones.

HF-01005 User to Terminal Interface Reference Model.

Modern telecommunications services are becoming increasingly complex, and this is resulting in services which are difficult to use and which are not consistent in their operation with other services. This recommendation will provide the designers of telecommunications terminals and services with a user-to-terminal interface reference model. Designers can base their designs on this reference model so that any operation can be carried out in a similar way on different services. An example concerns the user procedures for setting up a telephone call or for sending some electronic mail,. Operations should be carried out in a similar order for similar services, with appropriate feedback at each step. If users have to enter information in a different order in services with similar functions, or do not receive feedback that is normally expected, then they may become confused and stop using the service.

HF-01006 Human Factors in Videotelephony.

This recommendation provides guidelines for the design of videotelephony services, including the specification of controls and indications and user procedures for these services, The user procedures were originally specified in ETSI HF-TR 002-1 in 1991. The controls and indications will be specified in a separate report in 1993. An ETSI project team (PT16v) was set up in 1991 to evaluate the recommended user procedures, and to recommend two sets of procedures:

a) A set of generic user procedures that could be applied to any ISDN service throughout Europe. These include procedures for call set up, payment, identification, change, incoming call handling and call termination.

b) A set of user procedures for ISDN videotelephony. These include procedures for the selection of video or audio, incoming video indication, video pause, self view and switching between hands free and handset modes.

Evaluations were carried out in the UK, Germany, Italy and the Netherlands, with both objective and subjective measurement of results.

HF-01008 User Procedures for Multipoint Videotelephony.

This recommendation extends the guidelines for videotelephony given in HF-01006 to multipoint videotelephony. This covers simple or 'basic' videotelephony and does not include videophones incorporated as part of workstations or PCs. Multipoint videotelephony can be established in at least two modes in terms of video presentation and control:

- 'Switched' mode, where a single video signal from a terminal is transmitted to all the other terminals. Each user can then only see one participant's image at a time.
- 'Mixed' mode, where a mix of some or all of the video signals is transmitted to each terminal. This enables each user to have continuous visual presence of all participants' images.

Multipoint videotelephony services are more complicated and more difficult to use than point-to-point videotelephony. User procedures that are covered by this recommendation include procedures for:

- setting up a multipoint videophone call,
- switching the video signals in switched mode,
- controlling the 'mixture' of video signals in mixed mode.

So that multipoint user procedures are as easy to use as possible, users should be able to transfer knowledge gained from the use of point-to-point videotelephony services. Therefore a 'dynamic add-on procedure' is recommended. A 'side-call' can be made to a new user during any point-to-point or multipoint session who can then be brought in to the conference. This gives maximum flexibility to start off sessions as *either* point-to-point or multipoint sessions, and for any existing participant to introduce new participants.

To make it easier to start conferences from scratch if the participants are known in advance, an 'auto-originate procedure' is also recommended. The convener enters the videophone numbers of the participants and the start time of the conference. Each participant is then called automatically at the start of the conference. A 'call waiting' service is recommended in conjunction with this procedure, to make it easier for initially 'busy' participants to enter the conference at a later time.

HF-01010 Pictograms for Videotelephony Functions.

A pictogram or icon is a graphical representation of a function or element of a user interface. The use of pictograms can improve the usability of a telecommunications service provided the user can intuitively understand which function the pictogram represents. They can also reduce learning times, as standard pictograms for basic videotelephony functions would allow users to recognise these functions on any videophone without extra instruction. Pictograms can also be understood by illiterate users, and provide a degree of 'language independence' which makes them ideal for international use. The problem is that pictograms may mean different things in different cultures and communities, and so empirical tests are necessary to ensure that the optimum set of pictograms has been selected. Pictograms were evaluated for seven basic videotelephony functions:

- videophone/telephone
- camera on/off
- microphone on/off
- self view
- still picture
- document camera
- hands free mode

The report is in two parts:

1. The recommended methodology (the multiple index approach) for the evaluation of pictograms.
2. A report on the results of the evaluation of pictograms for point-to-point videotelephony and a set of recommended pictograms.

HF-01011 Human Factors Aspects of Pan-European Numbering Plan.

The CEC have asked ETSI to provide information on how it might be possible to move towards a telecommunications numbering plan that applied across the whole of Europe. At present the structure of the telephone number is different in each country, which makes it difficult to provide pan-European services.

Human Factors considerations applying to pan-European numbering include:

- an increase in consistency resulting from a single numbering plan throughout Europe,
- an increase in cost and decrease in usability during the change-over, as many users will have to adapt to a different numbering plan from that currently used.

HF-01012 Human Factors of Universal Personal Telecommunications (UPT).

UPT offers a set of services that will provide users with a degree of mobility by enabling them to use any telephone as if it was their own personal telephone. On registering as a UPT user at a given terminal, all incoming calls to their 'home' terminal will be diverted to the designated terminal, and all outgoing calls made by the UPT user from the designated terminal will be charged to the UPT user's account.

Human Factors considerations applying to UPT include the key sequences required to invoke the service, and feedback to the user on the status of the service.

HF-01014 Harmonisation of Code Schemes as Minimum Man Machine Interface for Telecommunication Terminals.

Many new telecommunications services are being introduced which use the telephone keypad as a 'minimum' MMI to invoke sets of functions. These can only use the digits 0 to 9 and * and #. The problem is that different services are being defined to use different key sequences for the same functions. For example, the UPT service uses a different set of keys to that used to invoke supplementary services for telephony. This will be confusing for users who regularly use a number of different services, as they will have to remember to use a different key for the same function in another service. This will lead to more mistakes by users, and a lower usage of the services.

This recommendation proposes a set of code sequences that should perform similar functions for any service using this 'minimum' MMI. This is based on the code scheme defined by CEPT in CEPT Recommendation T/CAC O 2 E. Adoption of this code scheme should allow users to use a wider range of services more easily, and will also make them more willing to try new services because they can apply their experience gained using one service to other services.

HF-01015 User Instructions for Public Telecommunications Terminals.

User instructions on how to operate public telecommunications terminals such as payphones are not always as easy to use as they could be. Content and presentation varies not only between countries but also for different types of terminal within a country. This is a problem for people wanting to use public terminals, and is exacerbated by the increase in international travel which requires people to use public terminals where the instructions are not in their own language.

Information on the layout and format of existing instructions has been collected and analysed with respect to the tasks that the user is trying to carry out. The next stage is to provide guidelines for simplifying the instructions to make them easier to understand and more consistent between countries. These guidelines will be produced during 1993.

HF-01016 Human Factors Aspects of Multimedia Telecommunications.

Multimedia telecommunications services are those which combine more than one medium; eg: voice and data or voice, video, text and graphics. These services can be used to carry out a wide variety of user tasks. However, there is the danger that, as the functionality is increased, these services will become too complex for users to use effectively. Multimedia services may require functions for stopping and fast-forwarding video sequences for example, which the user could very easily confuse with similar functions for moving through a document or scanning an image.

This report identifies issues in multimedia telecommunications that require human factors guidance, and provides this guidance where possible, based on work carried out in the RACE, ESPRIT and similar programmes. Guidance is given on the media and quality levels which are most effectively used to carry out different tasks, and the most effective way to integrate different media at the user interface.

HF-02004 Inductive Coupling of Telephone Earphones to Hearing Aids-Performance Requirements and Testing Methods.

The hearing impaired make up a significant proportion of users of telecommunications services. This standard will specify the ways of coupling telephone handsets to hearing aids, including performance requirements and methods of testing for adequate performance.

The standard will cover all public terminals, including those installed for safety reasons, for example: in lifts and on motorways.

HF-02007 Evaluation of Telephone Terminals for People with Special Needs.

This will extend DTR HF-3001 and HF-3002 to cover the requirements of people with special needs. A checklist will be produced for the evaluation of simple telephone terminals for use by disabled users. This will be completed by the end of 1993.

HF-03001 Guide on Usability Testing Methods for Telecommunication Terminals and Services.

This report extends the ISO 9241 definition of usability, and describes evaluation techniques that can be used to ensure that a product or service meets a specified level of usability. The ISO 9241 definition has been generally adopted in the report. However, the definition is enhanced in the following ways:

1. Usability is considered as a pure ergonomic concept not depending on costs of providing the work system. Usability and financial costs together form the concept of Utility. This means that an ergonomic highly usable system may have low utility for a particular user who considers the cost be too high in relation to the need for using the system.
2. Measures of usability are assumed to be of two kinds:
   a) Performance measures, which are 'objective' measures or observations of user behaviour and focussed on task performance, that is how well the user can achieve a given task.
   b) Attitude measures, which are 'subjective' measures of the user's opinion of working with the system, for example how much they like to use the system.

Based upon these measures, HF-3001 describes a generic process useful for undertaking criteria-related and comparative evaluations.

The method will be validated by application to a real case study in service development in 1993.

HF-03002 Usability Checklist for Telephones.

This is a checklist of items that can be used at an early stage in usability evaluation to obtain a rough estimate of the usability of a service or terminal. It covers the basic functions of setting up a call, receiving a call, carrying on a conversation, and terminating a call. It covers physical ergonomic aspects such as the weight, size and shape of the telephone handset, as well as the levels of tones and sidetone, and feedback to the user on telephone functions. This checklist is not a substitute for a more thorough usability evaluation. It is currently being extended to ISDN telephones, and a similar checklist will be be produced for videophones.

NA-12409 Base Document on Multimedia Services.

This is a report produced by ETSI STC NA1 *User Interfaces, Services and Charging*. ETSI NA1 is responsible for defining telecommunications services in Europe from the user viewpoint. This includes a textual description of the service and the feedback that will be available to the user at each stage. Other ETSI committees specify the signalling and protocols required to implement the service.

The most effective way of ensuring that all the enabling states that are required by a user to carry out a goal task exist is to provide multimedia services. These are flexible and can be configured to provide the media types and quality levels required by a user for any task. The basis for the provision of multimedia services is being defined NA-12409. This will define the types and quality levels of the media that can be used in a multimedia service and how these can be combined. It will therefore have an effect on how effectively and efficiently the users can carry out tasks. The way in which a multimedia service is provided will also affect the flexibility and therefore the usability of the service. Alternative ways of providing a multimedia service using ISDN or B-ISDN connections are shown in figure 4.

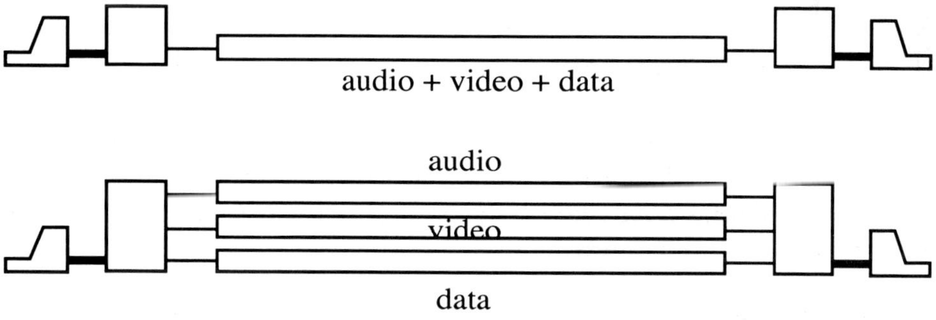

Figure 4. Alternative ways of providing a multimedia service.

In the first case, a single network connection is used. The media are multiplexed in the terminals (or terminal adaptors) and the network connection simply has to provide the information transfer rate and response times necessary to carry the media 'bundle'. This method will work but is not very flexible, as the quality levels of the individual media cannot be varied independently to meet the user's changing task requirements. The maximum likely information transfer rate must also be allocated at call set up, leading to a more expensive solution.

In the second case, separate network connections with different transfer capabilities are used for each medium. The quality levels of each medium can then be varied independently of the other media by varying the transfer capabilities of the connection. If the user requires higher quality video, to read an A4 document for example, then a standard TV quality service component can be deallocated and an HDTV service component allocated instead. This would require the connection for that medium to be released and a new connection with the appropriate capabilities allocated instead.

The second method is more flexible and should lead to more usable services. However, there may be problems in the short term before sufficient ISDN or B-ISDN network capacity becomes available. One problem is that synchronisation is often required between media, for example: to synchronise the audio and video streams so that lip synchronisation is maintained. Independent connections set up for each medium may take different routes from source to destination, and so end-to-end delays may be different requiring more buffering and hence an increase in overall delay. Also, when a connection is deallocated and a new one allocated, a new connection with the correct capabilities may not be available, or may have a different delay to the previously allocated connection.

5.2 European Committee for Standardisation (CEN)

CEN takes standards produced by ISO and applies them to EC and EFTA countries. TC 122 *Ergonomics* closely follows ISO TC 159, and is proposing to adopt ISO 9241 as the basis of standards for VDTs throughout the EC and EFTA. The CEN standard states that the machine and its associated elements (displays, controls etc) shall be suitable for the operator and the given task. An important issue is to what degree the human operator is needed in the system in order to accomplish a given task. The designer should consider if the planned selection of a system is in accordance with physical, psychological and social characteristics.

Six general design principles are given:

- suitability for the task,
- self-descriptiveness,
- controllability,
- conformity with user expectations,
- error tolerance,
- suitability for individualisation and learning.

Unfortunately, the recommended set of general principles in its present form is incomplete, is stated in a general and almost self-evident way, gives only minor help to designers in solving practical design problems, and gives no guidance on when and how to use the principles. This means that designers may have problems in applying the principles and may still result in a user interface which has a low usability for the target users.

6 National standards

The national standards bodies define standards applicable to their own countries. They also wish to ensure that national requirements are reflected in international standards. They therefore send representatives to monitor and contribute to the work done by ISO. The national standards bodies include:

- Afnor for France
- ANSI for USA
- BSI for UK
- DIN for Germany
- SIS for Sweden

The contributions of some of these bodies are described in more detail below.

6.1 British Standards Institution (BSI)

BSI sends representatives from UK companies and the Department of Trade and Industry to ISO. The important BSI committee for usability is PSM 39 *Applied Ergonomics*. This committee has produced standards such as BS 7179. This recommends items such as keyboard layouts and font sizes. PSM 39 tracks and contributes to ISO TC 159.

6.2 Swedish Standards Institution (SIS)

SIS engages people from Swedish companies, universities, research institutes and the Ministry of Industry and Commerce to ISO and CEN. A project group within SIS coordinates all usability issues in SIS four working groups (AGs) and tracks and contributes to ISO TC159 SC1 *Ergonomic guiding principles* and to CEN TC 122 *Ergonomics*.

6.3 Deutsches Institut für Normung eV (DIN)

This is the German national standards organisation. A standard produced by DIN is called a Deutsche Industrie Norm. DIN was very influential in the early days of producing standards for VDTs and DIN standards were produced for the physical ergonomics of VDTs. Consequently, DIN provided a major motivation for the production of ISO 9241, and Parts 3 and 4 of this were developed from DIN standards.

6.4 American National Standards Institute (ANSI)

ANSI does not itself prepare standards, but accredits other organisations to prepare standards on its behalf. So, for example, the IEEE is accredited to prepare electrical and electronic and communications standards, and T1, sponsored by the Exchange Carriers Standards Association (ECSA), is accredited to prepare telecommunications standards. In the usability area, ANSI has produced HFS Standard 100-1988 for the ergonomics of VDT workstations. This includes:

- Work-surface and display support height,
- Clearance under work surface for legs,
- Work-surface width and depth,
- Adjustability of heights and angles for chairs and work surfaces,
- Posture—seating depth and angle; back-rest height and lumbar support,
- Availability of arm rests, foot rests and palm rests,
- Use of chair casters.

These contribute to usability by making VDTs more comfortable to use over longer periods. This will ensure that the applications and services accessed by these terminals are more acceptable to users.

7 Pre-standardisation activities

Several pre-standards organisations have been set up to influence telecommunications standards or to try to get standards agreed more quickly. These can play an important role by:

- proposing new standards for products and services where they do not already exist,
- enhancing existing standards originally developed by CCITT or ISO.

The important pre-standards organisations concerned with usability are ECMA and RACE.

7.1 European Computer Manufacturers Association (ECMA)

ECMA was set up in 1960 to promulgate standards for data processing systems. Members include all major European computer companies, as well as U.S. and Japanese manufacturers through their European subsidiaries. It can therefore be said to represent the interests of computer companies worldwide. ECMA has developed over 250 standards and reports, many of which have been endorsed by ISO/IEC, ETSI and CENELEC. It currently has over 30 technical committees. Those directly relevant to usability are TC28 and TC35.

TC35 *User-System Interface* was set up in 1990 to develop an architectural reference model for the user system interface. So far TC35 has produced technical reports in the areas of User Interface Reference Architecture and User Interface Taxonomy. These describe user interface elements and concepts and their relationships and dependencies and identify necessary standards in this area. TC28 *Ergonomics of Workstations* completed its work in 1991. Work in this area is now done ETSI.

7.2 RACE

The RACE programme was set up by the CEC in 1985 to pave the way for the introduction of Integrated Broadband Communications (IBC) in Europe. IBC is a wider concept than B-ISDN, encompassing mobile and satellite networks as well as fixed networks. The programme will run until 1995, by which time the standards and technological infrastructure should be in place to enable the implementation of broadband networks throughout Europe. There are over 100 individual projects in the RACE programme, and over 800 standards contributions have been made to ETSI based on the work done. Material from RACE also forms a substantive part of many contributions to CCITT. Most of the contributions to ETSI have been made by the RACE Consensus Management project. This project was set up in 1988 as a joint Industry-Network Operators team in Brussels to technically steer the RACE programme, to promote a European consensus on technical issues, and to ensure that contributions based on this consensus are made to appropriate standards bodies. This consensus is generated in RACE Concertation Technical Workshops organised by the project, and recorded as the RACE Common Functional Specifications (CFS). These CFS act as pre-standards and contributions based on them are made to ETSI and EBU.

CFSs on usability are currently drafted in RACE Sub-Technical Group 1.3 *Usability Requirements*. Those agreed by the end of 1992 are listed below. It is also planned to produce CFSs on *Usability Requirements for IBC Terminal Hardware* covering the physical ergonomics of IBC terminals.

Usability results from the RACE programme are given elsewhere in this book, and so are not repeated in detail in this chapter. In particular, CFS P174 on the integration of telecommunications services is based on material described in detail in chapter 3.2.

P100 Introduction to Usability Requirements for IBC.

This CFS acts as an introduction to the usability work in RACE and to the usability CFS (P-series). It starts with a definition of an IBC service and the user's interaction with the service provider. It then identifies the interfaces for which usability is important.

P110 User characteristics relevant to the design and uptake of IBC services

This CFS describes the range of tasks that users will carry out using IBC services, and the range of user abilities, skills and limitations that should be catered for by designers of IBC services. It is clearly in designers' interests that their services should be able to be used by as wide a range of users as possible. Even if a service has been specified to be used by only a small group of users it is necessary to ensure that all their relevant characteristics are taken into account. Specific topics covered include intelligibility of speech, skills in keyboard use, literacy, and physical, visual, hearing and cognitive abilities. This CFS covers similar areas to ETSI HF2, and should contribute to a revision of ETSI ETR029 *Access to Telecommunications for People with Special Needs* to cover B-ISDN services.

P120 Methods for Determining Usability Requirements for IBC

This CFS provides a method of analysing and specifying the usability requirements for a telecommunications service. It begins by defining the criteria for usability and making the distinction between enabling tasks and goal tasks described in chapter 1.1. The analysis described in chapter 1.1 is then recommended to provide a set of operational requirements for the design of the service. The 12 usability design principles are then listed, with an example of how to apply them to a multimedia conferencing service. This CFS has contributed to ETSI DTR HF-3001 *Guide on Usability Testing Methods for Telecommunication Terminals and Services*.

P130 Methods for Assessing the Usability of IBC Services.

This CFS provides guidelines for designing and evaluating a telecommunications service, based on the RACE ISSUE project. It deals in detail with prototyping, and how to use prototypes effectively in the evaluation of services. This CFS has contributed to ETSI DTR HF-3001 *Guide on usability testing methods for telecommunication terminals and services.*

P150 The Relationship between Subjective QoS Appreciation and Objective Parameters

This CFS provides a framework within which usability can be seen as contributing to the usage of services offered by a broadband network. It explains in detail the relationships between issues contributing to usability and usage, as shown in figure 3.

P160 Generic Service Attributes

This CFS provides usability guidelines that apply to any telecommunications service. Generic service attributes include methods of providing feedback to the user on the status of the service, and generic call set up and clear down procedures that can apply to any service.

P170 Usability Aspects of IBC Services

This CFS introduces the more detailed series of usability recommendations P171—P174. It describes the division of telecommunications services into four types: conversational, distribution, retrieval and integrated. These types of service each have unique usability requirements and so are treated separately. More information is provided on the usability aspects of these services in chapter 3.2.

P171 Usability of Conversational Services

This CFS provides usability guidelines that apply to conversational services. Issues that are more important for conversational services include response time, delay and media synchronisation requirements. These are generally more stringent for conversational services than for distribution or retrieval services. This means that is is harder to synchronise different media at the receiving terminal as there is a lower overall delay specification over which the variability can be equalised, especially for a service offered over an ATM network or one employing low-speed codecs.

Taking an audiovisual service as an example, a retrieval service can simply delay both audio and video components so that they can be matched at the receiving terminal to maintain lip synchronisation. However, if the end-to-end delay for a conversational service is increased beyond about 0.25s then the participants will start to have turn-taking problems.

P172 Usability of Distribution Services

This CFS provides usability guidelines that apply to distribution (one to many) services. Issues that are more important for distributive services include feedback on the status of the service and on the information that has been sent. As information is not normally returned to the sending user as part of a distributive service, the user must have some way of knowing whether the information has been received and acted upon. This is a problem with many electronic mail systems at present, where there is often no guarantee that the information will be received by its intended recipient and no indication returned to the sending user as to whether it has been received (and acted upon) or not. More feedback would clearly be useful here.

P173 Usability of Retrieval Services

This CFS provides usability guidelines that apply to retrieval (many to one) services. Issues that are more important for retrieval services include compatibility and charging issues. It is important for the user to be able to determine the type, quality and delay requirements of the information to be retrieved (for example from a multimedia or hypermedia database). A user can then judge whether their terminal is capable of receiving and displaying the information in an acceptable timescale, and whether this will incur an acceptable cost.

An example is the retrieval of a high resolution image from a database. This may take 30 seconds or more to be received at the user's terminal, and when it arrives the terminal may not have the capabilities to display it effectively, and it may not contain information useful for the user's task. The retrieval will then have cost the user extra time in having to wait for the image to arrive, and extra time and access charges for the service.

P174 Usability of Integrated Services

This CFS provides usability guidelines applying to the integration of conversational, distributive and retrieval services at the user's terminal. The 12 generic usability principles covered in chapter 3.2 are described in detail, and are applied to the integration of conversational, distributive and retrieval services. The integration principles described in this CFS have been adopted in ETSI DTR HF-01013 *Human Factors Aspects of Multimedia Telecommunications*.

8 Proprietary user interface standards

All the standards previously described have been agreed and approved during some form of consensus making process. However, there are other standards, issued by an individual company, that are adopted by other organisations because of the benefits that they provide. The main standards in this category relevant to usability are those that define the 'look and feel' of a user interface.

Proprietary standards include:

- Apple Desktop
- Microsoft Windows
- Sun/AT&T Open Look
- OSF Motif

These define features of the user interface such as:

- window shape, size and features,
- button shapes, sizes and operation,
- menu style and layout,
- dialogue boxes.

The Apple Desktop user interface, although not the first graphical user interface (GUI), was largely responsible for establishing the use of GUI technology in the marketplace. Several proprietary standards have followed. One of the more accepted is MOTIF from the Open Software Foundation (OSF). This has attempted to give a 3-D feel to the buttons and dialogue boxes, as well as overcoming some of the interaction problems with the Apple Desktop user interface.

An example of proprietary user interface standards in the telecommunications management domain is TMOS (Telecommunications Management and Operations Support). The TMOS user interface is based on Open Look, and a user interface style guide has been produced by Swedish Telecom and Ericsson to show how a graphical user interface based on Open Look should be designed to meet customers requirements.

The adoption of one of these user interface standards for a new application or service should improve the usability of the application or service. Benefits include:

- *Consistency*
  Users will be able to transfer knowledge and skills more easily from one application to another. They will therefore be able to learn and use a larger number of applications and services.
- *Portability*
  User interfaces software will be able to be used across a range of applications, thus reducing costs for the developer and allowing faster implementation of new telecommunications services.

In this way, proprietary user interface standards have been of immense value in increasing the number of applications available for personal computers and the number of users able to use them. But would a single international standard 'look and feel' be of better value, and enable even *more* users to be able to use applications and services more easily? The answer to this is probably no. It is now too late. The existing standards are too entrenched and it would take many years to agree a new GUI standard. This could delay implementation of new applications and would increase costs for developers, thus ultimately increasing costs for users. Secondly, as with the adoption of any standard, there is a danger of stifling innovation and making new innovative, and possibly more usable, user interfaces less likely to evolve. Finally, there are many different types of users with different needs and capabilities, and a single user interface standard would not be optimal for all of them. For these reasons, proprietary user interface standards are likely to be around for many years yet and competition in the marketplace must be left to choose between these standards.

9 How designers use standards

How designers use standards depends on their level of expertise in human factors, and on their access to human factors professionals as part of the design process. Designers may simply refer to any standards they are already aware of or which are included in requirements specifications. For example, proprietary 'look and feel' standards may be specified as company policy or as part of the requirements specification for the product being designed.

In other cases, especially where government or open procurement policies apply, a certain level of usability may be specified in the offer to tender which will then also include references to relevant usability standards (such as the EC Directive on Display Screen Equipment). In these cases relevant usability standards should be available to the designer at the outset. However, these will usually be incomplete or insufficient to fully design the product or service to ensure optimum usability.

In general designers have to be successful in three interdependent tasks:

- *Finding the relevant standards.*
  The guidelines contained in standards and recommendations are identified and collected for application to a new product or service.
- *Specifying the guidelines.*
  The guidelines are specified and presented in a way that is easy to use by service designers.
- *Using the guidelines.*
  The resulting guidelines are used to design and evaluate products and services.

Significant problems that can occur during each of these tasks and the response of the human factors community to these problems are covered in more detail in the following sections. Although matters are improving, it is still extremely desirable to involve human factors professionals in the design of a product from the outset. They will at minimum be able to find and provide copies of relevant usability standards and so raise awareness amongst the design team, but preferably they will be part of the team throughout the design process, so that user interface designs based on an 'intelligent interpretation' of relevant standards can be evolved.

9.1 Finding the relevant standards

Usability experts making recommendations for a specific system or service can study well known guidelines (eg: Smith and Mosier), guidelines produced by the standards bodies covered earlier in this chapter, and relevant information available from research programmes such as RACE, ESPRIT and EURESCOM. However, although the 'database' of usability guidelines in the literature is extensive, an overview is lacking. It is too expensive in time and resources to carry out a new literature survey each time a practical problem is formulated by a designer or user, and even when time and money is spent on a literature search, it is hard to navigate through this and find the relevant sources for a given guideline. When a relevant guideline is found it is also hard to determine its practical value and whether it is applicable to the particular circumstance or not. Non-experts are even more confused by such limitations.

It is also not clear to users of guidelines (including Human Factors experts!) which guidelines are relevant at any point during the design process, and which are irrelevant or missing. One reason for this is the lack of methods which have been successful in integrating human factors and usability guidance into systems analysis, design and evaluation methods (Lim et.al.,1992). There is still no accepted method for service design. The problem of finding relevant usability standards could be solved by including references to these in a Design Support Package such as that described in the next section. This could provide automated references to relevant standards as part of the design information supplied.

9.2 Presenting the guidelines

Recommendations gathered are specified and produced in the form of guidelines that should be easy to understand by designers. Usability and human factors guidelines in their raw form are often seen by designers as difficult to read and understand, with large amounts of text full of vague definitions. The result of this is that, although guidelines are referred to, they are rarely used in the way intended.

The purpose of the usability expert at this stage is to re-specify and re-present the guidelines in a much easier to use form. A common example is the conversion of standards such as ISO 9241 to a checklist of categories to consider or even to a tick list with yes/no answers. The designer or evaluator can then simply tick the boxes to ensure that relevant guidelines have been followed. This approach is difficult to apply early in the design process but is often successfully used at the conformance testing stage when a design has been completed. It can easily be seen from the results where the problems lie and what aspect of the design needs to be improved further. However, in general, too little attention has been paid to the specification and presentation of usability guidelines. Therefore, this was one of the main goals of two usability projects in the RACE programme and these are described here.

The *IBC Systems and Services Usability Engineering* (ISSUE) project examined the usability of videocommunication and multimedia retrieval services. Guidelines are derived and presented in multimedia format,. Video sequences are used to show designers how to apply the guidelines. The guidelines were tested with designers of a multimedia retrieval system.

The *Generic Usability Information for the Design of Advanced Network Communications in Europe* (GUIDANCE) project examined multimedia, multiauthor document production. As has been described in chapters 3.1 and 3.2, a design support package was developed to support the use of the recommended design method and to provide access to the design principles in hypermedia format. The design principles could be accessed as a set of generic principles, with examples of their application to CSCW (Computer Supported Cooperative Work) and hypermedia services. For each of these domain specific principles, users can access examples which illustrate in textual (and sometimes in video) form the application of the selected principle. Alternatively, users can first select an application and then access a specific principle within that domain. The design support package was evaluated by designers involved in graphic design, human-computer interface design and software design of multimedia systems.

However, in the extreme, to provide adequate guidelines for designers of *new* systems and services is actually an impossible task without limiting the scope. For example, the range of possible broadband systems and services that could be offered to customers is so broad and in part unknown, that it will be impossible to specify reliable usability guidelines for all broadband services in the near future. For this reason, research within the RACE programme has been restricted to relatively well defined systems and services such as videocommunication and multimedia retrieval systems.

9.3 Using the guidelines

Finally, designers must apply standards and recommendations, either in their raw form or as guidelines or checklists. Some standards are used directly in the design of a product or service and some are used during evaluation.

The easiest standards to apply are the proprietary user interface style guides described in section 8. These are accompanied by lots of pictures and examples showing how the user interface should appear when the standards have been applied.

The current practice of usability evaluation relies heavily upon the experience and intuition of the evaluator and has been described as a craft (Long and Dowell,1989). Evaluators (experts as well as others) have different purposes and motivations for conducting evaluations, and testing against standards is only one of many objectives. Other purposes of evaluation that are often at least as important are the diagnosis of user problems and ways of improving the product (Whiteside et. al., 1988). For these reasons the benefit of using standards in usability evaluation is lower than expected.

The checklist approach attempts to estimate the usability of a system by checking conformance with available standards. It is assumed that it should capture some psychological principles which are sufficiently invariant across a range of contexts for them to be useful. Unfortunately human behaviour is contextually determined and so very complex. Therefore, it cannot be fully described even by a large number of simple independent statements or propositions about user behaviour. In fact, standards and recommendations may be true in one context but misleading in another.

10 Benefits of standardisation in usability

In Europe, EC Council decision 87/95/EEC requires reference to IT standards in public procurement. This means that any public sector organisation procuring IT products must demand conformance to the relevant IT standards. This only applies to purchases over Ecu 100,000 at present, but equipment purchased below this limit (and therefore currently exempt) must not 'prejudice the use of standards in later procurement over this threshold'. This means that, for example, telephone sets cannot be purchased that would not be able to interconnect with telephone switching equipment bought later to the appropriate standards.

EC Council decision 87/95/EEC is ultimately in the interests of the procurers themselves, because it encourages equipment to be manufactured to the appropriate international standards. The procurers will then have a wider range of equipment to choose from, and so are able to benefit from competition in the supply of this equipment. This is particularly relevant for VDTs, procurement of which is now subject to EC Directive 90/270. As well as leading to a wider range of suppliers for VDTs conforming to common standards, EC 90/270 will improve the usability of VDTs leading to a better quality of life and working environment for the employees of companies using them. This should also lead to increased productivity and a more effective work force.

Other benefits of usability standardisation include those of consistency and portability discussed in section 8. Consistency is not only required in the 'look and feel' of a user interface. It is also beneficial in the system functions, media types and quality levels available. If the functions and media quality levels expected by users are always available from a telecommunications service then the user will find the service more usable because the enabling states can always be reached. All standards, not just usability standards, increase portability, as components and functions will be able to be used across a wide range of systems and applications. This reduces costs for the developer and provides a potentially larger marker for products. For reasons of both consistency and portability, standards could be said to 'create the market' because only if the standards are in place will there be interworking possible between services offered by different suppliers. This will create a critical volume of demand as described in chapter 1.3 which could make the difference between a non-profitable and a profitable service.

But what of the danger of stifling innovation, also mentioned in section 8? It *can* sometimes be too early to introduce a standard in a particular technology area. A balance must be struck between introducing a standard too early, and thus reducing the functionality and therefore possibly the usability of a system, and leaving it too late when there may already be several entrenched proprietary systems using different 'standards'. Too much standardisation can also lead to an inability of manufacturers to effectively differentiate their products which can enable a single supplier to gain a dominant position in the marketplace and impose a de-facto monopoly. This is ultimately to the detriment of users by reducing competition and choice, making it less likely that a range of products to suit all users will be available.

What is the situation with standards affecting usability? As in the case of user interface 'look and feel' standards, there are many areas where standardisation has been left too late. An example is tones and indications in the telecommunications network. There are over 100 different tones used in different countries for the seven or so different status indications required to provide feedback to the user. It would apparently be sensible to standardise these and provide the same seven tones in all countries in the world. Customers would then recognise these, and be able to use telecommunications services more easily when travelling to other countries, which would increase overall usage of telecommunications services. But *which* seven tones should be standardised? To change the tones established in a particular country now would confuse customers and *reduce* the usability of the service. It would also be costly for both the administrations having to make the change and for the users in lost and misdialled calls during the transition. Therefore there is little prospect of actually achieving a standard range of tones in the foreseeable future.

The same is true for telecommunications numbering plans. It would seem to be a trivial change to the network to enable all customers anywhere in the world to use the same prefix for access to international services. In fact a recommendation has already been agreed in CCITT stating that '00' should be the prefix for international access worldwide. However, due to the cost to both users and network operators, this recommendation can only be implemented over a number of years, and even the EC will not have a uniform international access prefix until the year 2000.

The situation is more hopeful for services to be provided using ISDN and B-ISDN. Here the underlying standards are being defined *before* the services are implemented. This should in theory make future services more consistent than services on existing networks. However, the time it takes to fully define a network as complex as B-ISDN has meant that groups of manufacturers are not waiting for the standards to be defined before starting to implement solutions based on ATM techniques. This has led to the establishment of groups such as the ATM Forum which may push non-standard solutions. This must be resisted, or alternatively, standards eventually produced by CCITT must align with those used by the ATM Forum to ensure maximum usability of telecommunications services in the future.

11 Standards required for future integrated services

Standards are required that will enable integrated telecommunications services to be provided more flexibly to meet a wide range of user requirements. This will ensure that the enabling states required in the definition of usability in chapters 1.1 and 3.1 can be reached. A necessary element of this for an integrated broadband network will be to provide customers with multimedia services. These will enable a range of different media to be used during a call so that, for example, a multimedia document including speech, image, video and graphics can be retrieved from a multimedia database by a user using a multimedia service. 'Integration' implies that the user could also be holding a videophone conversation with another user at the same time using the same user interface. Work on the definition of multimedia services is being carried out in ETSI NA1 and CCITT SGXVIII as described previously in this chapter.

To implement multimedia services effectively requires standards for:

- User interface elements.
  It may not be possible to standardise the look and feel of a user interface, but it should be possible to standardise elements of the user interface so that it is clear to a user transferring from one environment to another which functions do what. Examples of these elements include user procedures and pictograms.
- User interface functions.
  It may also be possible to provide consistency and portability by standardising the user interface functions available to an application. All required input/output functions could then be carried out in some form on any user interface, from a simple telephone to a multimedia terminal.
- Service functions.
  The functions provided to an application by a service should come in a standard range. The actual functions themselves may not all be relevant or provided for a given service but the application designer should be able to anticipate what functions will be available. This will improve portability and consistency.
- Service media types and quality levels.
  There should be a standard range of media types and quality levels available to all services, so that applications can select the most appropriate to meet user requirements. A whole host of network capabilities is required to ensure that connections can be established to support these media types and quality levels.
- Service set up.
  Procedures will be required on service set up to ensure that the capabilities of the sending terminal are compatible with those of the receiving terminal and those of the intervening network(s). If the capabilities requested are not present, then fallback procedures should be provided to prevent mismatches.

12 Conclusions

Careful consideration of usability is essential to ensure maximum usage of telecommunications services by the general public. Standards for usability are an essential element of this and enable users to effectively use a much wider range of equipment and services than would otherwise be possible. In fact usability standards can be seen as a good way of 'institutionalising' good usability practise to make it more likely (but not certain) that products and services will be usable by a wide range of users. User interface standards also ensure portability of an application between systems, and can therefore reduce development costs for the application or service. However, commercial considerations will ensure that a single user interface standard will not arise in the foreseeable future so that differentiation between products and services remains possible. This will ensure that only 'minimum' user interfaces become standardised.

Even when all the relevant standards have been used, it is still possible to provide telecommunications services that are *not* easy to use for a variety of reasons. This may be, for example, because the target user population is not predicted correctly or because the application or service is used for tasks other than those for which it was designed. It is therefore always necessary to establish the target usability of an application or service at the outset as described in chapter 1.1, and to measure the actual usability with the target population of users using the methods described in chapter 3.4.

During the last decade graphical user interfaces have become the de-facto standard for most PC and workstation applications. New technical developments in input/output techniques, virtual reality and hypermedia systems will shape the design of future user interfaces. Systems are becoming ever more complex and there is an urgent need for standards and recommendations for good user interface design. These needs will continue to be met by a mixture of national, international and proprietary standards in the future.

These trends are also requiring closer cooperation between skilled designers, standards experts and active researchers to ensure that future standards meet the needs of designers and can be used effectively.

13 Glossary of standards abbreviations and terms

Aenor Asociación Española de Normalización y Certificación.
The Spanish national standards organisation.

Afnor Association française de normalisation.
The French national standards organisation.

ANSI The American National Standards Institute.
The primary standards setting body for the USA.

ATM Asynchronous Transfer Mode.
The preferred transfer mode for B-ISDN that can, given sufficient bandwidth, carry any type of network service.

ATM Forum Established in 1991 by telecommunications equipment manufacturers and service providers to accelerate the definition of ATM standards, so that equipment and services can be provided to common standards more quickly.

B-ISDN Broadband ISDN.
The extension of ISDN to enable the provision of services requiring more than 2 Mbit/s information transfer rate.

BSI British Standards Institution.
The UK national standards organisation.

CEC The Commission of the European Communities.
The 'civil service' of the European Community. The CEC issues European Directives which must be enacted in laws passed by member countries.

CEN The European Committee for Standardisation.
This covers both the EC and EFTA countries, and aims to eliminate trade barriers caused by differences in national standards.

CENELEC The European Committee for Electrotechnical Standardisation.
This also covers both the EC and EFTA countries.

CCIR International Radio Consultative Committee.
The part of the ITU responsible for radio communications.

CCITT International Telegraph and Telephone Consultative Committee.
The part of the ITU responsible for (non-mandatory) recommendations on public telecommunications services. The voting members of the CCITT are the national governments, but in practice membership is delegated to the national telecommunications administrations. CCITT publishes telecommunications recommendations in the form of books; the most recent is the Blue Book (1988).

CEPT European Committee for Posts and Telecommunications.
CEPT deals with legislative, regulatory and marketing issues. The members are the regulatory authorities of the European countries rather than the PNOs, although in some countries they are still the same body. The functions carried out by ETSI used to be part of CEPT.

CFS RACE Common Functional Specification.
These are pre-standards representing the consensus in the European RACE programme. There are over 100 CFS published by the RACE Consensus Management project.

CTR Common Technical Regulation.
A harmonised technical standard which is made mandatory through appropriate procedures to ensure compliance with essential requirements for telecommunications terminals. A CTR is usually based on part or the whole of an ETS.

CTS Conformance Testing Services programme.
An EC initiative to provide tools and facilities to meet the requirements for testing services.

DIN Deutsche Industrie Norm.
A standard produced by the German national standards organisation Deutsches Institut für Normung eV.

EBU European Broadcasting Union.

EC The European Community.
Covers the 12 member states: Belgium, Denmark, Eire, France, Germany, Greece, Holland, Italy, Luxembourg, Portugal, Spain and UK.

ECMA European Computer Manufacturers' Association.
An industrial body whose members include all major West European manufacturers, and many of the major US manufacturers through their European offices. Develops standards and technical reports to facilitate and standardize the use of information processing and telecommunications systems.

ECSA Exchange Carriers Standards Association.
The American network operators association which sponsors standards committee T1—telecommunications.

EFTA European Free Trade Association.
Covers Austria, Finland, Iceland, Norway, Sweden and Switzerland.

ETR ETSI Technical Report.

ETS ETSI Technical Standard.

ETSI European Telecommunications Standards Institute.
A non-profit making organisation setting telecommunications standards in Europe. ETSI has 12 technical committees (TCs) dealing with telecommunications, IT (in cooperation with CEN) and broadcasting (in cooperation with the EBU). ETSI has nearly 300 members including national administrations, public network operators, manufacturers, users and private service providers, research bodies and consulting companies.

EWOS European Workshop for Open Standards.
A pre-standardisation body set up to coordinate input to CEN/CENELEC and achieve convergence with contributions to ISO.

Functional Standard: This is shorthand for the sentence 'if you want to provide function X, then use standards A, B, C...'

GUI — Graphical User Interface.

IBC — Integrated Broadband Communications.
The RACE view of B-ISDN.

IEC — International Electrotechnical Commission.
A similar organisation to ISO. IEC sets worldwide standards for electrical and electronic engineering. Works closely with ISO through the Joint Technical Committee.

IEEE — Institute of Electrical and Electronics Engineers.
A professional society with over 300,000 members. Accredited by ANSI to prepare electrical and electronic and communications standards.

ISDN — Integrated Services Digital Network.

ISO — The International Organisation for Standardisation.
A federation of national standards bodies. ISO sets worldwide standards in any field not covered by a specialist standards body.

ITU — The International Telecommunications Union.
An agency of the United Nations based in Geneva. It is responsible for telecommunications standards worldwide and has 5 parts including CCITT and CCIR. On 1 March 1993 CCITT and CCIR were merged into a single part of ITU responsible for telecommunications standards.

MMI — Man-Machine Interface.

MOU — Memorandum of Understanding.

NET — Normes Européene de Télécommunications.
A technical specification which is given mandatory force according to the legal procedures established in Directive 86/361/EEC. NETs are being replaced by CTRs.

OSF — Open Software Foundation.
A consortium of computer manufacturers including IBM. Responsible for the MOTIF user interface standard.

OSI — Open Systems Interconnection.
A set of standards concerned with communication and interworking between computer systems of all types from any suppliers on the basis set out in the OSI Reference Model. This defines the framework within which the data communications standards are developed.

PNO — Public Network Operator.
An operator of public telecommunications services.

RACE — Research and Development in Advanced Communications Technologies in Europe.
A pre-competitive EC R&D programme for the development of advanced technologies in the field of telecommunications. RACE involves over 100 projects

	carried out in both EC and EFTA member countries. RACE acts as a regional pre-standardisation body through the production of the RACE Common Functional Specifications.
SIS	Swedish Standards Institution. The Swedish national standards organisation.
SPAG	Standards Promotion and Application Group. Established in 1983 by European manufacturers, with the aim of defining functional standards from the many permutations that are available from international standards, particularly those from ISO, so that multi-vendor working can be achieved.
T1	The committee defining telecommunications standards in the USA, established in 1984, sponsored by ECSA and accredited by ANSI.
TMOS	Telecommunications Management and Operations Support. A proprietary package for telecommunications network management from Swedish Telecom and Ericsson.
TTC	Telecommunications Technology Committee. A non-profit making private organisation which contributes to telecommunications network standards in Japan.
UNI	Ente Nazionale italiano di unificazione. The Italian national standards organisation.
UPT	Universal Personal Telecommunications.
VDT	Visual Display Terminal.

Standards documentation

Standardisation in Information Technology and Telecommunications, Commission of the European Communities DG XIII: 200 rue de la Loi, B-1049 Brussels, Belgium.

ANSI/HFS Standard No 100 (1988) *American National Standard for Human Factors Engineering of Visual Display Terminal Workstations,* Human Factors Society, Santa Monica, CA.

CCITT Recommendation E.134 *Generic Operating Procedures.*

CCITT Recommendation E.135 *Human Factors Aspects of Public Telecommunications Terminals for People with Special Needs.*

CCITT Framework Recommendation I.374 (1993) *Network Capabilities to support Multimedia Services.*

CCITT Recommendations Z.301—Z.341 *Man-Machine Language (MML)* Volume X, Fascicle X.7, Blue Book.

CEPT Telecom Recommendation T/CAC O 2 E (1992) *Subscriber control procedures for supplementary services in modern telecommunication systems.*

EC Directive 90/270/EEC (1990) *The VDT Directive* on Display Screen Equipment.

EC Directive 90/387/EEC (1990) *The ONP Directive* on the Establishment of the Internal Market for Telecommunications Services through the Implementation of Open Network Provision.

EC Directive 90/388/EEC (1990) *The Services Directive* on Competition in the Markets for Telecommunications Services.

EC Directive 92/400/EEC (1992) adopts ETSI as a European standards body so that ETSI standards have the same status as those adopted by CEN or CENELEC.

ETSI Technical Report ETR029 (1991) *Access to Telecommunications for People with Special Needs.*

ETSI Technical Report HF-2006 (1992) *European Standardisation Situation of Telecommunications Facilities for People with Special Needs—An Overview.*

ETSI Technical Report HF-3002 (1992) *Guide on Usability Testing Methods.*

ISO 9241 (1992) *Ergonomic Requirements for Office Work with Visual Display Terminals (VDTs),* ISO/TC159/SC4/N191 (draft CD 9241) Part 11: Guidance on Usability Specification and Measures.

IBM System Application Architecture (1987) *Common User Access: Panel Design and User Interaction.*

RACE Common Functional Specifications, Issue C (1992) published in 11 volumes by the RACE Consensus Management project,.

T1.203-1988 *American National Standard for Telecommunications—Operations, Administration, Maintenance and Provisioning (OAM&P)—Human-Machine Language.*

ISO standards can be obtained from the ISO Central Secretariat, 1, rue de Varembé, Case postale 56, CH-1211 Geneva 20, Switzerland.

CCITT Recommendations can be obtained from ITU Headquarters, Place des Nations, CH-1211, Geneva 20, Switzerland.

ETSI Technical Reports and ETSI Technical Standards can be obtained from the ETSI Secretariat, 06921 Sophia Antipolis Cedex, France.

EC directives can be obtained from the Commission of the European Communities, 200 rue de la Loi, B-1049 Brussels, Belgium.

RACE Common Functional Specifications can be obtained from the RACE Consensus Management Office, 165 boulevard du Souverain, B-1160 Brussels, Belgium.

Acknowledgements

Thanks are due to BT, Swedish Telecom and the RACE Industrial Consortium for their support and for permission to publish this chapter. Thanks are also due to colleagues in the RACE Consensus Management project for their helpful comments and suggestions, and to Phil Rushton, Vice Chairman of ETSI TCHF, for his help and advice on this chapter.

Acknowledgement is also made to the RACE ISSUE and GUIDANCE projects whose work has been referenced.

References

Lim, K.Y., Long, J.B. & Silcock, N. (1992) 'Integrating Human Factors with the Jackson System Development Method: an Illustrated Overview', *Ergonomics*, 36, 1135—1161.

Long, J.B. & Dowell, J. (1989) Conceptions of the Discipline of HCI: Craft, Applied Science and Engineering. In A. Sutcliffe & L. Macauley (eds) *People and Computers V*,. Cambridge University Press: Cambridge.

Smith, S.L, & Mosier, J.N. (1986) *Guidelines for Designing User Interface Software*, Report MTR-10090, The MITRE Corporation: Bedford, Massachusetts.

Whiteside, J, Benett, J. & Holtzbatt, K. (1988) Usability Engineering: Our Experience and Evolution. In M. Helande (ed) *Handbook of Human-Computer Interaction*. North-Holland Elsevier Science Publishers B.V.: Amsterdam, 757—789.

Afterword

The need to represent requirements in design

Jon May[†], Paul F. Byerley[*], and Philip J. Barnard[†]

[†] MRC—APU, 15 Chaucer Road, Cambridge, UK-CB2 2EF

[*] Alcatel-SEL AG, Lorenzstraße 10, D-7000 Stuttgart 40, Germany

Each of the three main sections of this book had a specific aim. The first section presented a structured view of requirements, tailored to the design of advanced communication services. It elaborated the needs for requirements from the point of view of the providers of IBC services, the marketing of those services, and their design. Without structuring the elicitation and presentation of requirements in some way, there can be no assurance that appropriate products can be designed, provided, or marketed.

The second section described some basic research on the identification of specific aspects of requirements. Three chapters respectively covered the cognitive requirements of the human end-users, an approach to the formal modelling of system specifications, and a framework for representing the interaction between users and systems. Two chapters dealt with representational techniques for structuring the design space to make explicit the decision making process and the trade-offs involved in meeting all types of requirements. These approaches to requirements come from different theoretical backgrounds, even from different research disciplines, but they have a common aim in providing a comprehensive and well-grounded way in dealing with specific areas of the problem. Clearly they do not cover all of the ground possible, but taken together they go some way to providing the new concepts and ideas that are essential for the structured representation of requirements presented in section one.

The first three chapters in the third section dealt with one particular way of applying a structured representation of requirements in design, the 'enabling states approach'. This method, while it is developed from the eight-step method described in the first section, is not intended to be seen as a 'stand-alone' or complete 'design method'. What it does attempt to do is to be explicit about the sequence in which the requirements of the 'internal perspective' of the first section must be dealt with in designing integrated services. It does this by adopting a set of 'specification tables' that can be completed with the support of 'usability principles'.

The information in these principles, the grounds for the trading-off that must be carried out in their application, and the recording of the decision-making process, can be derived from the research reported in the second section. The section concludes with recommendations for the successful evaluation of designs, and the way in which standardisation of usability information is carried out.

The introduction to each section contained a figure linking the chapters together. We can now assemble these figures to summarise the content of the book as a whole:

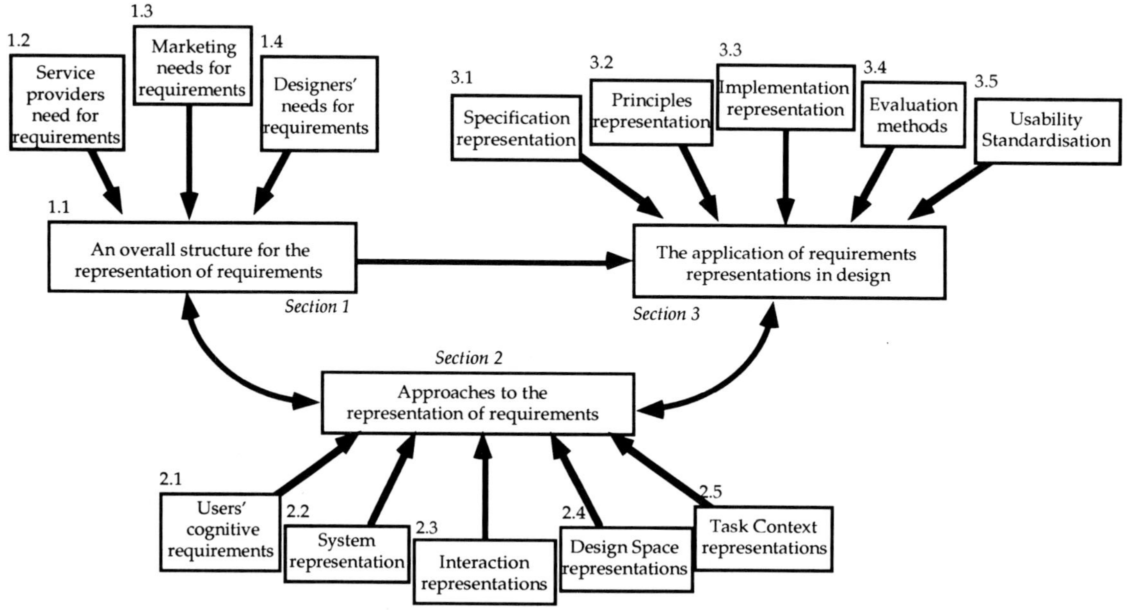

The relationship indicated between the topic areas of sections 1 and 3 is essentially that of providing, in the future, for consistency between specific design and evaluation methods within an overall structure. This has been illustrated in this book by the work reported in the chapters of section 3 which are consistent, one with another, within the framework provided by chapter 1.1.

The relationship between the topic areas of section 1 and section 2 provides for the migration of precise system property definitions to the overall framework. These properties can include the usability properties investigated in the AMODEUS project, but must also include properties relevant to, for example, the 'external' perspective described in chapter 1.1. While it is clear that some aspects of the CO-SITUE frames can be related to the external perspective, most of the work in this area has been concerned with the internal perspective. The RACE 1077 'Usage Reference Model' is an exception in that it explicitly examines the properties of the external perspective.

Many readers will undoubtedly raise important issues about the transferability of basic techniques to practical contexts. It is our view that any such difficulties are worth striving to overcome. It seems unlikely that basic research will lead to modelling techniques that will be so good that usability testing becomes redundant. However, as technologies evolve to more integrated systems, the design issues they raise become more complex and inter-related. Resolving answers to such inter-related questions may require a deep understanding of interactions that will only emerge when the different disciplines involved develop both adequate basic concepts and practical methods of working together in a range of design and development contexts.

Obviously, basic research of the type presented in section 2 should also, over time, have an increasing impact in the application of requirements of the types described in section 3. That impact may come in a number of different ways. Naturally, some techniques developed within basic research may require large amounts of effort and be rather impractical in some design contexts. Nevertheless, individual principles dealing with user cognition, systems, or their interactions may be absorbed into practitioners evolving methods. The principles may even be re-represented within new methods to support design decision making. Techniques such as the Questions Options and Criteria notation of chapter 2.4 or the COSITUE frames of chapter 2.5 are themselves potential means of migrating information from research to design and development contexts. In this sense, these chapters are themselves one form of important bridge between fundamental research and a downstream impact for modelling techniques.

We hope that our general argument is clear from this figure. There is a need for the appropriate level of requirement to be considered at the appropriate point in design, and there are ways of obtaining and representing these requirements so that they are available and interpretable. We are not making any claims about the exclusive validity of any of the approaches described in this book: they are not intended to be exclusive. They do intend, however, to make the need for requirements explicit.

One of the problems that has become apparent in the promotion of formal design methods is that even the adherents of a method adapt it to their own ends, neglecting some elements and adding others, with the risk that any implicit connection between the requirements and design is weakened or lost. Requirement gathering and validation can be particularly susceptible to being bypassed, given the pressure for concrete design 'results', since (to a harassed designer) they may not seem to make a direct contribution to the creative aspects of design. This may be exacerbated by the division of the design process into discrete steps or stages, with ever more reports to deliver and deadlines to be achieved. As section 1 of this book has emphasised, accurate requirements remain crucial if the right product or service is to be designed.

One way of avoiding this problem is to explicitly separate requirements from design, so that whatever method of design is taken up, the requirements effort can continue alongside the design task, obtaining the relevant information and providing it in an accessible manner. If this option is taken up, though, the continuity between the design process and the requirements provision must be made clear, or requirements may be ignored altogether. As reported in section 3, the 'specification tables' that are used in the Enabling States Approach are a straightforward way of representing both requirements and design decisions in a common format, tailored to the needs of those fulfilling the human factors, design and implementation roles.

It is, in our opinion, not necessary to settle on one particular way to 'do design'. What is necessary is for designers to have access to the requirements for their design, so that they know what they are to design, and why. The requirements must be structured so that the relevant information is available at the relevant point in design, however the design process is structured. The representation of these requirements should be linked to the presentation of human-factors and usability information so that principled design decisions can be made and justified. There must be a way of evaluating designs so that they are tested against the requirements that they are meant to fulfil.

The structure of the figure represents the rationale underlying the various research projects whose work is covered in this book. These projects are drawn from different areas of the European Communities 'Framework' research programme. The figure indicates that projects from different streams of activity can, nevertheless, be drawn together effectively. It is to be hoped that successor programmes can not only build upon the individual strands of research but also further capitalise upon their inter-relationships.

We hope that this book will be of value to those people involved in the provision and marketing of IBC services. They should be able to draw upon the research material to create more successful products. This should benefit us all, since we are all 'end-users' to some extent. Equally, we hope that the book will encourage those involved in basic research to consider more aspects of requirements elicitation and representation. Finally, and most importantly, we hope that the research will be appreciated by designers and implementers, and that it will help them to argue for adequate specification of usability requirements and the systematic provision of human-factors support throughout the design process.